Pitman Research Notes in Mathematics Series

Submission of proposals for consideration
Suggestions for publication, in the form of outlines and representative samples, are invited by the Editorial Board for assessment. Intending authors should approach one of the main editors or another member of the Editorial Board, citing the relevant AMS subject classifications. Alternatively, outlines may be sent directly to the publisher's offices. Refereeing is by members of the board and other mathematical authorities in the topic concerned, throughout the world.

Preparation of accepted manuscripts
On acceptance of a proposal, the publisher will supply full instructions for the preparation of manuscripts in a form suitable for direct photo-lithographic reproduction. Specially printed grid sheets can be provided and a contribution is offered by the publisher towards the cost of typing. Word processor output, subject to the publisher's approval, is also acceptable.

Illustrations should be prepared by the authors, ready for direct reproduction without further improvement. The use of hand-drawn symbols should be avoided wherever possible, in order to maintain maximum clarity of the text.

The publisher will be pleased to give any guidance necessary during the preparation of a typescript, and will be happy to answer any queries.

Important note
In order to avoid later retyping, intending authors are strongly urged not to begin final preparation of a typescript before receiving the publisher's guidelines. In this way it is hoped to preserve the uniform appearance of the series.

Addison Wesley Longman Ltd
Edinburgh Gate
Harlow, Essex, CM20 2JE
UK
(Telephone (0) 1279 623623)

Titles in this series. A full list is available from the publisher on request.

251 Stability of stochastic differential equations with respect to semimartingales
X Mao
252 Fixed point theory and applications
J Baillon and M Théra
253 Nonlinear hyperbolic equations and field theory
M K V Murthy and S Spagnolo
254 Ordinary and partial differential equations. Volume III
B D Sleeman and R J Jarvis
255 Harmonic maps into homogeneous spaces
M Black
256 Boundary value and initial value problems in complex analysis: studies in complex analysis and its applications to PDEs 1
R Kühnau and W Tutschke
257 Geometric function theory and applications of complex analysis in mechanics: studies in complex analysis and its applications to PDEs 2
R Kühnau and W Tutschke
258 The development of statistics: recent contributions from China
X R Chen, K T Fang and C C Yang
259 Multiplication of distributions and applications to partial differential equations
M Oberguggenberger
260 Numerical analysis 1991
D F Griffiths and G A Watson
261 Schur's algorithm and several applications
M Bakonyi and T Constantinescu
262 Partial differential equations with complex analysis
H Begehr and A Jeffrey
263 Partial differential equations with real analysis
H Begehr and A Jeffrey
264 Solvability and bifurcations of nonlinear equations
P Drábek
265 Orientational averaging in mechanics of solids
A Lagzdins, V Tamuzs, G Teters and A Kregers
266 Progress in partial differential equations: elliptic and parabolic problems
C Bandle, J Bemelmans, M Chipot, M Grüter and J Saint Jean Paulin
267 Progress in partial differential equations: calculus of variations, applications
C Bandle, J Bemelmans, M Chipot, M Grüter and J Saint Jean Paulin
268 Stochastic partial differential equations and applications
G Da Prato and L Tubaro
269 Partial differential equations and related subjects
M Miranda
270 Operator algebras and topology
W B Arveson, A S Mishchenko, M Putinar, M A Rieffel and S Stratila
271 Operator algebras and operator theory
W B Arveson, A S Mishchenko, M Putinar, M A Rieffel and S Stratila
272 Ordinary and delay differential equations
J Wiener and J K Hale
273 Partial differential equations
J Wiener and J K Hale
274 Mathematical topics in fluid mechanics
J F Rodrigues and A Sequeira
275 Green functions for second order parabolic integro-differential problems
M G Garroni and J F Menaldi
276 Riemann waves and their applications
M W Kalinowski
277 Banach C(K)-modules and operators preserving disjointness
Y A Abramovich, E L Arenson and A K Kitover
278 Limit algebras: an introduction to subalgebras of C*-algebras
S C Power
279 Abstract evolution equations, periodic problems and applications
D Daners and P Koch Medina
280 Emerging applications in free boundary problems
J Chadam and H Rasmussen
281 Free boundary problems involving solids
J Chadam and H Rasmussen
282 Free boundary problems in fluid flow with applications
J Chadam and H Rasmussen
283 Asymptotic problems in probability theory: stochastic models and diffusions on fractals
K D Elworthy and N Ikeda
284 Asymptotic problems in probability theory: Wiener functionals and asymptotics
K D Elworthy and N Ikeda
285 Dynamical systems
R Bamon, R Labarca, J Lewowicz and J Palis
286 Models of hysteresis
A Visintin
287 Moments in probability and approximation theory
G A Anastassiou
288 Mathematical aspects of penetrative convection
B Straughan
289 Ordinary and partial differential equations. Volume IV
B D Sleeman and R J Jarvis
290 *K*-theory for real *C**-algebras
H Schröder
291 Recent developments in theoretical fluid mechanics
G P Galdi and J Necas
292 Propagation of a curved shock and nonlinear ray theory
P Prasad
293 Non-classical elastic solids
M Ciarletta and D Ieșan
294 Multigrid methods
J Bramble
295 Entropy and partial differential equations
W A Day
296 Progress in partial differential equations: the Metz surveys 2
M Chipot
297 Nonstandard methods in the calculus of variations
C Tuckey
298 Barrelledness, Baire-like- and (LF)-spaces
M Kunzinger
299 Nonlinear partial differential equations and their applications. Collège de France Seminar. Volume XI
H Brezis and J L Lions

300 Introduction to operator theory
T Yoshino
301 Generalized fractional calculus and applications
V Kiryakova
302 Nonlinear partial differential equations and their applications. Collège de France Seminar Volume XII
H Brezis and J L Lions
303 Numerical analysis 1993
D F Griffiths and G A Watson
304 Topics in abstract differential equations
S Zaidman
305 Complex analysis and its applications
C C Yang, G C Wen, K Y Li and Y M Chiang
306 Computational methods for fluid-structure interaction
J M Crolet and R Ohayon
307 Random geometrically graph directed self-similar multifractals
L Olsen
308 Progress in theoretical and computational fluid mechanics
G P Galdi, J Málek and J Necas
309 Variational methods in Lorentzian geometry
A Masiello
310 Stochastic analysis on infinite dimensional spaces
H Kunita and H-H Kuo
311 Representations of Lie groups and quantum groups
V Baldoni and M Picardello
312 Common zeros of polynomials in several variables and higher dimensional quadrature
Y Xu
313 Extending modules
N V Dung, D van Huynh, P F Smith and R Wisbauer
314 Progress in partial differential equations: the Metz surveys 3
M Chipot, J Saint Jean Paulin and I Shafrir
315 Refined large deviation limit theorems
V Vinogradov
316 Topological vector spaces, algebras and related areas
A Lau and I Tweddle
317 Integral methods in science and engineering
C Constanda
318 A method for computing unsteady flows in porous media
R Raghavan and E Ozkan
319 Asymptotic theories for plates and shells
R P Gilbert and K Hackl
320 Nonlinear variational problems and partial differential equations
A Marino and M K V Murthy
321 Topics in abstract differential equations II
S Zaidman
322 Diffraction by wedges
B Budaev
323 Free boundary problems: theory and applications
J I Diaz, M A Herrero, A Liñan and J L Vazquez
324 Recent developments in evolution equations
A C McBride and G F Roach
325 Elliptic and parabolic problems: Pont-à-Mousson 1994
C Bandle, J Bemelmans, M Chipot, J Saint Jean Paulin and I Shafrir
326 Calculus of variations, applications and computations: Pont-à-Mousson 1994
C Bandle, J Bemelmans, M Chipot, J Saint Jean Paulin and I Shafrir
327 Conjugate gradient type methods for ill-posed problems
M Hanke
328 A survey of preconditioned iterative methods
A M Bruaset
329 A generalized Taylor's formula for functions of several variables and certain of its applications
J-A Riestra
330 Semigroups of operators and spectral theory
S Kantorovitz
331 Boundary-field equation methods for a class of nonlinear problems
G N Gatica and G C Hsiao
332 Metrizable barrelled spaces
J C Ferrando, M López Pellicer and L M Sánchez Ruiz
333 Real and complex singularities
W L Marar
334 Hyperbolic sets, shadowing and persistence for noninvertible mappings in Banach spaces
B Lani-Wayda
335 Nonlinear dynamics and pattern formation in the natural environment
A Doelman and A van Harten
336 Developments in nonstandard mathematics
N J Cutland, V Neves, F Oliveira and J Sousa-Pinto
337 Topological circle planes and topological quadrangles
A E Schroth
338 Graph dynamics
E Prisner
339 Localization and sheaves: a relative point of view
P Jara, A Verschoren and C Vidal
340 Mathematical problems in semiconductor physics
P Marcati, P A Markowich and R Natalini
341 Surveying a dynamical system: a study of the Gray–Scott reaction in a two-phase reactor
K Alhumaizi and R Aris
342 Solution sets of differential equations in abstract spaces
R Dragoni, J W Macki, P Nistri and P Zecca
343 Nonlinear partial differential equations
A Benkirane and J-P Gossez
344 Numerical analysis 1995
D F Griffiths and G A Watson
345 Progress in partial differential equations: the Metz surveys 4
M Chipot and I Shafrir
346 Rings and radicals
B J Gardner, Liu Shaoxue and R Wiegandt
347 Complex analysis, harmonic analysis and applications
R Deville, J Esterle, V Petkov, A Sebbar and A Yger
348 The theory of quantaloids
K I Rosenthal
349 General theory of partial differential equations and microlocal analysis
Qi Min-you and L Rodino

350 Progress in elliptic and parabolic partial differential equations
A Alvino, P Buonocore, V Ferone, E Giarrusso, S Matarasso, R Toscano and G Trombetti
351 Integral representations for spatial models of mathematical physics
V V Kravchenko and M V Shapiro
352 Dynamics of nonlinear waves in dissipative systems: reduction, bifurcation and stability
G Dangelmayr, B Fiedler, K Kirchgässner and A Mielke
353 Singularities of solutions of second order quasilinear equations
L Véron
354 Mathematical theory in fluid mechanics
G P Galdi, J Málek and J Necas
355 Eigenfunction expansions, operator algebras and symmetric spaces
R M Kauffman
356 Lectures on bifurcations, dynamics and symmetry
M Field
357 Noncoercive variational problems and related results
D Goeleven
358 Generalised optimal stopping problems and financial markets
D Wong
359 Topics in pseudo-differential operators
S Zaidman
360 The Dirichlet problem for the Laplacian in bounded and unbounded domains
C G Simader and H Sohr
361 Direct and inverse electromagnetic scattering
A H Serbest and S R Cloude
362 International conference on dynamical systems
F Ledrappier, J Lewowicz and S Newhouse
363 Free boundary problems, theory and applications
M Niezgódka and P Strzelecki
364 Backward stochastic differential equations
N El Karoui and L Mazliak
365 Topological and variational methods for nonlinear boundary value problems
P Drábek
366 Complex analysis and geometry
V Ancona, E Ballico, R M Mirò-Roig and A Silva
367 Integral expansions related to Mehler–Fock type transforms
B N Mandal and N Mandal
368 Elliptic boundary value problems with indefinite weights: variational formulations of the principal eigenvalue and applications
F Belgacem
369 Integral transforms, reproducing kernels and their applications
S Saitoh
370 Ordinary and partial differential equations. Volume V
P D Smith and R J Jarvis
371 Numerical methods in mechanics
C Conca and G N Gatica
372 Generalized manifolds
K-G Schlesinger
373 Independent axioms for Minkowski space-time
J W Schutz
374 Integral methods in science and engineering Volume one: analytic methods
C Constanda, J Saranen and S Seikkala
375 Integral methods in science and engineering Volume two: approximation methods
C Constanda, J Saranen and S Seikkala
376 Inner product spaces and applications
T M Rassias

Themistocles M Rassias (Editor)

National Technical University of Athens

Inner product spaces and applications

LONGMAN

Addison Wesley Longman Limited
Edinburgh Gate, Harlow
Essex CM20 2JE, England
and Associated Companies throughout the world.

Published in the United States of America
by Addison Wesley Longman Inc.

First published 1997

AMS Subject Classifications (Main) 46Cxx
(Subsidiary) 46Bxx

ISSN 0269-3674

ISBN 0 582 31711 8

British Library Cataloguing in Publication Data

A catalogue record for this book is
available from the British Library

Printed and bound in Great Britain
by Biddles Ltd, Guildford and King's Lynn

Contents

Preface

Javier Alonso
Some properties of Birkhoff and Isosceles orthogonality in normed linear spaces 1

Ali R Amir-Moez
Quadratic norms and inner products 12

Jacek Chmielińksi
On the stability of isometric operators for Hilbert spaces 15

B Danković, G V Milovanović and S LJ Rančić
Malmquist and Müntz orthogonal systems and applications 22

J-L Garcia-Roig
On the group of isometries and a system of functional equations of a real normed plane 42

Samuel S Holland, Jr
The eigenvalues of the sum of two projections 54

PL Kannappan
On inner product spaces IV 65

Manfred Kracht and Erwin Kreyszig
On Hilbert space theory and kernel functions 70

Gradimir V Milovanović
Orthogonal polynomial systems and some applications 115

M Aslam Noor
Multivalued variational inequalities 183

Jean-Paul Penot
What is a subgradient in a space which is not an inner product space? 208

Themistocles M Rassias
Stability of the generalized orthogonality functional equation 219

Parfeny P Saworotnow
Hilbert algebras and Hilbert modules 241

K Sundaresan
Geometry of spaces of quadratic forms 261

László Székelyhidi
Cosine operator functions on inner product spaces 266

Preface

In this volume the contributing authors deal primarily with the interaction among problems of analysis and geometry in the context of inner product spaces. They present new and old characterizations of inner product spaces among normed linear spaces and the use of such spaces in various research problems of pure and applied mathematics. The methods employed are accessible to students familiar with normed linear spaces. We remark here that some of the theorems presented are at the same time simple and challenging.

In a real inner product space the orthogonality of two vectors can be expressed in different ways. This gives rise to several definitions of generalized orthogonality. Some of the topics dealt with in this volume include: recent progress in the field of orthogonal systems (orthogonal algebraic and trigonometric polynomials, orthogonal Müntz polynomials, orthogonal rational functions, etc.), Hilbert algebras and Hilbert modules, Hilbert space theory and kernel functions, functional equations involving norms of the real plane, cosine operator functions on inner product spaces, the Wigner equation connected with the stability of isometric operators for Hilbert spaces, the generalized orthogonality equation and some of its characteristic stability properties, the eigenvalues of the sum of two projections, the question of what is a subgradient in a space which is not an inner product space, and geometry of spaces of quadratic forms.

I wish to express my deep appreciation to all the scientists who contributed to this volume, to Professor Haïm Brezis for his help and encouragement and, of course, to Addison Wesley Longman for their expert support.

August 1997 Themistocles M Rassias

JAVIER ALONSO

Some properties of Birkhoff and isosceles orthogonality in normed linear spaces

In a real inner product space the orthogonality of two vectors can be expressed in different ways by means of the norm induced by the scalar product. That is, if $(H, \langle\cdot,\cdot\rangle)$ is a real inner product space with induced norm $\|\cdot\|$ and $x, y \in H$, then $\langle x, y\rangle = 0$ if and only if, for example, $\|x + y\| = \|x - y\|$. This, and similar properties, make it possible to transfer the concept of orthogonality to the wider frame of normed linear spaces, giving rise to several definitions of generalized orthogonalities [1, 2]. Among them, in this paper we shall pay attention to Birkhoff and isosceles orthogonalities introduced, respectively, by G. Birkhoff [4] and R.C. James [6].

Let E be a real normed linear space of dimension at least two, and let $x, y \in E$. Then, x is *Birkhoff orthogonal* to y, $(x \perp_B y)$, if $\|x + \lambda y\| \geq \|x\|$, for every $\lambda \in \mathbb{R}$, and x is *isosceles orthogonal* to y, $(x \perp_I y)$, if $\|x + y\| = \|x - y\|$.

It is well known that if H is a real inner product space, then the relation of orthogonality has the following properties:

Nondegeneracy: If $\lambda, \mu \in \mathbb{R}$ and $x \in H$, then $\lambda x \perp \mu x$ if and only if either $\lambda x = 0$ or $\mu x = 0$.

Simplification: If $x \perp y$, then $\lambda x \perp \lambda y$ for every $\lambda \in \mathbb{R}$.

Continuity: If $\{x_n\}$ and $\{y_n\}$ are sequences in H such that $x_n \to x$, $y_n \to y$ and $x_n \perp y_n$, $(n \in \mathbb{N})$, then $x \perp y$.

Existence: For every $x, y \in H$ there exists $\alpha \in \mathbb{R}$ such that $x \perp \alpha x + y$.

Uniqueness: In the above property, the scalar α is unique if $x \neq 0$.

Symmetry: If $x \perp y$, then $y \perp x$.

Homogeneity: If $x \perp y$, then $x \perp \lambda y$ for every $\lambda \in \mathbb{R}$.

Additivity: If $x \perp y$ and $x \perp z$ then $x \perp y + z$.

Extension: If $x \perp y$, then there exists a closed maximal subspace $P \subset H$ such that $y \in P$ and $x \perp P$.

Nevertheless, as far as we know, if we change the inner product space H into an arbitrary real normed linear space E, then there is no known generalized orthogonality with all the listed properties. In fact, only the properties of *nondegeneracy*, *simplification* and *continuity* are common to every known generalized orthogonality, whereas some of the other properties for some orthogonalities are characteristic of inner product spaces. With regard to Birkhoff and isosceles orthogonalities, in addition to the above three properties, they have the following ones [3] ... [11].

Properties of Birkhoff orthogonality

B.1 It is not necessarily symmetric. If $\dim E \geq 3$ then B-orthogonality is symmetric if and only if E is an inner product space, whereas if $\dim E = 2$ it is symmetric if and only if E has a *Radom norm* (e.g., a norm whose spheres are regular hexagons).

B.2 It has the properties of *right* and *left-existence*, i.e., for every $x, y \in E$ there exist $\alpha, \beta \in \mathbb{R}$ such that $x \perp_B \alpha x + y$ and $\beta x + y \perp_B x$.

B.3 It is *right (left)-unique* if and only if the space is smooth (strictly convex).

B.4 It is always *homogeneous.*

B.5 It is *right-additive* if and only if the space is smooth. If $\dim E = 2$, then it is *left-additive* if and only if E is strictly convex; and if $\dim E \geq 3$, then it is left-additive if and only if E is an inner product space.

B.6 It can always be *extended by the right*: if $x \perp_B y$, then there exists a closed hyperplane $P \subset E$ such that $y \in P$ and $x \perp_B P$.

B.7 If $\dim E = 2$ the B-orthogonality can always be *extended by the left* because it is homogeneous. If $\dim E \geq 3$, then this property is characteristic of inner product spaces and can be stated, equivalently, as follows: E is an inner product space if and only if for every $x \in E$ there exists a closed hyperplane $P \subset E$ such that $P \perp_B x$.

Properties of isosceles orthogonality

I.1 It is obviously *symmetric.*

I.2 It has the *existence* property.

I.3 It has the *uniqueness* property if and only if E is strictly convex. But any normed linear space E has the following properties:

I.3.1 For every $x, y \in E$, with $0 < \|y\| \leq \|x\|$, there exists a unique $\alpha \in \mathbb{R}$ such that $x \perp_I \alpha x + y$.

I.3.2 Let $x \in L \subset E$, where $\dim L = 2$. Then, for every $0 \leq \gamma \leq \|x\|$, there exists a unique (except in the sign) $y \in L$, with $\|y\| = \gamma$, such that $x \perp_I y$.

I.4 It is either *homogeneous* or *additive* only in inner product spaces.

I.5 *Extension* property implies homogeneity. Therefore, I-orthogonality can be extended only in inner product spaces. Also the following is a characteristic property of inner product spaces: for every $x \in S$ there exists a closed hyperplane $P \subset E$ such that $P \cap S \perp_I x$, where S denotes the unit sphere of E.

In the light of the above properties it is plain that in arbitrary normed linear spaces the behaviour of Birkhoff and isosceles orthogonality are very different. In fact, both orthogonalities coincide only in inner product spaces. More precisely, we have,

B-I E is an inner product space if and only if one of the next properties holds:

B-I.1 If $x, y \in S$ and $x \perp_B y$, then $x \perp_I y$.

B-I.2 If $x, y \in S$ and $x \perp_I y$, then $x \perp_B y$.

New properties in spaces of three or more dimensions

G. Marino and P. Pietramala [12] proved that a normed linear space of three or more dimensions, strictly convex and smooth, is an inner product space if and only if Birkhoff orthogonality is *left-additive for biorthogonal pairs of vectors (l.a.b.)*, namely,

$$\text{if } x \perp_B y,\ y \perp_B x,\ x \perp_B z \text{ and } y \perp_B z, \text{ then } x + y \perp_B z.$$

The next proposition removes the hypothesis of strict convexity and smoothness from Marino-Pietramala's result, and the proof follows an essentially different technique.

Proposition 1 *Let E be a real normed linear space of three or more dimensions. Then E is an inner product space if and only if Birkhoff orthogonality is left-additive for biorthogonal pairs of vectors.*

Proof. Obviously, if E is an inner product space, B-orthogonality is additive. Conversely, taking into account that a normed linear space of dimension greater or equal than three is an inner product space if and only if every three-dimensional linear subspace is an inner product space, we can assume that E is $\mathbb{R}^3$ endowed with a norm. We shall prove that for every $z \in E$ there exists a plane through the origin B-orthogonal to z. This is by property B.7 a characterization of inner product spaces.

Before going on to the proof, we must recall some easy results. For every $x, y, z \in \mathbb{R}^3$, linearly independent, we denote by $[x, y, z]$ the parallelepiped centred at 0 and with $\pm x$, $\pm y$ and $\pm z$ as centres of its faces, i.e.,

$$[x, y, z] = \{\lambda_1 x + \lambda_2 y + \lambda_3 z : \ |\lambda_i| \leq 1, \ i = 1, 2, 3\}.$$

Let $\{e_1, e_2, e_3\}$ be a basis of $\mathbb{R}^3$. If we assume that the volume of $[e_1, e_2, e_3]$ is 2^3, then

$$\text{vol}[x, y, z] = 2^3 |\det(x, y, z)|,$$

where $\det(x, y, z)$ denotes the determinant of the matrix of the coordinates of x, y, z, in the basis $\{e_1, e_2, e_3\}$.

Let S be the unit sphere of E. Then $S \subset [x, y, z]$ if and only if

$$\|\lambda_1 x + \lambda_2 y + \lambda_3 z\| \geq \max\{|\lambda_1|, |\lambda_2|, |\lambda_3|\},$$

for every $\lambda_1, \lambda_2, \lambda_3 \in \mathbb{R}$.

For every $x, y \in E$, linearly independent, we denote by $P(x, y)$ the two-dimensional linear subspace spanned by x and y.

Let us turn to the proof. Fix $z \in E$ and let $x, y \in E$ be such that $S \subset [x, y, \lambda z]$ for some $\lambda > 0$ and

$$\text{vol}[x, y, z] = \inf\{\text{vol}[x', y', z] : \ x', y' \in E, \ S \subset [x', y', \lambda' z] \text{ for some } \lambda' > 0\}.$$

For every $\mu \in \mathbb{R}$, $\|x + \mu z\| \geq 1$, because $S \subset [x, y, \lambda z]$. Let $\mu_0 \in \mathbb{R}$ be such that

$$\|x + \mu_0 z\| = \inf_{\mu \in \mathbb{R}} \|x + \mu z\| \geq 1,$$

and let $x_0 = (x + \mu_0 z)/\|x + \mu_0 z\|$. Then, $x_0 \perp_B z$ and, by property B.6, there exists $\alpha \in \mathbb{R}$ such that $x_0 \perp_B P(\alpha x + y, z)$.

Next, we shall show that $S \subset [x_0, \alpha x + y, \delta z]$ for some $\delta > 0$. Let $u \in S$ and let $\lambda_1, \lambda_2, \lambda_3 \in \mathbb{R}$ be such that $u = \lambda_1 x_0 + \lambda_2(\alpha x + y) + \lambda_3 z$. We only need to see that $|\lambda_1| \leq 1$ and $|\lambda_2| \leq 1$, and to observe that $|\lambda_3|$ is bounded. Taking into account that $x_0 \perp_B P(\alpha x + y, z)$, it follows that

$$1 = \|\lambda_1 x_0 + \lambda_2(\alpha x + y) + \lambda_3 z\| \geq |\lambda_1|.$$

On the other hand,

$$u = \left(\frac{\lambda_1}{\|x + \mu_0 z\|} + \lambda_2 \alpha\right) x + \lambda_2 y + \left(\frac{\lambda_1 \mu_0}{\|x + \mu_0 z\|} + \lambda_3\right) z,$$

and then $|\lambda_2| \leq 1$.

Therefore,

$$\text{vol}[x, y, z] \leq \text{vol}[x_0, \alpha x + y, z] = \frac{1}{\|x + \mu_0 z\|} \text{vol}[x, y, z],$$

and then $\|x + \mu_0 z\| = 1$.

In the same way as above, it can be shown that there also exists $\gamma_0 \in \mathbb{R}$ such that $\|y + \gamma_0 z\| = 1$ and $y + \gamma_0 z \perp_B z$.

Taking into account that for every $s, t \in \mathbb{R} \setminus \{0\}$,

$$\frac{s(x + \mu_0 z) + t(y + \gamma_0 z)}{\|s(x + \mu_0 z) + t(y + \gamma_0 z)\|} \in S \subset [x, y, \lambda z],$$

we get

$$\|s(x + \mu_0 z) + t(y + \gamma_0 z)\| \geq \max\{|s|, |t|\},$$

that is, $x + \mu_0 z \perp_B y + \gamma_0 z$ and $y + \gamma_0 z \perp_B x + \mu_0 z$.

Finally, from Birkhoff homogeneity and the hypothesis of l.a.b., it follows that $P(x + \mu_0 z, y + \gamma_0 z) \perp_B z$. □

Birkhoff orthogonality is in general non-symmetric. Therefore we can also consider the property of *right additivity for biorthogonal pairs of vectors (r.a.b.)*. This orthogonality is right-additive if and only if the space is smooth. Therefore in every smooth normed linear space B-orthogonality is r.a.b. As we shall see in the next section, there are non-smooth two-dimensional spaces in which B-orthogonality is r.a.b. But we do not know if in spaces of dimension equal or greater than three r.a.b. property for this orthogonality implies smoothness. As an example of three-dimensional space in which B-orthogonality is not r.a.b. we can consider $\mathbb{R}^3$ endowed with the norm $\|(x_1, x_2, x_3)\| = |x_1| + |x_2| + |x_3|$. The unit vectors $x = (1/2, 1/2, 0)$, $y = (1/2, -1/2, 0)$ and $z = (1/2, 0, 1/2)$ are such that $x \perp_B y$, $y \perp_B x$, $z \perp_B x$ and $z \perp_B y$, but z is not B-orthogonal to $x + y$. From this example we can easily imagine also unit spheres strictly convex in which B-orthogonality is not r.a.b.

Because isosceles orthogonality is symmetric, for this orthogonality the property of *additivity for biorthogonal pairs of vectors (a.b.)* is stated as follows:

if $x \perp_I y$, $x \perp_I z$ and $y \perp_I z$, then $x \perp_I y + z$, $y \perp_I x + z$ and $z \perp_I x + y$.

Lemma 1 *Let E be a real normed linear space of three or more dimensions and with unit sphere S. Let $x, y \in S$ be such that $x \perp_I y$. Then, for every three-dimensional subspace $F \subset E$, such that $x, y \in F$, there exists $z \in F \cap S$ such that $x \perp_I z$ and $y \perp_I z$.*

Proof. We can assume without loss of generality that E is $\mathbb{R}^3$ endowed with a norm and that $x = (0, 0, 1)$ and $y = (1, 0, 0)$. Let

$$u(\theta, \psi) = \frac{(\cos\theta \cos\psi,\ \sin\theta\cos\psi,\ \sin\psi)}{\|(\cos\theta\cos\psi,\ \sin\theta\cos\psi,\ \sin\psi)\|}, \quad 0 \le \theta \le 2\pi,\ -\frac{\pi}{2} \le \psi \le \frac{\pi}{2},$$

be a parametrization of S by polar coordinates. By property I.3.2, for every $0 \le \theta \le 2\pi$ there exists a unique $-\pi/2 < \psi(\theta) < \pi/2$ such that $x \perp_I u(\theta, \psi(\theta))$. Thus, $y = u(0, \psi(0))$, because we are assuming $x \perp_I y$. Next, we shall see that $\psi(\theta)$ is a continuous function. Let $0 \le \bar{\theta} \le 2\pi$ and let $\{\theta_n\}$ be a sequence in $[0, 2\pi]$ such that $\theta_n \to \bar{\theta}$. The sequence $\{\psi(\theta_n)\}$ is bounded and, therefore, there exists $\bar{\psi} \in [-\pi/2, \pi/2]$ and a subsequence $\{\theta_{n_k}\}$ such that $\psi(\theta_{n_k}) \to \bar{\psi}$. Then, $u(\theta_{n_k}, \psi(\theta_{n_k})) \to u(\bar{\theta}, \bar{\psi})$, because u is continuous. Since I-orthogonality is continuous and $x \perp_I u(\theta_{n_k}, \psi(\theta_{n_k}))$, then $x \perp_I u(\bar{\theta}, \bar{\psi})$. But also $x \perp_I u(\bar{\theta}, \psi(\bar{\theta}))$. By uniqueness property I.3.2, we get $\bar{\psi} = \psi(\bar{\theta})$ and therefore $\psi(\theta)$ is continuous in $\bar{\theta}$.

Now, the function

$$f(\theta) = \|y + u(\theta, \psi(\theta))\| - \|y - u(\theta, \psi(\theta))\|, \quad 0 \le \theta \le 2\pi,$$

is continuous and $f(0) = -f(\pi) = 2$. Therefore, there exists $0 < \tilde{\theta} < \pi$ such that $f(\tilde{\theta}) = 0$ and $z = u(\tilde{\theta}, \psi(\tilde{\theta}))$ is such that $x \perp_I z$ and $y \perp_I z$. □

Proposition 2 *Let E be a real normed linear space of three or more dimensions. Then E is an inner product space if and only if isosceles orthogonality is additive for biorthogonal pairs of vectors.*

Proof. We can assume that E is $\mathbb{R}^3$ endowed with a norm. Let S be the unit sphere. We shall prove that if I-orthogonality is a.b. then property B-I.2 holds.

Let $x, y \in S$ be such that $x \perp_I y$. By Lemma 1 there exists $z \in S$ such that $x \perp_I z$ and $y \perp_I z$. Next, we shall see by mathematical induction that $y \perp_I 2^n x$ and $y \perp_I 2^n z$ for every $n \in \mathbb{N}$. For $n = 0$ we have $y \perp_I x$ and $y \perp_I z$. Assume that $y \perp_I 2^{n-1}x$ and $y \perp_I 2^{n-1}z$. Taking into account that I-orthogonality has simplification property and that if $u \perp_I v$ then $u \perp_I -v$, it follows from the hypothesis that $y \perp_I 2^{n-1}x + 2^{n-1}z$ and $y \perp_I 2^{n-1}x - 2^{n-1}z$. We also have that $2^{n-1}x + 2^{n-1}z \perp_I 2^{n-1}x - 2^{n-1}z$, because $\|x\| = \|z\|$. Thus, $y \perp_I 2^n x$ and $y \perp_I 2^n z$.

The function $f(\lambda) = \|x + \lambda y\|$, $(\lambda \in \mathbb{R})$, is convex and $\|x + 2^{-n}y\| = \|x - 2^{-n}y\|$ for every $n \in \mathbb{N}$. Therefore,

$$\|x\| = f(0) = f\left(\frac{1}{2}(2^{-n}) + \frac{1}{2}(-2^{-n})\right) \le$$
$$\frac{1}{2}f(2^{-n}) + \frac{1}{2}f(-2^{-n}) = \|x + 2^{-n}y\|, \quad n \in \mathbb{N},$$

which implies that $x \perp_B y$. □

What happens in two-dimensional spaces?

By property B.3 we know that B-orthogonality is right-unique in smooth spaces. Therefore, in two-dimensional smooth spaces B-orthogonality is trivially l.a.b. Marino and Pietramala [12] also show that there are two-dimensional spaces in which B-orthogonality is not l.a.b. Such is the case of $\mathbb{R}^2$ endowed with the sup norm. On the other hand, in two-dimensional spaces, left-additivity of B-orthogonality is equivalent to be the space strictly convex. Therefore in any two-dimensional strictly convex space B-orthogonality is l.a.b. The next example shows that there are two-dimensional non-smooth and non-strictly convex spaces in which B-orthogonality is l.a.b.

Example 1 Let us consider the polygon of vertexes $\pm P_1, \pm P_2, \pm P_3$ and $\pm P_7$, where

$$P_j = \left(\cos\frac{j\pi}{8}, \sin\frac{j\pi}{8}\right).$$

Now, polish lightly the vertexes, except $\pm P_2$, to make the curve smooth except at these two points. In this way, we get a curve which is the unit sphere of a non-smooth and non-strictly convex norm in $\mathbb{R}^2$. In this space the B-orthogonality is l.a.b. because the only two points of the unit sphere in which the norm is non-smooth do not belong to any pair of B-biorthogonal vectors.

A normed linear space is strictly convex if and only if B-orthogonality is left-unique. Therefore B-orthogonality is r.a.b in every two-dimensional strictly convex space. On the other hand, assume that E is a two-dimensional normed linear space and that there exist $x, y, z \in E \setminus \{0\}$ such that $x \perp_B y$, $y \perp_B x$, $z \perp_B x$ and $z \perp_B y$. Then we can assume that $\|x\| = \|y\| = \|z\| = 1$. Let $\alpha, \beta \in \mathbb{R} \setminus \{0\}$ be such that $z = \alpha x + \beta y$. Then

$$1 = \|\alpha x + \beta y\| = |\alpha| \left\| x + \frac{\beta}{\alpha} y \right\| \geq |\alpha| = \|z - \beta y\| \geq 1.$$

Therefore, $|\alpha| = 1$. In the same way, we get $|\beta| = 1$ and then we can assume that $z = x + y$. But this implies by convexity that $\|x + \lambda y\| = \|y + \lambda x\| = 1$ for $0 \leq \lambda \leq 1$. Hence, in every two-dimensional space in which there is not a couple of points x, y in the above situation, B-orthogonality is r.a.b.; and of course we can find such spaces non-strictly convex and non-smooth. As an example of two-dimensional space in which B-orthogonality is not r.a.b. we can consider $\mathbb{R}^2$ endowed with the sup norm.

Proposition 3 shall show that with isosceles orthogonality the property of additivity for biorthogonal pairs of vectors has no sense in two-dimensional spaces, because in such spaces there are not three vectors different from zero, two by two I-orthogonal.

Lemma 2 [3] *Let S be the unit sphere of a real normed linear space of dimension at least two. Let $u, v \in S$ and $\alpha, \beta \in \mathbb{R}$ be such that $u \perp_I v$ and $u \perp_I \alpha u + \beta v$. Then,*

$$|\alpha| \leq \begin{cases} \min\left\{|\beta|(\|u+v\| - 1), \frac{(1-|\beta|)(2-\|u+v\|)}{\|u+v\|}\right\}, & \text{if } |\beta| \leq 1, \\ \max\left\{|\beta|(2 - \|u+v\|) - 1, \frac{(|\beta|-1)(2-\|u+v\|)}{\|u+v\|}\right\}, & \text{if } |\beta| \geq 1, \end{cases}$$

and the estimations are sharp.

Lemma 3 *Let S be the unit sphere of a real normed linear space of dimension at least two. If $u, v \in S$ and $\alpha, \beta \in \mathbb{R}$ are such that $u \perp_I v$, $u \perp_I \alpha u + \beta v$ and $\|\alpha u + \beta v\| \leq 1$, then $|\beta| \leq 1$.*

Proof. We can assume without loss of generality that $\alpha \geq 0$ and $\beta \geq 0$. Then,

$$\alpha + \beta = \|\alpha u + \beta v + \alpha(v - u)\| \leq \|\alpha u + \beta v\| + \alpha\|u + v\| \leq 1 + \alpha\|u + v\|,$$

and therefore,

$$\beta \leq \alpha(\|u + v\| - 1) + 1.$$

If $\alpha = 0$, we get $\beta \leq 1$. On the other hand, if $\alpha > 0$, then, taking into account that since $u \perp_I v$, then $1 \leq \|u+v\| \leq 2$, and that for every $t \in \mathbb{R}$, $t^2 - 2t + 2 > 0$ and $t^2 - 3t + 3 > 0$, it follows

$$\beta(2 - \|u+v\|) \leq [\alpha(\|u+v\| - 1) + 1](2 - \|u+v\|) < \min\{2 + (\alpha - 1)\|u+v\|, \alpha + 1\}.$$

Thus,

$$\alpha > \max\left\{\beta(2 - \|u+v\|) - 1, \frac{(\beta - 1)(2 - \|u+v\|)}{\|u+v\|}\right\}$$

and, from Lemma 2, we get $\beta < 1$. □

Proposition 3 *Let E be a real normed linear space of dimension two. If $x, y, z \in E$ are such that $x \perp_I y$, $x \perp_I z$ and $y \perp_I z$, then $\|x\|\|y\|\|z\| = 0$.*

Proof. We shall suppose that x, y and z are different from zero and we shall get a contradiction. We can assume without loss of generality that $\|x\| = 1$, $\|y\| \leq 1$ and $\|z\| \leq 1$. By property I.3.2, there exists $v \in E$ with $\|v\| = 1$ such that $x \perp_I v$. Then, x and v are linearly independent and there exist $\alpha_i, \beta_i \in \mathbb{R}$, $i = 1, 2$, such that $y = \alpha_1 x + \beta_1 v$ and $z = \alpha_2 x + \beta_2 v$. By Lemma 3 we know that $|\beta_i| \leq 1$, $i = 1, 2$. We can assume that $0 < \beta_2 \leq \beta_1$. We shall assume also that $\alpha_1 \geq 0$. The proof in the case $\alpha_1 < 0$ is completely similar. It follows from Lemma 2 that if $\|x+v\| = 1$ or $\|x+v\| = 2$, then $\alpha_1 = \alpha_2 = 0$, which is an absurd because $y \perp_I z$. Therefore, $1 < \|x+v\| < 2$. Again, by Lemma 2, we get $\alpha_1 < \beta_1$ and $\alpha_1 + \beta_1 \leq 1$. Because $y \perp_I z$ we have also that $\|y - z\| \geq \max\{\|y\|, \|z\|\}$.

From Lemma 2 and the identity

$$\beta_1(x+v) = (\beta_1 - \alpha_1)x + y$$

it follows

$$\|y\| \geq \alpha_1 + \beta_1(\|x+v\| - 1). \tag{1}$$

Now, we must consider several cases.

Case 1. Assume that $\alpha_1\beta_2 - \alpha_2\beta_1 \geq 0$. From the identity

$$y - z = \frac{1}{\beta_1}\big((\beta_1 - \beta_2)y + (\alpha_1\beta_2 - \alpha_2\beta_1)x\big),$$

Lemma 2 and (1), it follows

$$\begin{aligned}
\|y\| \leq \|y - z\| &\leq \frac{1}{\beta_1}\big((\beta_1 - \beta_2)\|y\| + \alpha_1\beta_2 - \alpha_2\beta_1\big)\\
&\leq \frac{1}{\beta_1}\big((\beta_1 - \beta_2)\|y\| + \alpha_1\beta_2 + \beta_1\beta_2(\|x+v\| - 1)\big)\\
&= \frac{1}{\beta_1}\Big(\beta_1\|y\| + \beta_2\big(\alpha_1 + \beta_1(\|x+v\| - 1) - \|y\|\big)\Big)\\
&\leq \|y\|.
\end{aligned}$$

Therefore,

$$\|y\| = \alpha_1 + \beta_1(\|x+v\| - 1). \tag{2}$$

Let us consider the convex function

$$f(\lambda) = \left\| x + \lambda \left(\frac{x+v}{\|x+v\|} - x \right) \right\|, \quad \lambda \in \mathbb{R}.$$

Then $f(0) = f(1) = 1$ and, by (2),

$$\begin{aligned} f\left(\frac{\beta_1\|x+v\|}{\|y\|}\right) &= \frac{1}{\|y\|}\left\|\big(\|y\| + \beta_1(1 - \|x+v\|)\big)x + \beta_1 v\right\| \\ &= \frac{1}{\|y\|}\left\|\alpha_1 x + \beta_1 v\right\| = 1. \end{aligned}$$

But, $\beta_1\|x+v\|/\|y\| = 1 + (\beta_1 - \alpha_1)/\|y\| > 1$. Therefore, $f(\lambda) = 1$ for $0 \leq \lambda \leq \beta_1\|x+v\|/\|y\|$. Then,

$$\begin{aligned} 1 - \alpha_1 + \beta_1(\|x+v\| - 1) &\geq \left\|(1 - \alpha_1 - \beta_1)x + \beta_1(x - v)\right\| \\ &= \|x - y\| = \|x + y\| \\ &= \big(1 + \alpha_1 + \beta_1(\|x+v\| - 1)\big) f\left(\frac{\beta_1\|x+v\|}{1 + \alpha_1 + \beta_1(\|x+v\| - 1)}\right) \\ &= 1 + \alpha_1 + \beta_1(\|x+v\| - 1), \end{aligned}$$

because

$$0 < \frac{\beta_1\|x+v\|}{1 + \alpha_1 + \beta_1(\|x+v\| - 1)} \leq 1.$$

Hence $\alpha_1 = 0$, because we are assuming $\alpha_1 \geq 0$. Finally, from (2) we get $\beta_1 = \|y\| = \beta_1(\|x+v\| - 1) < \beta_1$, which is absurd.

Case 2. Assume now that $\alpha_1\beta_2 - \alpha_2\beta_1 < 0$. Then $\alpha_2 > 0$. From the identity

$$(\alpha_1 + \beta_1)v = \alpha_1(v - x) + y,$$

it follows

$$\|y\| \geq \beta_1 + \alpha_1(1 - \|x+v\|). \tag{3}$$

Case 2.1. Assume that $0 < \alpha_2 \leq \alpha_1$. Then

$$\begin{aligned} \|y\| \leq \|y - z\| &= \frac{1}{\alpha_1}\left\|(\alpha_2\beta_1 - \alpha_1\beta_2)v + (\alpha_1 - \alpha_2)y\right\| \\ &\leq \frac{1}{\alpha_1}\big(\alpha_2\beta_1 - \alpha_1\beta_2 + (\alpha_1 - \alpha_2)\|y\|\big) \\ &= \frac{\alpha_2\beta_1 - \alpha_1\beta_2 - \alpha_2\|y\|}{\alpha_1} + \|y\|. \end{aligned}$$

and, we get from (3)

$$\begin{aligned} 0 \le \alpha_2\beta_1 - \alpha_1\beta_2 - \alpha_2\|y\| &\le \alpha_2\beta_1 - \alpha_1\beta_2 - \alpha_2\big(\beta_1 + \alpha_1(1 - \|x+v\|)\big) \\ &= -\alpha_1\big(\beta_2 + \alpha_2(1 - \|x+v\|)\big). \end{aligned}$$

From Lemma 2 it follows

$$0 \ge \beta_2 + \alpha_2(1 - \|x+v\|) \ge \frac{\alpha_2}{\|x+v\| - 1}\big(1 - (\|x+v\| - 1)^2\big) > 0,$$

which is a contradiction.

Case 2.2. Assume now that $\alpha_2 > \alpha_1 \ge 0$. If $\beta_1 - \beta_2 + \alpha_1 - \alpha_2 \ge 0$, then from (3) it follows

$$\begin{aligned} \|y\| \le \|y - z\| &= \|(\alpha_2 - \alpha_1)(v - x) + (\beta_1 - \beta_2 + \alpha_1 - \alpha_2)v\| \\ &\le (\alpha_2 - \alpha_1)\|x+v\| + \beta_1 - \beta_2 + \alpha_1 - \alpha_2 \\ &= \beta_1 + \alpha_1(1 - \|x+v\|) + \alpha_2(\|x+v\| - 1) - \beta_2 \\ &\le \|y\| + \alpha_2(\|x+v\| - 1) - \beta_2, \end{aligned}$$

and from Lemma 2, we get the contradiction

$$\beta_2 \le \alpha_2(\|x+v\| - 1) \le \beta_2(\|x+v\| - 1)^2 < \beta_2.$$

On the other hand, if $\beta_1 - \beta_2 + \alpha_1 - \alpha_2 < 0$, then by Lemma 2 and (1) it follows

$$\begin{aligned} \|y\| \le \|y - z\| &= \|(\beta_1 - \beta_2)(v - x) + (\beta_1 - \beta_2 + \alpha_1 - \alpha_2)x\| \\ &\le (\beta_1 - \beta_2)\|x+v\| + \beta_2 - \beta_1 + \alpha_2 - \alpha_1 \\ &= \beta_1(\|x+v\| - 1) + \beta_2(1 - \|x+v\|) + \alpha_2 - \alpha_1 \\ &\le \beta_1(\|x+v\| - 1) - \alpha_1 \\ &\le \|y\| - 2\alpha_1. \end{aligned}$$

Therefore, $\alpha_1 \le 0$. But we are assuming $\alpha_1 \ge 0$. Hence $\alpha_1 = 0$ and the above inequalities become identities. Then $\|y\| = \beta_1(\|x+v\| - 1)$, and we get a contradiction as at the end of Case 1. □

References

[1] ALONSO, J.; BENÍTEZ, C., *Orthogonality in normed linear spaces: A survey. Part I: Main properties.* Extracta Math. **3** (1) (1988), 1–15.

[2] ALONSO, J.; BENÍTEZ, C., *Orthogonality in normed linear spaces: A survey. Part II: Relations between main orthogonalities.* Extracta Math. **4** (3) (1989), 121–131.

[3] ALONSO, J., *Uniqueness properties of isosceles orthogonality in normed linear spaces.* Ann. Sci. Math. Québec **18** (1994), no. 1, 25–38.

[4] BIRKHOFF, G., *Orthogonality in linear metric spaces.* Duke Math. Jour. **1** (1935), 169–172.

[5] DAY, M.M., *Some characterizations of inner product spaces.* Trans. Amer. Math. Soc. **62** (1947), 320–337.

[6] JAMES, R.C., *Orthogonality in normed linear spaces.* Duke Math. Jour. **12** (1945), 291–301.

[7] JAMES, R.C., *Inner products in normed linear spaces.* Bull. Amer. Math. Soc. **53** (1947), 559–566.

[8] JAMES, R.C., *Orthogonality and linear functionals in normed linear spaces.* Trans. Amer. Math. Soc. **61** (1947), 265–292.

[9] KAPOOR, O.P.; PRASAD, J.; *Orthogonality and characterizations of inner product spaces.* Bull. Aust. Math. Soc. **19** (1978), 403–416.

[10] KAPOOR, O.P.; MATHUR, S.B., *Some geometric characterizations of inner product spaces.* Bull. Aust. Math. Soc. **24** (1981), 239–246.

[11] LIN, PEI-KEE, *A remark on the Singer-orthogonality in normed linear spaces.* Math. Nachr. **160** (1993), 325–328.

[12] MARINO, G.; PIETRAMALA, P., *A note about inner products on Banach Spaces.* Bollettino U.M.I. (7) **1**-A (1987), 425–427.

Dpto. Matemáticas, Univ. de Extremadura, 06071 Badajoz, Spain
E-mail: jalonso@ba.unex.es Fax: 34-24-272911

ALI R AMIR-MOEZ

Quadratic norms and inner products

1. NOTATION: A unitary space of dimension n will be denoted by E_n. Vectors and scolars will be denoted by Greek and Latin letters respectively. The inner product of ξ and η is represented by (ξ, η). Thus the Euclidean norm of ξ will be defined by $\|\xi\|^2 = (\xi, \xi)$. Matrices will be denoted by capital letters. In what follows all bases (coordinate systems) are orthonormal [2].

2. QUADRATIC NORMS: Let A be positive definite linear transformation on E_n whose matrix with respect to an orthonormal basis is

$$A = \begin{pmatrix} a_{11} & \dots & a_{1n} \\ \dots & & \dots \\ a_{n1} & \dots & a_{nn} \end{pmatrix}. \tag{1}$$

Let $\xi = (x_1, \dots, x_n)$. Then

$$(x \dots x_n) \begin{pmatrix} a_{11} & \dots & a_{1n} \\ \dots & & \dots \\ a_{n1} & \dots & a_{nn} \end{pmatrix} \begin{pmatrix} x_1^* \\ \vdots \\ x_n^* \end{pmatrix} = (a). \tag{2}$$

is a matrix of one positive element. We define the quadratic norm of ξ by

$$\|\xi\|_Q = a^{\frac{1}{2}}. \tag{3}$$

One can easily show that

$$\|\xi\|_Q^2 = (A\xi, \xi). \tag{4}$$

If we replace A by a Hermitian transformation, then (3) may be imaginary. In this case (3) may be called pseudo norm.

3. THEOREM (Apollonius): Let α and β be any two vectors in E_n. Then

$$\|\alpha + \beta\|^2 + \|\alpha - \beta\|^2 = 2\left[\|\alpha\|^2 + \|\beta\|^2\right]. \tag{5}$$

This actually means that in any parallelogram the sum of the squares of the diagonals is equal to the sum of the squares of its four sides.

The vector proof of (5) is quite simple and will be omitted.

4. THEOREM (P. Jordan and J. Von Neumann): Let S be a normed linear space. Let $\alpha, \beta \in S$ which satisfy

$$\|\alpha+\beta\|^2+\|\alpha-\beta\|^2=2\left[\|\alpha\|^2+\|\beta\|^2\right]. \tag{6}$$

Then S is an inner product space. (This is the converse of Apollonius Theorem).

For the proof one may refer to [3].

Let (6) be satisfied in S. Then the inner product can be defined by

$$(\alpha,\beta)=\frac{1}{4}\left[\|\alpha+\beta\|^2-\|\alpha-\beta\|^2\right]. \tag{7}$$

5. THEOREM (Quadratic inner product): The quadratic norm satisfies (6). This means that the quadratic norm induces an inner product.

Proof: We shall test the Appolonius, Jordan, Von Neuman property of this norm [3]. Let $\xi,\eta\in E_n$. Then

$$\|\xi+\eta\|_Q^2+\|\xi-\eta\|_Q^2=(A[\xi+\eta],\xi+\eta)+(A[\xi-\eta],\xi-\eta)=2\left[(A\xi,\xi)+(A\xi,\eta)\right].$$

Therefore one can define the quadratic inner product as

$$(\xi,\eta)_Q=\frac{1}{4}\left[\|\xi+\eta\|_Q^2-\|\xi-\eta\|_Q^2\right]=(A\xi,\eta). \tag{8}$$

So we may define $\{\xi,\eta\}$ as quadratic orthogonal if $(A\xi,\eta)=0$. In what follows we shall use the word orthogonal for quadratic orthogonal. We shall give an example for application of this orthogonality.

If A is Hermitian, then $(A\xi,\xi)$ will be a pseudo norm. In this case (8) is also satisfied and $(A\xi,\xi)$ will be called pseudo inner product.

6. SOME APPLICATIONS: Through the quadratic norm one can study some properties of quadric hypersurfaces. These surfaces are very similar to spheres. In what follows we shall choose A to be a Hermitian matrix with real entries.

7. THEOREM: Let $\{\alpha_1,\ldots,\alpha_{n-1}\}$ be linearly independent in E_n, and let $\alpha_n=-(\alpha_1+\ldots+\alpha_{n-1})$. We have chosen the origin at the centroid of the convex hull of $\{\alpha_1,\ldots,\alpha_n\}$. Let ξ vary so that

$$\sum_{i=1}^{n}\|\xi-\alpha_i\|_Q^2=k^2. \tag{9}$$

Then ξ describes a quadric hypersurface.

Proof: Note that (9), by (4), can be written as

$$\sum_{i=1}^{n}(A[\xi-\alpha_i],\xi-\alpha_i)=k^2. \tag{10}$$

We shall expand (10) and get

$$n(A\xi,\xi) + \sum_{i=1}^{n}(A\alpha_i,\alpha_i) - 2\left(A\xi, \sum_{i=1}^{n}\alpha_i\right) = k^2. \tag{11}$$

Since $\sum_{i=1}^{n}\alpha_i = 0$, (11) will become $(A\xi,\xi) = \frac{1}{n}[k^2 - \sum_{i=1}^{n}(A\alpha_i,\alpha_i)]$, which is a central quadric hypersurface in E_n, with the center at the origin.

8. ORTHOGONALITY: By (4), the quadratic norm is defined by $\|\xi\|_Q^2 = (A\xi,\xi)$, where A is a Hermitian matrix with real entries. (Note that this may be a pseudo-norm).

9. THEOREM: Let $\{\alpha_1,\ldots,\alpha_{n-1}\}$ be linearly independent and $\alpha_n = -(\alpha_1+,\ldots,+\alpha_{n-1})$. Let ξ vary such that

$$(A[\xi-\alpha_i],\xi-\alpha_j) = 0, i<j, i=1,\ldots,n-1. \tag{12}$$

Then ξ describes a quadric surface.

Proof: Adding the equations of (12), we obtain

$$\frac{n(n-1)}{2}(A\xi,\xi) + \sum_{i<j}(A\alpha_i,\alpha_j) = 0.$$

This implies that

$$\frac{n(n-1)}{2}(A\xi,\xi) = \frac{1}{2}\sum_{i=1}^{n}(A\alpha_i,\alpha_i).$$

Consequently the equation of the quadric is

$$(A\xi,\xi) = \frac{1}{n(n-1)}\sum_{i=1}^{n}(A\alpha_i,\alpha_i).$$

References

[1] A.R. Amir-Moéz, Vector Geometry of Spheres, *Journal of Science*, Vol. 40, No. 1, February 1988.

[2] A.R. Amir-Moéz and A.L. Fass, *Elements of Linear Spaces*, Pergamon Press, Oxford, England, 1962.

[3] P. Jordan and J. Von Neumann, On Inner Products in Linear Metric Spaces, *Ann. of Math.*, (2) **36** (1935), pp. 719–723.

[4] Themistocles M. Rassias, *New Characterizations of Inner Product spaces*, Bull.Sc. math, 2^eserie,108,1984,p.95-99.

Texas Tech University, Lubbock, TX 79409-1042 (USA)

JACEK CHMIELIŃKSI

On the stability of isometric operators for Hilbert spaces

Introduction

Let E_1 and E_2 be (real or complex) inner product spaces. Let $\langle\cdot|\cdot\rangle_1$ and $\langle\cdot|\cdot\rangle_2$ denote inner products in E_1 and E_2, respectively, and $\|\cdot\|_1$, $\|\cdot\|_2$ norms associated with them. A mapping $T : E_1 \to E_2$ is called *isometric* if it is a solution of the *orthogonality equation*:

$$\langle T(x)|T(y)\rangle_2 = \langle x|y\rangle_1 \quad \text{for } x, y \in E_1. \tag{1}$$

It is well known that an isometric operator T is linear and injective (cf. [5]). If, additionally, it is surjective then it is called *unitary* mapping (*unitary isomorphism*) from E_1 onto E_2 and the spaces E_1 and E_2 are called *isomorphic* (or *equivalent*). Isometries for Hilbert spaces are widely investigated in [6].

In the present paper we investigate a class of approximate solutions of functional equation (1) (*approximate isometric operators*). This class is defined (following definitions of approximate homomorphisms and approximate isometries given by S.M Ulam and D.H. Hyers – cf. [3], [4]) by the functional inequality

$$|\langle T(x)|T(y)\rangle_2 - \langle x|y\rangle_1| \le \varepsilon \quad \text{for } x, y \in E_1, \tag{2}$$

where $T : E_1 \to E_2$ and ε is a nonnegative constant.

We are going to prove the stability of equation (1) with respect to the class of approximate solutions defined by (2) (Hyers-Ulam stability) in the case where the target space E_2 is a Hilbert space. In the second part of the paper we prove that the existence of an (not necessarily linear) approximate isometric and "almost surjective", in a sense, mapping implies that E_1 and E_2 are isomorphic Hilbert spaces.

The stability of equation (1) as well as of its generalization (the Wigner equation – cf. [6])

$$|\langle T(x)|T(y)\rangle| = |\langle x|y\rangle| \quad \text{for } x, y \in E$$

in the case where $T : E \to E$ for E being a real Hilbert space was investigated in [1] and [2].

Stability

For a mapping $T : E_1 \to E_2$ we define a (Hyers') sequence of mappings $T_n : E_1 \to E_2$ ($n = 0, 1, 2, \ldots$) by

$$T_n(x) := 2^{-n}T(2^n x) \quad \text{for } x \in E_1. \tag{3}$$

AMS(1991) Subject classification: Primary 39B72, Secondary 47H19.

If T satisfies (2) we can establish some properties of the sequence T_n. Namely, we have the following two lemmas. Here and subsequently, $\Re(z)$ stands for the real part of z.

Lemma 1. *If $T : E_1 \to E_2$ satisfies (2) and $T_n : E_1 \to E_2$ $(n = 0, 1, 2, \ldots)$ are defined by (3), then for each $x \in E_1$ and arbitrary $m, n \in \mathbf{N}$:*

$$\|x\|_1^2 - \frac{\varepsilon}{2^{n+m}} \le \Re\left(\langle T_n(x)|T_m(x)\rangle_2\right) \le \|x\|_1^2 + \frac{\varepsilon}{2^{n+m}}, \tag{4}$$

$$\|x\|_1^2 - \frac{\varepsilon}{4^n} \le \|T_n(x)\|_2^2 \le \|x\|_1^2 + \frac{\varepsilon}{4^n}, \tag{5}$$

$$\|x\|_1^2 - \varepsilon \le \|T(x)\|_2^2 \le \|x\|_1^2 + \varepsilon. \tag{6}$$

Proof: From (2) we have

$$|\Re\left(\langle T(x)|T(y)\rangle_2 - \langle x|y\rangle_1\right)| \le |\langle T(x)|T(y)\rangle_2 - \langle x|y\rangle_1| \le \varepsilon$$

for $x, y \in E_1$. Putting $2^n x$ and $2^m x$ in place of x and y, respectively, we obtain

$$\left|\Re\left(\langle T_n(x)|T_m(x)\rangle_2\right) - \|x\|_1^2\right| \le \frac{\varepsilon}{2^{n+m}},$$

whence (4) follows. Putting there $n = m$ and then $n = m = 0$, one obtains (5) and (6). □

Lemma 2. *If $T : E_1 \to E_2$ satisfies (2) and $T_n : E_1 \to E_2$ $(n = 0, 1, 2, \ldots)$ are defined by (3), then for arbitrarily fixed $x \in E_1$ the sequence $\{T_n(x)\}_{n=1}^{\infty}$ is a Cauchy sequence in E_2.*

Proof: For $x \in E_1$ and $n, m \in \mathbf{N}$ there is

$$\|T_n(x) - T_m(x)\|_2^2 = \|T_n(x)\|_2^2 + \|T_m(x)\|_2^2 - 2\Re\left(\langle T_n(x)|T_m(x)\rangle_2\right).$$

Applying (4) and (5) one gets

$$\|T_n(x) - T_m(x)\|_2 \le \sqrt{\frac{\varepsilon}{2^n} + \frac{\varepsilon}{2^m} + \frac{2\varepsilon}{2^{n+m}}} \le \sqrt{\frac{2\varepsilon}{2^{N-1}}} \quad \text{for } n, m \ge N,$$

which means that $\{T_n(x)\}$ is a Cauchy sequence. □

We can now formulate the main result in this section. This theorem expresses the Hyers-Ulam stability of equation (1).

Theorem 1. *Let E_1 be an inner product space and let E_2 be a Hilbert space. Then, for an approximate isometric mapping $T : E_1 \to E_2$ (a solution of (2)) there exists a unique linear isometric mapping $T_* : E_1 \to E_2$ (a solution of (1)) such that*

$$\|T(x) - T_*(x)\|_2 \le \sqrt{\varepsilon} \quad \text{for } x \in E_1. \tag{7}$$

Proof: We define

$$T_*(x) := \lim_{n\to\infty} T_n(x) \quad \text{for } x \in E_1$$

(it is Lemma 2 and the completness of E_2 that makes the definition allowable). Let $x, y \in E_1$ and $n \in \mathbf{N}$. Replacing in (2) x and y by $2^n x$ and $2^n y$, respectively, we get

$$|\langle T_n(x)|T_n(y)\rangle_2 - \langle x|y\rangle_1| \le \frac{\varepsilon}{4^n},$$

whence, on letting $n \to \infty$, we obtain

$$|\langle T_*(x)|T_*(y)\rangle_2 - \langle x|y\rangle_1| = 0.$$

Thus T_* satisfies (1) and consequently it is a linear isometry. For $x \in E_1$ and $n \in \mathbf{N}$ we have also

$$\|T_n(x) - T(x)\|_2^2 = \|T_n(x)\|_2^2 + \|T(x)\|_2^2 - 2\Re\left(\langle T_n(x)|T(x)\rangle_2\right),$$

whence, using (5), (6) and (4) with $m = 0$

$$\|T_n(x) - T(x)\|_2^2 \le \frac{\varepsilon}{4^n} + \varepsilon + \frac{\varepsilon}{2^n}.$$

Letting $n \to \infty$ we obtain

$$\|T_*(x) - T(x)\|_2 \le \sqrt{\varepsilon}$$

and (7) is proved.

Finally, assume that there exists a (linear) mapping $T'_* : E_1 \to E_2$ satisfying (1) and such that (with some $\delta \ge 0$) $\|T'_*(x) - T(x)\|_2 \le \delta$ for $x \in E_1$. Replacing in the last inequality x by $2^n x$ we obtain

$$\|T'_*(x) - T_n(x)\|_2 \le \frac{\delta}{2^n},$$

which implies

$$T'_*(x) = \lim_{n\to\infty} T_n(x) = T_*(x)$$

and proves uniqueness in the assertion of the theorem. □

Remark 1. *The constant $\sqrt{\varepsilon}$ is optimal (it cannot be replaced by a smaller one).*

This remark is based on the following example.

Example 1. Let $E_1 = E_2 = l^2$ (l^2 is a Hilbert space of infinite sequences $x = (t_1, t_2, \ldots)$ such that $\sum_{i=1}^{\infty} |t_i|^2 < \infty$ with the inner product defined by $\langle x|y\rangle := \sum_{i=1}^{\infty} t_i \overline{s_i}$ for $x = (t_1, t_2, \ldots)$, $y = (s_1, s_2, \ldots) \in l^2$). Define a mapping $T : l^2 \to l^2$ by the formula

$$T(t_1, t_2, t_3, \ldots) = (\sqrt{\varepsilon}, t_1, t_2, t_3, \ldots).$$

T satisfies (2). Indeed,

$$\langle T(x)|T(y)\rangle_{l^2} = \varepsilon + \sum_{i=1}^{\infty} t_i \overline{s_i} = \langle x|y\rangle_{l^2} + \varepsilon.$$

On the other hand for a T_* satisfying (1) we have $T_*(0) = 0$ and hence

$$\|T(0) - T_*(0)\|_{l^2} = \|(\sqrt{\varepsilon}, 0, 0, \ldots)\|_{l^2} = \sqrt{\varepsilon}.$$

Thus $\sqrt{\varepsilon}$ is an optimal constant.

Corollary 1. *Under assumptions of Theorem 1 if $T : E_1 \to E_2$ satisfies* (2) *then T is an approximate additive mapping.*

Proof: Let us take the operator T_* from the assertion of the Theorem 1. T_* is additive; thus we have

$$\|T(x+y)-T(x)-T(y)\|_2 = \|T(x+y)-T_*(x+y)-T(x)+T_*(x)-T(y)+T_*(y)\| \leq 3\sqrt{\varepsilon}.$$

□

The same assertion can be obtained without the assumption of the compactness of E_2.

Equivalence of Hilbert spaces

The second part of the paper deals with the connection between the existence of some approximate unitary operator between two Hilbert spaces and the equivalence of them. We will discuss some kind of surjectivity. We begin with a result which has a more general setting. Recall that for normed spaces X and Y a mapping $f : X \to Y$ is called an *isometry* iff $\|f(x) - f(y)\|_Y = \|x - y\|_X$ for all $x, y \in X$. A mapping f is *δ-surjective* ($\delta > 0$) iff for each $y \in Y$ there exists an $x \in X$ such that $\|f(x) - y\| \leq \delta$.

Proposition 1. *Suppose that X is a Banach space and Y is a strictly convex space. If an isometry $f : X \to Y$ is δ-surjective (with some $\delta \geq 0$), then it is surjective.*

Proof: 1. At first we prove that

$$f(\mathrm{conv}\,\{x_1, x_2\}) = \mathrm{conv}\{f(x_1), f(x_2)\} \quad \text{for } x_1, x_2 \in X. \tag{8}$$

Here conv $\{p, q\}$ ($p, q \in X$ or $p, q \in Y$) denotes a segment $\{\lambda p + (1-\lambda)p : \lambda \in [0, 1]\}$ in X or Y.
Let $y_1 := f(x_1)$ and $y_2 := f(x_2)$ and let $\lambda \in (0, 1)$. If $x := \lambda x_1 + (1-\lambda)x_2$ and $y := f(x)$ then $x - x_1 = (1-\lambda)(x_2 - x_1)$ and $x - x_2 = \lambda(x_1 - x_2)$, whence $\|x - x_1\| = (1-\lambda)\|x_2 - x_1\|$ and $\|x - x_2\| = \lambda\|x_1 - x_2\|$. From the fact that f is an isometry we have also

$$\|y - y_1\| = \|x - x_1\|, \quad \|y - y_2\| = \|x - x_2\|, \quad \|y_1 - y_2\| = \|x_1 - x_2\|,$$

which yields

$$\|y - y_1\| = (1-\lambda)\|y_2 - y_1\| \quad \text{and} \quad \|y - y_2\| = \lambda\|y_2 - y_1\| \tag{9}$$

and then

$$\|y_1 - y\| + \|y - y_2\| = \|y_1 - y_2\|.$$

As Y is a strictly convex space we get from the above equality

$$y_1 - y = \mu(y - y_2) \quad \text{for some } \mu > 0. \tag{10}$$

From (9) and (10) we have, respectively, $\|y - y_1\| = \frac{1-\lambda}{\lambda}\|y - y_2\|$ and $\|y - y_1\| = \mu\|y - y_2\|$. Thus $\mu = \frac{1-\lambda}{\lambda}$ and (10) becomes $y_1 - y = \frac{1-\lambda}{\lambda}(y - y_2)$, whence $y = \lambda y_1 + (1-\lambda)y_2$ and (8) is proved.

2. Suppose, contrary to our claim, that there exists $y_0 \in Y \setminus f(X)$. If it were true that

$$\forall \eta > 0 \ \exists x_\eta \in X \text{ such that } \|f(x_\eta) - y_0\| \leq \eta \tag{11}$$

then, in particular, there would exist a sequence $\{x_n\}$ in X such that $\|f(x_n) - y_0\| \leq \frac{1}{n}$. It would mean that

$$\|x_n - x_m\| = \|f(x_n) - f(x_m)\| \leq \|f(x_n) - y_0\| + \|f(x_m) - y_0\| \leq \frac{1}{n} + \frac{1}{m},$$

i.e., $\{x_n\}$ is a Cauchy sequence, convergent to some $x_0 \in X$. From the continuity of f we derive that

$$y_0 = \lim_{n\to\infty} f(x_n) = f(x_0),$$

whence $y_0 \in f(X)$, a contradiction. It means that (11) fails and we have

$$\text{if } y_0 \in Y \setminus f(X), \text{ then } \exists \delta_0 > 0 \text{ such that } \forall x \in X \ \|f(x) - y_0\| \geq \delta_0. \tag{12}$$

3. As f is δ-surjective, δ_0 has to be less than or equal to δ. From δ-surjectivity of f we obtain the existence of an $x_1 \in X$ such that for $y_1 := f(x_1)$ there is $\|y_1 - y_0\| \leq \delta$. Let N be such that $N > \frac{\delta}{\delta_0} \geq 1$ and let $y_2 := y_1 + N(y_0 - y_1)$. Thus there exists an $x_3 \in X$ such that for $y_3 := f(x_3)$ there is $\|y_3 - y_2\| \leq \delta$.
Now, if $y := y_1 + \frac{1}{N}(y_3 - y_1) = (1 - \frac{1}{N})y_1 + \frac{1}{N}y_3$, then from what we have proved in the first step of the proof, there exists an $x \in \operatorname{conv}\{x_1, x_2\}$ such that $f(x) = y$. Consequently we have

$$y - y_0 = (y_1 - y_0) + (y - y_1) = \frac{1}{N}(y_1 - y_2) + \frac{1}{N}(y_3 - y_1) = \frac{1}{N}(y_3 - y_2),$$

whence

$$\|y - y_0\| = \frac{1}{N}\|y_3 - y_2\| \leq \frac{1}{N}\delta < \frac{\delta_0}{\delta} \cdot \delta = \delta_0$$

and we get that there exists an $y \in f(X)$ such that $\|y - y_0\| < \delta_0$, which contradicts (12). Thus $f(X) = Y$.

□

Now we apply the above result as well as Theorem 1 to prove the following

Theorem 2. *Let E_1, E_2 be Hilbert spaces. E_1 and E_2 are equivalent (isomorphic) if and only if there exists a δ-surjective (with some $\delta \geq 0$) mapping $T : E_1 \to E_2$ satisfying* (2) *(with any $\varepsilon \geq 0$).*

Proof: If E_1 and E_2 are isomorphic then, obviously, there exists a surjective mapping $T : E_1 \to E_2$ satisfying, in fact, (1). Thus what we need to prove is the 'if' part of the assertion. If there exists a mapping $T : E_1 \to E_2$ satisfying (2) then, according to Theorem 1, there exists a linear isometry $T_* : E_1 \to E_2$ satisfying (1) and such that $\|T_*(x) - T(x)\|_2 \leq \sqrt{\varepsilon}$ for $x \in E_1$. From the δ-surjectivity of T we get that T_* is $(\sqrt{\varepsilon} + \delta)$-surjective. Indeed, for a $y \in E_2$ there exists an $x \in E_1$ such that $\|T(x) - y\|_2 \leq \delta$, whence

$$\|T_*(x) - y\|_2 \leq \|T_*(x) - T(x)\|_2 + \|T(x) - y\|_2 \leq \delta + \sqrt{\varepsilon}.$$

T_* is a $(\delta + \sqrt{\varepsilon})$-surjective isometry, thus – according to Proposition 1 (a Hilbert space E_2 is, by all means, strictly convex) – T_* is surjective. To sum up, T_* is a linear isomorphism from E_1 onto E_2, satisfying (1) and that is exactly what we need to state the equivalence of E_1 and E_2.

□

Remark 2. *The assumption of δ-surjectivity is not superfluous.*

The above remark is based on the final example.

Example 2. Let $E_1 = l_n^2$ (n-dimensional Euclidean space) and $E_2 = l^2$. Obviously the two Hilbert spaces are not isomorphic. However, an embedding $T : l_n^2 \to l^2$ defined by

$$T(t_1, t_2, \ldots, t_n) := (t_1, t_2, \ldots, t_n, 0, 0, \ldots)$$

satisfies (2) (actually, it satisfies (1)). Notice that if $y_k = (0, \ldots, 0 \underset{\text{place}}{\overset{(n+1)-st}{,}} -k,\ 0, \ldots) \in l^2$ then for any $x = (t_1, t_2, \ldots, t_n) \in l_n^2$,

$$\|T(x) - y_k\|_{l^2} = \|(t_1, t_2, \ldots, t_n, k, 0, 0, \ldots)\|_{l^2} = \sqrt{k^2 + \sum_{i=1}^{n} t_i^2} \geq k.$$

Thus T cannot be a δ-surjective mapping.

References

[1] Chmieliński, J., *On the Hyers–Ulam stability of the generalized orthogonality equation in real Hilbert spaces*, in: *Stability of Mappings of Hyers-Ulam Type*, Th. M. Rassias, J. Tabor (eds.), Hadronic Press, Palm Harbor, 1994, 31-41.

[2] Chmieliński, J., *On the superstability of the generalized orthogonality equation in Euclidean space*, Ann. Math. Sil. **8**(1994), 127-140.

[3] Hyers, D. H., *On the stability of the linear functional equation*, Proc. Nat. Acad. Sci. U.S.A. **27**(1941), 222–224.

[4] Hyers, D. H. and Ulam, S. M. *On approximate isometries*, Bull. Amer. Math. Soc. **51**(1945), 288–292.

[5] Maurin, K., *Methods of Hilbert spaces*, Monografie Matematyczne vol. 45, Polish Scientific Publishers (PWN), Warszawa 1972.

[6] Rassias, Th. M. and Sharma C. S., *Properties of Isometries*, J. Natural Geometry **3** (1993), No. 1, 1-38.

INSTITUTE OF MATHEMATICS
PEDAGOGICAL UNIVERSITY
PODCHORĄŻYCH 2
PL-30-084 KRAKÓW, POLAND
E-MAIL: smchmiel@cyf-kr.edu.pl

B DANKOVIĆ, G V MILOVANOVIĆ AND S LJ RANČIĆ

Malmquist and Müntz orthogonal systems and applications(*)

1. Introduction

The last fifty years have seen a great deal of progress in the field of orthogonal systems (orthogonal algebraic and trigonometric polynomials, orthogonal Müntz polynomials, orthogonal rational functions, etc.), as well as many their applications in mathematics, physics and other computational and applied sciences (electronics and communication, control system theory, process identification, etc.). The orthogonal polynomial systems, especially classical orthogonal polynomials (cf. Szegő [53], Chihara [11], Freud [16], Suetin [50], Geronimus [23–24], Nevai [43], Milovanović, Mitrinović, Rassias [36]), play very important role in many problems in approximation theory and numerical analysis. Some of them found very important applications in applied sciences and have become the main tool in several methods and procedures. Also, there are many non-classical orthogonal polynomials on the real line (cf. Gautschi [18] and [17] for software), as well as several classes of non-standard orthogonal polynomials (cf. Milovanović [32–33] and [35]). For example, the Szegő class on the unit circle (cf. Szegő [51–52] and [53, pp. 287–295], Geronimus [22–23], Nevai [44]), orthogonal polynomials on the semicircle and on a circular arc (see Gautschi and Milovanović [20–21], Gautschi, Landau, Milovanović [19], de Bruin [7], and also papers [29–30], [37–39]), orthogonal polynomials on the radial rays in the complex plane (see [31], [33–34] and [40]), etc.

In the recent years, the other classes of orthogonal systems, as Malmquist systems (cf. Walsh [55], Müntz systems (cf. Borwein and Erdélyi [5]), and others, have also taken a significant role in applications. A great progress in the theoretical sense for such orthogonal systems was made too. The first papers for orthogonal rational functions on the unit circle whose poles are fixed were given by M.M. Djrbashian [12–13]. These rational functions generalize the orthogonal polynomials of Szegő [53, pp. 287–295]. Recently, the paper of Djrbashian [13], which originally appeared in two parts, has been translated to English by K. Müller and A. Bultheel [42]. A survey on the theory of orthogonal systems and some open problems was written also by Djrbashian [14]. Several papers in this direction have been appeared in the last period ([8–10], [26], [46–49]).

The orthogonal Müntz systems were considered first by Armanian mathematicians Badalyan [3] and Taslakyan [54]. Recently, it was investigated by McCarthy, Sayre

(*) This work was supported in part by the Serbian Scientific Foundation.

and Shawyer [27] and more completely by Borwein, Erdélyi, and Zhang [6] (see also the recent book [5]).

This paper is organized as follows. In Section 2 we use the orthogonality relation for Jacobi polynomials in order to obtain rational functions (via Laplace transform), which can be applied in many technical fields. In the same section, we mention some of these important applications. Section 3 is devoted to the orthogonal Müntz systems which represent an important extension of the orthogonal polynomial systems. The Malmquist systems of orthogonal rational functions are considered in Section 4. Finally, in Section 5 we give some connections between the previous mentioned orthogonal systems using one new nonstandard inner product.

2. Applications of Orthogonal Systems

The applications of the classical orthogonal polynomials in technical fields as *electrical network synthesis, electronics and telecommunication, signal processing theory, control system theory* and *process identification* are well known. The Legendre, Laguerre, Chebyshev and Jacobi polynomials are very useful for design and construction of electrical network, transfer functions, orthogonal filters, adaptive control, telecommunication systems, etc. These applications are based on the least squares polynomial approximations. The orthogonality of these polynomials enable the construction of optimal network and optimal filters. Moreover, the Laplace transforms of the classical polynomials, or their modifications are rational functions, which can be easy factorized. This property is very convenient in constructing simple procedures for forming signal generators, adaptive controllers, and practical realizations of the transfer functions. For instance, Chebyshev orthogonal polynomials are used very much for technical applications, so there exist Chebyshev orthogonal filters. Many papers were published about Chebyshev polynomials and their applications in technics.

For designing orthogonal filters and optimal transfer functions may be used some modifications of the Legendre or Jacobi polynomials, which are orthogonal on the interval $(-1,1)$. On the other side, technical systems which are designed using orthogonal polynomials work in the real time, so we need the corresponding approximations on the interval $(0,\infty)$. For example, by substituting $x = 2e^{-at} - 1$ $(a > 0)$ in a polynomial orthogonal on $(-1,1)$, we obtain an exponential polynomial orthogonal on the $(0,\infty)$. For polynomials orthogonal on $(0,1)$ we use the substitution $x = e^{-at}$ $(a > 0)$.

Applying the Laplace transform on the exponential orthogonal polynomials we obtain orthogonal rational functions. In the sequel we will consider some applications of the classical orthogonal polynomials in technics, using their Laplace transforms and corresponding orthogonal functions.

Starting from the orthogonality relation for Jacobi polynomials $P_n^{(\alpha,\beta)}(x)$ with the parameters $\alpha, \beta > -1$, i.e.,

$$\int_{-1}^{1} (1-x)^{\alpha}(1+x)^{\beta} P_n^{(\alpha,\beta)}(x) P_m^{(\alpha,\beta)}(x)\, dx = \|P_n^{(\alpha,\beta)}\|^2 \delta_{nm}, \tag{2.1}$$

where

$$\|P_n^{(\alpha,\beta)}\|^2 = \frac{2^{\alpha+\beta+1}\Gamma(n+\alpha+1)\Gamma(n+\beta+1)}{n!(2n+\alpha+\beta+1)\Gamma(n+\alpha+\beta+1)}$$

and putting $x+1 = 2e^{-2t}$, we obtain

$$\int_0^{\infty} \varphi_n(t)\varphi_m(t)\, dt = \delta_{nm},$$

where

$$\varphi_n(t) = \frac{2^{(\alpha+\beta+2)/2}}{\|P_n^{(\alpha,\beta)}\|} e^{-(\beta+1)t}(1-e^{-2t})^{\alpha/2} P_n^{(\alpha,\beta)}(2e^{-2t}-1).$$

Then, the Laplace transform of φ_n can be expressed in the form

$$\begin{aligned}
W_n(s) = \mathcal{L}[\varphi_n(t)] &= \int_0^{\infty} e^{-st}\varphi_n(t)\, dt \\
&= \frac{2^{-(s+1)/2}}{\|P_n^{(\alpha,\beta)}\|} \int_{-1}^{1} (1-x)^{\alpha/2}(1+x)^{(s+\beta-1)/2} P_n^{(\alpha,\beta)}(x)\, dx \\
&= \frac{2^{(\alpha+\beta)/2}}{\|P_n^{(\alpha,\beta)}\|} \cdot \frac{\Gamma\left(\frac{\alpha}{2}+1\right)\Gamma\left(\frac{1}{2}(s+\beta+1)\right)}{\Gamma\left(\frac{1}{2}(\alpha+\beta+s+3)\right)} \times \\
&\quad \times {}_3F_2\left(-n, \alpha+\beta+n+1, \tfrac{1}{2}\alpha+1; \alpha+1, \tfrac{1}{2}(\alpha+\beta+s+3); 1\right),
\end{aligned}$$

where the hypergeometric function ${}_3F_2$ is reduces to

$$\begin{aligned}
Q_n^{(\alpha,\beta)}(s) &= {}_3F_2\left(-n, \alpha+\beta+n+1, \tfrac{1}{2}\alpha+1; \alpha+1, \tfrac{1}{2}(\alpha+\beta+s+3); 1\right) \\
&= \sum_{k=0}^{\infty} \frac{(-n)_k(\alpha+\beta+n+1)_k\left(\frac{1}{2}\alpha+1\right)_k}{(\alpha+1)_k\left(\frac{1}{2}(\alpha+\beta+s+3)\right)_k k!} \\
&= \sum_{k=0}^{\infty} (-1)^k \binom{n}{k} \frac{(\alpha+\beta+n+1)_k\left(\frac{1}{2}\alpha+1\right)_k}{(\alpha+1)_k\left(\frac{1}{2}(\alpha+\beta+s+3)\right)_k}.
\end{aligned}$$

In a simpler case when $\alpha = 0$, the function ${}_3F_2$ can be reduced to ${}_2F_1$. Namely,

$$\begin{aligned}
Q_n^{(0,\beta)}(s) &= {}_3F_2\left(-n, \beta+n+1, 1; 1, \tfrac{1}{2}(\beta+s+3); 1\right) \\
&= {}_2F_1\left(-n, \beta+n+1; \tfrac{1}{2}(\beta+s+3); 1\right).
\end{aligned}$$

Using a representation of the Jacobi polynomials, with parameters

$$\hat{\alpha} = \frac{s+\beta+1}{2} \quad \text{and} \quad \hat{\beta} = -\frac{s-\beta+1}{2},$$

in terms of hypergeometric function (see Bateman and Erdélyi [2])

$${}_2F_1\left(-n, n+\hat{\alpha}+\hat{\beta}+1; \hat{\alpha}+1; \tfrac{1}{2}(1-x)\right) = \frac{P_n^{(\hat{\alpha},\hat{\beta})}(x)}{\binom{n+\hat{\alpha}}{n}},$$

we find

$$Q_n^{(0,\beta)}(s) = \frac{P_n^{(\hat{\alpha},\hat{\beta})}(-1)}{\binom{n+\hat{\alpha}}{n}} = (-1)^n \frac{\binom{n+\hat{\beta}}{n}}{\binom{n+\hat{\alpha}}{n}},$$

because of $n+\hat{\alpha}+\hat{\beta}+1 = n+\beta+1$ and $\hat{\alpha}+1 = (s+\beta+3)/2$.

Thus, in this case we get

$$\begin{aligned} W_n(s) &= \mathfrak{L}\left[\frac{2^{\beta/2+1}}{\|P_n^{(0,\beta)}\|} e^{-(\beta+1)t} P_n^{(0,\beta)}(2e^{-2t}-1)\right] \\ &= \frac{2^{\beta/2}}{\|P_n^{(0,\beta)}\|} \cdot \frac{\Gamma\left(\frac{1}{2}(s+\beta+1)\right)}{\Gamma\left(\frac{1}{2}(s+\beta+3)\right)} \cdot \frac{(-1)^n(n+\hat{\beta})(n-1+\hat{\beta})\cdots(1+\hat{\beta})}{(n+\hat{\alpha})(n-1+\hat{\alpha})\cdots(1+\hat{\alpha})}, \end{aligned}$$

i.e.,

$$W_n(s) = \sqrt{2(2n+\beta+1)}\, \frac{\prod\limits_{k=0}^{n-1}(s-(2k+1+\beta))}{\prod\limits_{k=0}^{n}(s+(2k+1+\beta))}. \tag{2.2}$$

Sometimes, we put $x+1 = 2e^{-t}$ in (2.1). Then the previous rational functions become

$$\widetilde{W}_n(s) = \sqrt{2}\, W_n(2s) = \sqrt{2n+\beta+1}\, \frac{\prod\limits_{k=0}^{n-1}\left(s - \frac{2k+1+\beta}{2}\right)}{\prod\limits_{k=0}^{n}\left(s + \frac{2k+1+\beta}{2}\right)}. \tag{2.3}$$

In a degenerating case $\beta = -1$, when the Jacobi polynomial $P_n^{(0,-1)}(x)$ can be expressed in the form

$$P_n^{(0,-1)}(x) = \frac{1+x}{2}\, P_{n-1}^{(0,1)}(x),$$

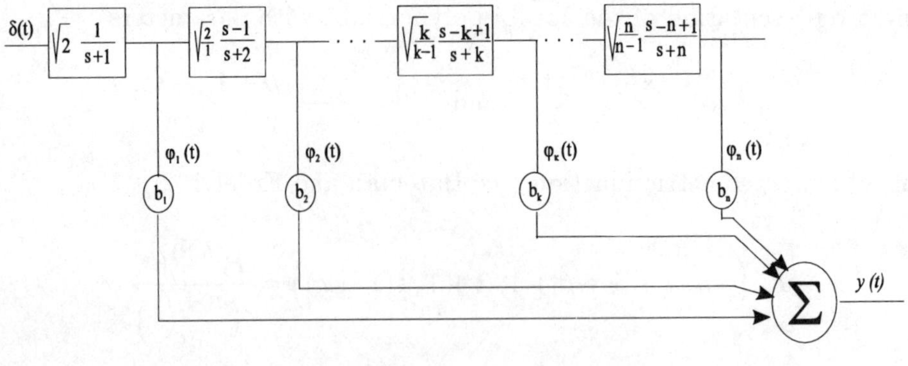

Fig. 2.1

from (2.3) we obtain

$$\widetilde{W}_n(s) = \sqrt{2n}\,\frac{\prod_{k=1}^{n-1}(s-k)}{\prod_{k=1}^{n}(s+k)}\,. \tag{2.4}$$

Using (2.4) we can obtain an orthogonal filter (see [1, pp. 42–71]), which is given in Fig. 2.1. In this case the corresponding function $\widetilde{\varphi}(t)$ can be expressed in the form

$$\widetilde{\varphi}(t) = \sqrt{2n}\sum_{k=1}^{n} a_{n,k}e^{-kt},$$

where

$$a_{n,k} = (-1)^k\binom{n+k-1}{k-1}\binom{n}{k}.$$

This Jacobi filter has been used in signal theory for obtaining the orthogonal signals $\phi(t) = \widetilde{\varphi}(t)$. Also, this filter is useful in the adaptive control in construction of adjustable models. The application of this filter is very often in the process identification as well. In this case, the network, obtained by Jacobi's filter, is connected with an unknown process as it is shown in Fig. 2.2.

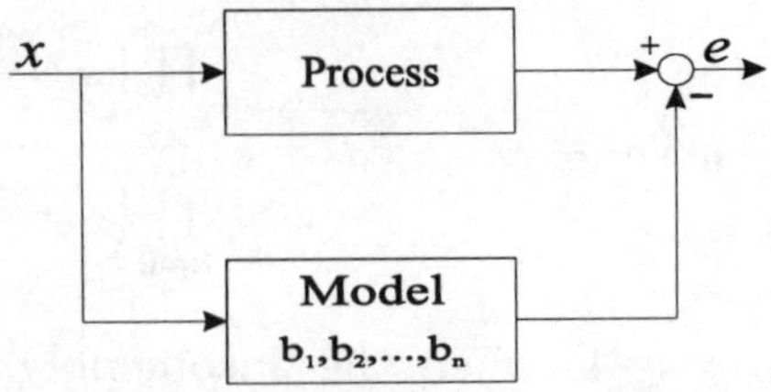

Fig. 2.2

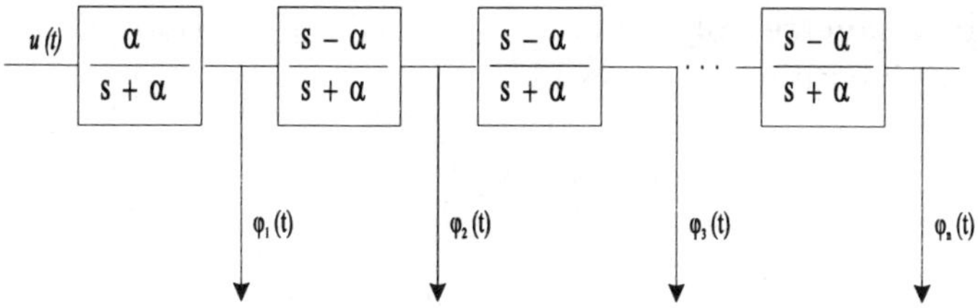

Fig. 2.3

By adjusting the parameters b_k, so that the error $e(t)$ is minimal (or zero), one can derive the model of an unknown process.

An application of the Legendre polynomials is also given in [1]. Namely, using (2.2) with $\beta = 0$, one can obtain the following sequence of the rational functions

$$W_n(s) = \sqrt{2(2n+1)}\,\frac{\prod_{k=0}^{n-1}(s-(2k+1))}{\prod_{k=0}^{n}(s+2k+1)}.$$

These functions can be used in practical realization in the same way as the previous functions of the Jacobi type.

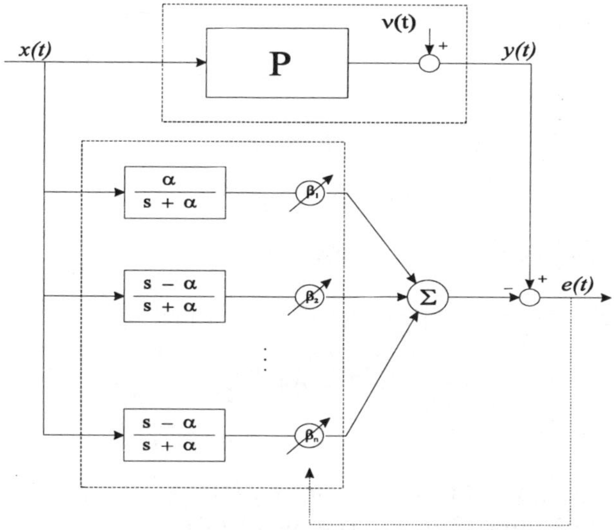

Fig. 2.4

In [15] it was also presented a method for designing of orthogonal filters, based on the classical orthogonal polynomials. In construction of some real filters has been used the Hermite polynomials. A method for process identification was derived in [4] and based on the Laguerre polynomials. It is well known that the Laplace transform of the Laguerre polynomials is given by

$$\mathcal{L}[L_n(x)] = \frac{(-1)^n}{n!}\frac{(s-1)^n}{s^{n+1}}\,.$$

In order to obtain the Laguerre orthogonal filters, we can take the modified Laguerre functions $\phi_{n,\alpha}(x)$, which Laplace transform is given by

$$\mathcal{L}[\phi_{n,\alpha}(x)] = \frac{\sqrt{2\alpha}}{s+\alpha}\left(\frac{s-\alpha}{s+\alpha}\right)^n. \tag{2.5}$$

The orthogonal filter obtained by (2.5) is displayed in Fig. 2.3.

In this way, we can obtain one base of the orthogonal functions $\phi_{n,\alpha}(x)$, useful for process identification. Such a system is presented in Fig. 2.4.

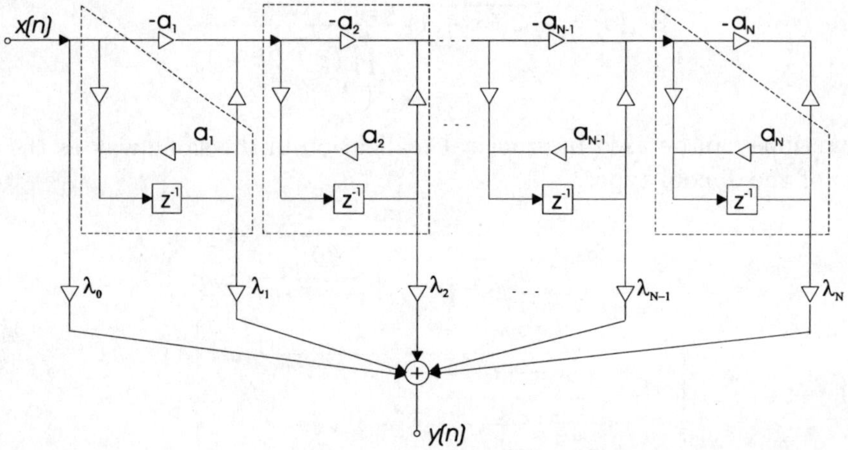

Fig. 2.5

Using the complex rational functions $F_n(s)$ defined by

$$F_n(s) = \frac{\sqrt{1-\alpha_n^2}}{s-\alpha_n}\prod_{k=0}^{n-1}\frac{1-\bar{\alpha}_k s}{s-\alpha_k}$$

(see Section 4), some applications in obtaining orthogonal filters are given in [25] and [45]. Such an orthogonal filter is presented in Fig. 2.5.

3. Müntz Orthogonal Systems

Let $\Lambda = \{\lambda_0, \lambda_1, \dots\}$ be a given sequence of complex numbers. Taking the following definition for x^λ:

$$x^\lambda = e^{\lambda \log x}, \qquad x \in (0, +\infty),\ \lambda \in \mathbb{C},$$

and the value at $x = 0$ is defined to be the limit of x^λ as $x \to 0$ from $(0, +\infty)$ whenever the limits exists, we will consider orthogonal Müntz polynomials as linear combinations of the Müntz system $\{x^{\lambda_0}, x^{\lambda_1}, \dots, x^{\lambda_n}\}$ (see [5–6]). The set of all such polynomials we will denote by $M_n(\Lambda)$, i.e.,

$$M_n(\Lambda) = \operatorname{span}\{x^{\lambda_0}, x^{\lambda_1}, \dots, x^{\lambda_n}\},$$

where the linear span is over the complex numbers $\mathbb{C}$ in general. The union of all $M_n(\Lambda)$ is denoted by $M(\Lambda)$.

The first considerations of orthogonal Müntz systems were made by Badalyan [3] and Taslakyan [54]. Recently, it was investigated by McCarthy, Sayre and Shawyer [27] and more completely by Borwein, Erdélyi, and Zhang [6].

Supposing that $\mathrm{Re}(\lambda_k) > -1/2$ $(k = 0, 1, \dots)$ we can give the following definition of the *Müntz-Legendre polynomials* on $(0, 1]$ (see [54], [6]):

Definition 3.1. The nth Müntz-Legendre polynomial on $(0, 1]$ is given by

$$P_n(\lambda_0, \dots, \lambda_n; x) = \frac{1}{2\pi i} \oint_\Gamma \prod_{\nu=0}^{n-1} \frac{s + \bar{\lambda}_\nu + 1}{s - \lambda_\nu} \frac{x^s \, ds}{s - \lambda_n} \qquad (n = 0, 1, \dots), \tag{3.1}$$

where the simple contour Γ surrounds all the zeros of the denominator in the integrand.

For polynomials $P_n(x) \equiv P_n(\lambda_0, \dots, \lambda_n; x)$ one can prove an orthogonality relation on $(0, 1)$:

Theorem 3.1. *Let the polynomials $P_n(x)$ be defined by (3.1). Then*

$$\int_0^1 P_n(x) \overline{P_m(x)} \, dx = \delta_{n,m} / (1 + \lambda_n + \bar{\lambda}_n)$$

holds for every $n, m = 0, 1, \dots$.

Evidently, that the polynomials $P_n^*(x) = (1 + \lambda_n + \bar{\lambda}_n)^{1/2} P_n(x)$ are *orthonormal.* Supposing that $\lambda_\nu \neq \lambda_\mu$ $(\nu \neq \mu)$ it is easy to show that polynomials $P_n(x)$ can be expressed in the form

$$P_n(x) = \sum_{k=0}^{n} c_{n,k} x^{\lambda_k},$$

where

$$c_{n,k} = \frac{\prod_{\nu=0}^{n-1} (1 + \lambda_k + \bar{\lambda}_\nu)}{\prod_{\substack{\nu=0 \\ \nu \neq k}}^{n} (\lambda_k - \lambda_\nu)}.$$

In a limit case when $\lambda_0 = \lambda_1 = \cdots = \lambda_n = \lambda$, the polynomials (3.1) reduce to

$$P_n(\lambda_0, \ldots, \lambda_n; x) = x^\lambda L_n(-(1 + \lambda + \bar{\lambda}) \log x),$$

where $L_n(x)$ is the Laguerre polynomial orthogonal on the $(0, \infty)$ with respect to the exponential weight e^{-x}, and for which $L_n(0) = 1$.

There is also a generalized Rodrigues formula for the Müntz-Legendre polynomials (see [27]). Some recurrence relations also hold:

Theorem 3.2. *Let the polynomials $P_n(x)$ be defined by* (3.1). *Then*

$$x(P_n'(x) - P_{n-1}'(x)) = \lambda_n P_n(x) + (1 + \bar{\lambda}_{n-1}) P_{n-1}(x),$$

$$x P_n'(x) = \lambda_n P_n(x) + \sum_{\nu=0}^{n-1} (1 + \lambda_\nu + \bar{\lambda}_\nu) P_\nu(x),$$

$$x P_n''(x) = (\lambda_n - 1) P_n'(x) + \sum_{\nu=0}^{n-1} (1 + \lambda_\nu + \bar{\lambda}_\nu) P_\nu'(x).$$

Similar to the Legendre polynomials and here also $P_n(1) = 1$ for each n.

An interesting question is connected by the zero distribution of the Müntz-Legendre polynomials. A nice proof of the following result was given in [6].

Theorem 3.3. *For real numbers $\lambda_\nu > -1/2$ $(\nu = 0, 1, \ldots)$ the Müntz-Legendre polynomial $P_n(x)$ has exactly n distinct zeros in $(0, 1)$, and it changes sign at each of these zeros.*

Some inequalities for Müntz polynomials were also investigated in [6].

4. Malmquist Orthogonal Systems

Let $A = \{a_0, a_1, \ldots\}$ be an arbitrary sequence of complex numbers in the unit circle $(|a_\nu| < 1)$. The Malmquist system of rational functions (see [55], [12–14]) is defined in the following way

$$\phi_n(s) = \frac{(1 - |a_n|^2)^{1/2}}{1 - \bar{a}_n s} \prod_{\nu=0}^{n-1} \frac{a_\nu - s}{1 - \bar{a}_\nu s} \cdot \frac{|a_\nu|}{a_\nu} \qquad (n = 0, 1, \ldots), \tag{4.1}$$

where for $a_\nu = 0$ we put $|a_\nu|/a_\nu = \bar{a}_\nu/|a_\nu| = -1$. Such system of functions was intensively investigated in several papers by Djrbashian [12–14], Bultheel, González-Vera, Hendriksen, and Njåstad [8–10], Pan [46–49], etc.

In this section we want to prove some auxiliary results in order to connect this system of orthogonal functions with some Müntz system of polynomials.

Excluding the normalization constants, the system (4.1) can be represented in the form

$$W_n(s) = \frac{\prod_{\nu=0}^{n-1}(s-a_\nu)}{\prod_{\nu=0}^{n}(s-a_\nu^*)}, \tag{4.2}$$

where $a_\nu^* = 1/\bar{a}_\nu$. For $a_\nu = 0$ we put only s instead of $(s-a_\nu)/(s-a_\nu^*)$.

Suppose now that $a_\nu \neq a_\mu$ for $\nu \neq \mu$. Then (4.2) can be written in the form

$$W_n(s) = \sum_{k=0}^{n} \frac{A_{n,k}}{s-a_k^*}, \tag{4.3}$$

where

$$A_{n,k} = \frac{\prod_{\nu=0}^{n-1}(a_k^*-a_\nu)}{\prod_{\substack{\nu=0\\ \nu\neq k}}^{n}(a_k^*-a_\nu^*)} \qquad (k=0,1,\ldots,n). \tag{4.4}$$

The case when $a_\nu = a_\mu$ can be considered as a limiting process $a_\nu \to a_\mu$.

Alternatively, for $|a_\nu| < 1$, (4.4) can be reduced to

$$A_{n,k} = \frac{\prod_{\nu=0}^{n-1}\left(\dfrac{1}{\bar{a}_k}-a_\nu\right)}{\prod_{\substack{\nu=0\\ \nu\neq k}}^{n}\left(\dfrac{1}{\bar{a}_k}-\dfrac{1}{\bar{a}_\nu}\right)} = \frac{\bar{a}_0\cdots\bar{a}_n}{\bar{a}_k}\cdot\frac{\prod_{\nu=0}^{n-1}(\bar{a}_k a_\nu - 1)}{\prod_{\substack{\nu=0\\ \nu\neq k}}^{n}(\bar{a}_k-\bar{a}_\nu)}. \tag{4.5}$$

It is well-known that system of functions (4.1) is orthonormal on the unit circle $|s| = 1$ with respect to the inner product

$$(u,v) = \frac{1}{2\pi i}\oint_{|s|=1} u(s)\overline{v(s)}\,\frac{ds}{s} = \frac{1}{2\pi}\int_{-\pi}^{\pi} u(e^{i\theta})\overline{v(e^{i\theta})}\,d\theta. \tag{4.6}$$

Namely, $(\phi_n, \phi_m) = \delta_{nm}$ $(n, m = 0, 1, \ldots)$.

We note also that

$$(u,v) = \frac{1}{2\pi i}\oint_{|s|=1} u(s)\overline{v(s^*)}\,\frac{ds}{s},$$

where $s^* = 1/\bar{s}$ on the unit circle $|s| = 1$.

Because of completeness we prove the following result:

Theorem 4.1. *If the system of rational functions* $\{W_n\}_{n=0}^{+\infty}$ *defined by* (4.2) *and the inner product* $(.,.)$ *by* (4.6), *then*

$$(W_n, W_m) = \|W_n\|^2 \delta_{nm},$$

where

$$\|W_n\|^2 = \frac{|a_0 a_1 \cdots a_n|^2}{1 - |a_n|^2}.$$

Proof. It is enough to prove that $(W_n, W_m) = 0$ for $m < n$, because of the property $(u, v) = \overline{(v, u)}$. Supposing that $m < n$, we have

$$\begin{aligned}
(W_n, W_m) &= \frac{1}{2\pi i} \oint\limits_{|s|=1} \frac{\prod_{\nu=0}^{n-1} (s - a_\nu)}{\prod_{\nu=0}^{n} (s - a_\nu^*)} \cdot \frac{\prod_{\nu=0}^{m-1} (\bar{s} - \bar{a}_\nu)}{\prod_{\nu=0}^{m} (\bar{s} - \bar{a}_\nu^*)} \cdot \frac{ds}{s} \\
&= \frac{1}{2\pi i} \oint\limits_{|s|=1} \frac{\prod_{\nu=0}^{n-1} (s - a_\nu)}{\prod_{\nu=0}^{n} (s - a_\nu^*)} \cdot \frac{\prod_{\nu=0}^{m-1} \left(\frac{1}{s} - \frac{1}{a_\nu^*}\right)}{\prod_{\nu=0}^{m} \left(\frac{1}{s} - \frac{1}{a_\nu}\right)} \cdot \frac{ds}{s} \\
&= \frac{1}{2\pi i} \oint\limits_{|s|=1} \frac{\prod_{\nu=0}^{n-1} (s - a_\nu)}{\prod_{\nu=0}^{n} (s - a_\nu^*)} \cdot \frac{\prod_{\nu=0}^{m-1} (s - a_\nu^*)}{\prod_{\nu=0}^{m} (s - a_\nu)} \cdot \frac{(-1) a_0 \cdots a_m}{a_0^* \cdots a_{m-1}^*} \, ds.
\end{aligned}$$

Evidently, for $|a_\nu| < 1$, i.e., $|a_\nu^*| > 1$, and $m < n$, Cauchy's theorem gives $(W_n, W_m) = 0$. For $m = n$ we see that

$$\begin{aligned}
\|W_n\|^2 &= -\frac{a_0 \cdots a_n}{a_0^* \cdots a_{n-1}^*} \cdot \frac{1}{2\pi i} \oint\limits_{|s|=1} \frac{ds}{(s - a_n^*)(s - a_n)} \\
&= -\frac{a_0 \cdots a_n}{(1/\bar{a}_0) \cdots (1/\bar{a}_{n-1})} \cdot \frac{1}{a_n - a_n^*} = \frac{|a_0 a_1 \cdots a_n|^2}{1 - |a_n|^2}. \quad \square
\end{aligned}$$

Lemma 4.1. *Let* $|z| \le 1$ *and let* F *be defined by*

$$F(z) = \frac{1}{2\pi i} \oint\limits_{|s|=1} W_n(s) \overline{W_m(\bar{z} s)} \, \frac{ds}{s}, \tag{4.7}$$

where the system functions $W_n(s)$ *is defined by* (4.2) *with mutually different numbers* a_ν $(\nu = 0, 1, \dots)$ *in the unit circle* $|s| = 1$. *Then*

$$F(z) = \sum_{i=0}^{n} \sum_{j=0}^{m} \frac{A_{n,i} \bar{A}_{m,j}}{a_i^* \bar{a}_j^* - z}, \tag{4.8}$$

where the numbers $A_{n,k}$ are given in (4.4).

Proof. For $|z| \le 1$ we conclude that the function

$$F(z) = \frac{1}{2\pi i} \oint_{|s|=1} \frac{\prod_{\nu=0}^{n-1} (s - a_\nu)}{\prod_{\nu=0}^{n} (s - a_\nu^*)} \cdot \frac{\prod_{\nu=0}^{m-1} \left(\frac{z}{s} - \frac{1}{a_\nu^*}\right)}{\prod_{\nu=0}^{m} \left(\frac{z}{s} - \frac{1}{a_\nu}\right)} \cdot \frac{ds}{s}$$

$$= \frac{(-1)a_0 \cdots a_m}{a_0^* \cdots a_{m-1}^*} \cdot \frac{1}{2\pi i} \oint_{|s|=1} \frac{\prod_{\nu=0}^{n-1} (s - a_\nu)}{\prod_{\nu=0}^{n} (s - a_\nu^*)} \cdot \frac{\prod_{\nu=0}^{m-1} (s - a_\nu^* z)}{\prod_{\nu=0}^{m} (s - a_\nu z)} \, ds$$

has $(m+1)$ poles in the circle $|s| = 1$: $a_j z$ $(j = 0, 1, \ldots, m)$. By Cauchy's residue theorem we find that

$$F(z) = \frac{(-1)a_0 \cdots a_m}{a_0^* \cdots a_{m-1}^*} \sum_{j=0}^{m} \frac{\prod_{\nu=0}^{n-1} (a_j z - a_\nu)}{\prod_{\nu=0}^{n} (a_j z - a_\nu^*)} \cdot \frac{\prod_{\nu=0}^{m-1} (a_j z - a_\nu^* z)}{\prod_{\substack{\nu=0 \\ \nu \ne j}}^{m} (a_j z - a_\nu z)}.$$

Since

$$G_{m,j} = \frac{\prod_{\nu=0}^{m-1} (a_j z - a_\nu^* z)}{\prod_{\substack{\nu=0 \\ \nu \ne j}}^{m} (a_j z - a_\nu z)} = \frac{\prod_{\nu=0}^{m-1} (a_j \bar{a}_\nu - 1)}{\prod_{\substack{\nu=0 \\ \nu \ne j}}^{m} (a_j - a_\nu)} \cdot \frac{1}{\bar{a}_0 \cdots \bar{a}_{m-1}},$$

because of (4.5), we have

$$G_{m,j} = \frac{a_j}{a_0 \cdots a_m} \cdot \bar{A}_{m,j} \cdot \frac{1}{\bar{a}_0 \cdots \bar{a}_{m-1}} = \frac{a_j \bar{a}_m \bar{A}_{m,j}}{|\bar{a}_0 \cdots \bar{a}_m|^2}.$$

Thus,

$$\frac{a_0 \cdots a_m}{a_0^* \cdots a_{m-1}^*} G_{m,j} = a_j \bar{A}_{m,j},$$

and we obtain

$$F(z) = -\sum_{j=0}^{m} \bar{A}_{m,j} \frac{\prod_{\nu=0}^{n-1} \left(z - \frac{a_\nu}{a_j}\right)}{\prod_{\nu=0}^{n} \left(z - \frac{a_\nu^*}{a_j}\right)}.$$

On the other hand, expanding $Q_j(z) = \prod_{\nu=0}^{n-1}(z - a_\nu/a_j) \Big/ \prod_{\nu=0}^{n}(z - a_\nu^*/a_j)$ in partial fractions, we get

$$Q_j(z) = \sum_{i=0}^{n} \frac{\prod_{\nu=0}^{n-1}\left(\dfrac{a_i^*}{a_j} - \dfrac{a_\nu}{a_j}\right)}{\prod_{\substack{\nu=0\\ \nu\neq i}}^{n}\left(\dfrac{a_i^*}{a_j} - \dfrac{a_\nu^*}{a_j}\right)} \cdot \frac{1}{z - \dfrac{a_i^*}{a_j}} = \sum_{i=0}^{n} \frac{\prod_{\nu=0}^{n-1}(a_i^* - a_\nu)}{\prod_{\substack{\nu=0\\ \nu\neq i}}^{n}(a_i^* - a_\nu^*)} \cdot \frac{1}{z - a_i^*\bar{a}_j^*}.$$

Because of (4.4), the right hand side of the last equality becomes

$$\sum_{i=0}^{n} \frac{A_{n,i}}{z - a_i^*\bar{a}_j^*},$$

so that we obtain

$$F(z) = \sum_{j=0}^{m} \bar{A}_{m,j} \sum_{i=0}^{n} A_{n,i} \frac{1}{a_i^*\bar{a}_j^* - z},$$

i.e., (4.8). □

For $z = 1$, from (4.7) we see that $F(1) = (W_n, W_m)$. Thus,

$$(W_n, W_m) = \sum_{i=0}^{n}\sum_{j=0}^{m} \frac{A_{n,i}\bar{A}_{m,j}}{a_i^*\bar{a}_j^* - 1}. \tag{4.9}$$

5. A Connection Between Malmquist and Müntz Systems

In this section we give a connection between the Malmquist system of rational functions (4.2) and a Müntz system, which is orthogonal with respect to a new inner product.

Using the numbers a_ν^* which appear in (4.2), i.e., $A^* = \{a_0^*, a_1^*, \dots\}$, we form the Müntz system $\{x^{a_0^*}, x^{a_1^*}, \dots, x^{a_n^*}\}$. Notice that all such numbers are outside the unit circle. In order to short our notation and to be consistent to Section 3, we write $a_\nu^* = \lambda_\nu$ $(\nu = 0, 1, \dots)$.

In our consideration we need a new operation:

Definition 5.1. For $\alpha, \beta \in \mathbb{C}$ we have

$$x^\alpha \odot x^\beta = x^{\alpha\beta} \qquad (x \in (0, \infty)). \tag{5.1}$$

Using (5.1) we can introduce an external operation for the Müntz polynomials from $M(A^*) = M(\Lambda)$.

Definition 5.2. Let $P \in M_n(\Lambda)$ and $Q \in M_m(\Lambda)$, i.e.,

$$P(x) = \sum_{i=0}^{n} p_i x^{\lambda_i} \quad \text{and} \quad Q(x) = \sum_{j=0}^{m} q_j x^{\lambda_j}, \tag{5.2}$$

then

$$(P \odot Q)(x) = \sum_{i=0}^{n} \sum_{j=0}^{m} p_i q_j x^{\lambda_i \lambda_j}. \tag{5.3}$$

Under restrictions that for each i and j we have

$$|\lambda_i| > 1, \qquad \text{Re}(\lambda_i \bar{\lambda}_j - 1) > 0, \tag{5.4}$$

then we can introduce a new inner product for Müntz polynomials:

Definition 5.3. Let the conditions (5.4) be satisfied for the Müntz polynomials $P(x)$ and $Q(x)$ given by (5.2). Their inner product $[P, Q]$ is defined by

$$[P, Q] = \int_0^1 (P \odot \overline{Q})(x) \, \frac{dx}{x^2}, \tag{5.5}$$

where $(P \odot Q)(x)$ is determined by (5.3).

It is not clear immediately that (5.5) represents an inner product. Therefore, we prove the following result:

Theorem 5.1. *Let* $\Lambda = \{\lambda_0, \lambda_1, \dots\}$ *be a sequence of the complex numbers such that the conditions* (5.4) *hold. Then*

(i) $[P, P] \geq 0$;
(ii) $[P, P] = 0 \quad \Longleftrightarrow \quad P(x) \equiv 0$;
(iii) $[P + Q, R] = [P, R] + [Q, R]$;
(iv) $[cP, Q] = c[P, Q]$;
(v) $[P, Q] = \overline{[Q, P]}$

for each $P, Q, R \in M(\Lambda)$ *and each* $c \in \mathbb{C}$.

Proof. Let $P(x)$ and $Q(x)$ be given by (5.2). Using (5.5) and (5.3) we have

$$[P, Q] = \int_0^1 (P \odot Q)(x) \frac{dx}{x^2} = \sum_{i=0}^{n} \sum_{j=0}^{m} p_i \bar{q}_j \int_0^1 x^{\lambda_i \bar{\lambda}_j - 2} dx.$$

Because of (5.4), we find that $\int_0^1 x^{\lambda_i \bar{\lambda}_j - 2} dx = 1/(\lambda_i \bar{\lambda}_j - 1)$ for each i and j, so that we get

$$[P, Q] = \sum_{i=0}^{n} \sum_{j=0}^{m} \frac{p_i \bar{q}_j}{\lambda_i \bar{\lambda}_j - 1}. \tag{5.6}$$

In order to prove (i) and (ii) it is enough to conclude that the quadratic form

$$[P,P] = \sum_{i=0}^{n}\sum_{j=0}^{n} \frac{1}{\lambda_i\bar{\lambda}_j - 1} p_i\bar{p}_j,$$

i.e., its matrix $H_n = \left[1/(\lambda_i\bar{\lambda}_j - 1)\right]_{i,j=0}^{n}$, is positive definite. Therefore, we use the Sylvester's necessary and sufficient conditions (cf. [28, p. 214])

$$D_k = \det H_k = \det\left[1/(\lambda_i\bar{\lambda}_j - 1)\right]_{i,j=0}^{k} > 0 \qquad (k = 0, 1, \dots, n).$$

In order to evaluate the determinants D_k, we use Cauchy's formula (see Muir [41, p. 345])

$$\det\left[\frac{1}{a_i + b_j}\right]_{i,j=0}^{k} = \frac{\prod\limits_{i>j=0}^{k} (a_i - a_j)(b_i - b_j)}{\prod\limits_{i,j=0}^{k} (a_i + b_j)}$$

with $a_i = \lambda_i$ and $b_j = -1/\bar{\lambda}_j$. Thus, we obtain

$$D_k = \frac{1}{\prod\limits_{j=0}^{k} \bar{\lambda}_j} \det\left[\frac{1}{\lambda_i - 1/\bar{\lambda}_j}\right]_{i,j=0}^{k} = \frac{1}{\prod\limits_{j=0}^{k} \bar{\lambda}_j} \cdot \frac{\prod\limits_{i>j=0}^{k} (\lambda_i - \lambda_j)\left(-\frac{1}{\bar{\lambda}_i} + \frac{1}{\bar{\lambda}_j}\right)}{\prod\limits_{i,j=0}^{k} \left(\lambda_i - \frac{1}{\bar{\lambda}_j}\right)},$$

i.e.,

$$D_k = \frac{1}{\prod\limits_{j=0}^{k} \bar{\lambda}_j} \cdot \frac{\prod\limits_{i>j=0}^{k} \frac{|\lambda_i - \lambda_j|^2}{\bar{\lambda}_i\bar{\lambda}_j}}{\prod\limits_{i,j=0}^{k} \frac{\lambda_i\bar{\lambda}_j - 1}{\bar{\lambda}_j}}.$$

Since $D_0 = 1/(|\lambda_0|^2 - 1) > 0$ (because of (5.4)) and

$$D_k = \frac{D_{k-1}}{|\lambda_k|^2 - 1} \prod_{i=0}^{k-1} \frac{|\lambda_k - \lambda_i|^2}{|\lambda_i\bar{\lambda}_k - 1|^2},$$

by induction we conclude that $D_k > 0$ for all k.

The properties (iii)–(v) follow directly from (5.5) or (5.6). □

Now we are ready to define the Müntz polynomials

$$Q_n(x) \equiv Q_n(\lambda_0, \lambda_1, \dots, \lambda_n; x), \qquad n = 0, 1, \dots, \tag{5.7}$$

orthogonal with respect to the inner product (5.5).

Consider again the Malmquist system of rational functions (4.2), i.e.,

$$W_n(s) = \frac{\prod_{\nu=0}^{n-1} (s - 1/\bar{\lambda}_\nu)}{\prod_{\nu=0}^{n} (s - \lambda_\nu)}, \tag{5.8}$$

where $\Lambda = \{\lambda_0, \lambda_1, \dots\}$ is a complex sequence such that (5.4) holds.

Definition 5.4. The nth Müntz polynomial $Q_n(x)$, associated to the Malmquist function (5.8), is defined by

$$Q_n(x) = \frac{1}{2\pi i} \oint_\Gamma W_n(s) x^s \, ds, \tag{5.9}$$

where the simple contour Γ surrounds all the points λ_ν $(\nu = 0, 1, \dots, n)$.

Using (5.9), (4.3), (4.4) and Cauchy's residue theorem we get, a representation of (5.7) in the form

$$Q_n(x) = \sum_{k=0}^{n} A_{n,k} x^{\lambda_k}, \tag{5.10}$$

where

$$A_{n,k} = \frac{\prod_{\nu=0}^{n-1} (\lambda_k - 1/\bar{\lambda}_\nu)}{\prod_{\substack{\nu=0 \\ \nu\neq k}}^{n} (\lambda_k - \lambda_\nu)} \qquad (k = 0, 1, \dots, n), \tag{5.11}$$

and where we assumed that $\lambda_i \neq \lambda_j$ $(i \neq j)$.

Finally, we prove an orthogonality relation for polynomials $Q_n(x)$. Other results in this direction will be published elsewhere.

Theorem 5.2. *Under previous conditions on the sequence* Λ, *the Müntz polynomials* $Q_n(x)$, $n = 0, 1, \dots,$ *defined by* (5.9), *are orthogonal with respect to the inner product* (5.5). *i.e.,*

$$[Q_n, Q_m] = \frac{1}{(|\lambda_n|^2 - 1)|\lambda_0 \lambda_1 \cdots \lambda_{n-1}|^2} \delta_{n,m}. \tag{5.12}$$

Proof. According to (5.6) and (5.10), we have

$$[Q_n, Q_m] = \sum_{i=0}^{n} \sum_{j=0}^{m} \frac{A_{n,i} \bar{A}_{m,j}}{\lambda_i \bar{\lambda}_j - 1},$$

where $A_{n,k}$ is given by (5.11).

Using Lemma 4.1 with $z = 1$, i.e., equality (4.9), we conclude that

$$[Q_n, Q_m] = (W_n, W_m),$$

where $W_n(s)$ is determined by (5.8).

Since $a_\nu = 1/\bar{\lambda}_\nu$ ($\nu = 0, 1, \ldots$), from Theorem 4.1 it follows (5.12). □

For the norm of polynomial $Q_n(x)$ we obtain

$$\|Q_n\| = \sqrt{[Q_n, Q_n]} = \frac{1}{|\lambda_0\lambda_1 \cdots \lambda_{n-1}|} \cdot \frac{1}{\sqrt{|\lambda_n|^2 - 1}},$$

where the complex numbers satisfy the condition (5.4).

One particular result can be interesting:

Corollary 5.3. *Let $Q_n(x)$ be defined by* (5.9) *and let $\lambda_0 = \lambda_1 = \cdots = \lambda_n = \lambda$. Then*

$$Q_n(x) = x^\lambda L_n\bigl(-(\lambda - 1/\bar{\lambda}) \log x\bigr), \tag{5.13}$$

where $L_n(x)$ is the Laguerre polynomial orthogonal with respect to e^{-x} *on* $[0, \infty)$ *and such that $L_n(0) = 1$.*

Proof. For $\lambda_0 = \lambda_1 = \cdots = \lambda_n = \lambda$, (5.9) reduces to

$$Q_n(x) = \frac{1}{2\pi i} \oint_\Gamma \frac{(s - 1/\bar{\lambda})^n}{(s - \lambda)^{n+1}} x^s \, ds,$$

where $\lambda \in \operatorname{int} \Gamma$. Since

$$\operatorname*{Res}_{z=\lambda} \left[\frac{(s - 1/\bar{\lambda})^n}{(s - \lambda)^{n+1}} x^s\right] = \frac{1}{n!} \lim_{z \to \lambda} \frac{d^n}{ds^n} \left[(s - 1/\bar{\lambda})^n x^s\right]$$

we obtain by Cauchy's residue theorem

$$\begin{aligned} Q_n(x) &= \frac{1}{n!} \sum_{k=0}^{n} \binom{n}{k} n(n-1) \cdots (k+1)(\lambda - 1/\bar{\lambda})^k x^\lambda (\log x)^k \\ &= x^\lambda \sum_{k=0}^{n} \frac{1}{k!} \binom{n}{k} (\lambda - 1/\bar{\lambda})^k (\log x)^k, \end{aligned}$$

which gives (5.13). □

References

[1] M.M. Aĭzinov, *Selected Topics in Signal Theory and Circuit Theory*, Izdat. Svyaz', Moscow, 1971. (Russian)

[2] H. Bateman and A. Erdélyi, *Higher Transcendental Functions*, Vol. 2, McGraw–Hill, New York, 1953.

[3] G.V. Badalyan, *Generalization of Legendre polynomials and some of their applications*, Akad. Nauk. Armyan. SSR Izv. Ser. Fiz.-Mat. Estest. Tekhn. Nauk **8** (5) (1955), 1–28 and **9** (1) (1956), 3–22. (Russian, Armanian summary)

[4] A.A. Bessonov, Yu.V. Zagashvili, and A.S. Markelov, *Method and tools for Dynamic System Identification*, È nergoizdat, Leningrad, 1989. (Russian)

[5] P. Borwein and T. Erdélyi, *Polynomials and Polynomial Inequalities*, Springer Verlag, New York – Berlin – Heidelberg, 1995.

[6] P. Borwein, T. Erdélyi, and J. Zhang, *Müntz systems and orthogonal Müntz-Legendre polynomials*, Trans. Amer. Math. Soc. **342** (1994), 523–542.

[7] M.G. de Bruin, *Polynomials orthogonal on a circular arc*, J. Comput. Appl. Math. **31** (1990), 253–266.

[8] A. Bultheel, P. González-Vera, E. Hendriksen, and O. Njåstad, *The computation of orthogonal rational functions and their interpolating properties*, Numer. Algorithms **2** (1992), 85–114.

[9] A. Bultheel, P. González-Vera, E. Hendriksen, and O. Njåstad, *A Favard theorem for orthogonal rational functions on the unit circle*, Numer. Algorithms **3** (1992), 81–89.

[10] A. Bultheel, P. González-Vera, E. Hendriksen, and O. Njåstad, *Moment problems and orthogonal functions*, J. Comput. Appl. Math. **48** (1993), 49–68.

[11] T. S. Chihara, *An Introduction to Orthogonal Polynomials*, Gordon and Breach, New York, 1978.

[12] M. M. Djrbashian, *Orthogonal systems of rational functions on the circle with given set of poles*, Dokl. Akad. Nauk SSSR **147** (1962), 1278–1281.

[13] M. M. Djrbashian, *Orthogonal systems of rational functions on the circle*, Akad. Nauk. Armyan. SSR Izv. Ser. Mat. **1** (1) (1966), 3–24 and **1** (2) (1966), 106–125. (Russian)

[14] M. M. Djrbashian, *A survey on the theory of orthogonal systems and some open problems*, Orthogonal Polynomials – Theory and Practice (P. Nevai, ed.), NATO ASI Series, Series C; Mathematical and Physical Sciences, Vol. 294, Kluwer, Dordrecht, 1990, pp. 135–146.

[15] L.G. Evlanov, *Dynamic System Control*, Nauka, Moscow, 1972. (Russian)

[16] G. Freud, *Orthogonal Polynomials*, Akadémiai Kiadó/Pergamon, Budapest, 1971.

[17] W. Gautschi, *Algorithm 726: ORTHPOL – a package of routines for generating orthogonal polynomials and Gauss-type quadrature rules*, ACM Trans. Math. Software **20** (1994), 21–62.

[18] W. Gautschi, *Orthogonal polynomials: applications and computation*, Acta Numerica (Cambridge) (1996), 45–119.

[19] W. Gautschi, H. Landau, and G.V. Milovanović, *Polynomials orthogonal on the semicircle, II*, Constr. Approx. **3** (1987), 389–404.

[20] W. Gautschi and G.V. Milovanović, *Polynomials orthogonal on the semicircle*, Rend. Sem. Mat. Univ. Politec. Torino (Special Functions: Theory and Computation) (1985), 179–185.

[21] W. Gautschi and G.V. Milovanović, *Polynomials orthogonal on the semicircle*, J. Approx. Theory **46** (1986), 230–250.

[22] Ya. L. Geronimus, *Polynomials orthogonal on a circle and their applications*, Zap. Nauč.-issled. Inst. Mat. Mech. HMO **19** (1948), 35–120. (Russian)

[23] Ya. L. Geronimus, *Polynomials orthogonal on a circle and interval*, Gos. Izd. Fiz.-Mat. Lit., Moscow, 1958. (Russian)

[24] Ya. L. Geronimus, *Orthogonal polynomials*, Amer. Math. Soc. Transl. **108** (1977), 37–130.

[25] A.H. Gray and J.D. Markel, *A normalized digital filter structure*, IEEE Trans. **ASSP-23** (1975), 268–277.

[26] S. Kuś, *On the completeness of certain systems in* $L^2(|z|=1)$, Demonstratio Math. **18** (1985), 395–408.

[27] P.C. McCarthy, J.E. Sayre, and B.L.R. Shawyer, *Generalized Legendre polynomials*, J. Math. Anal. Appl. **177** (1993), 530–537.

[28] G.V. Milovanović, *Numerical Analysis,* Part I, Naučna Knjiga, Belgrade, 1988. (Sebian)

[29] G.V. Milovanović, *Complex orthogonality on the semicircle with respect to Gegenbauer weight: Theory and applications*, Topics in Mathematical Analysis (Th.M. Rassias, ed.), World Scientific Publ., Singapore, 1989, pp. 695–722.

[30] G.V. Milovanović, *On polynomials orthogonal on the semicircle and applications*, J. Comput. Appl. Math. **49** (1993), 193–199.

[31] G.V. Milovanović, *Generalized Hermite polynomials on the radial rays in the complex plane*, Theory of Functions and Applications (H. B. Nersessian, ed.), Collection of Works Dedicated to the Memory of M. M. Djrbashian, Louys Publishing House, Èrevan, 1995, pp. 125–129.

[32] G. V. Milovanović, *Orthogonal polynomials and applications*, Scientific Review **14** (1995), 5–28.

[33] G. V. Milovanović, *Some nonstandard types of orthogonality (A survey)*, FILOMAT (Formerly: Zb. Rad.) **9** (1995), 517–542.

[34] G. V. Milovanović, *A class of orthogonal polynomials on the radial rays in the complex plane*, J. Math. Anal. Appl. **206** (1997), 121–139.

[35] G.V. Milovanović, *Orthogonal polynomial systems and some applications*, This Book.

[36] G.V. Milovanović, D.S. Mitrinović and Th.M. Rassias, *Topics in Polynomials: Extremal Problems, Inequalities, Zeros*, World Scientific, Singapore – New Jersey – London – Hong Kong, 1994.

[37] G.V. Milovanović and P.M. Rajković, *Geronimus concept of orthogonality for polynomials orthogonal on a circular arc*, Rend. di Matematica, Serie VII (Roma) **10** (1990), 383–390.

[38] G.V. Milovanović and P.M. Rajković, *On polynomials orthogonal on a circular arc*, J. Comput. Appl. Math. **51** (1994), 1–13.

[39] G.V. Milovanović and P.M. Rajković, *Complex orthogonal polynomials with the Hermite weight*, Univ. Beograd Publ. Elektrotehn. Fak. Ser. Mat. **6** (1995), 63–73.

[40] G.V. Milovanović, P.M. Rajković and Z.M. Marjanović, *A class of orthogonal polynomials on the radial rays in the complex plane, II*, Facta Univ. Ser. Math. Inform. **11** (1996), 29–47.

[41] T. Muir, *The Theory of Determinats,* Vol. I, Macmillan & Co., London, 1906.

[42] K. Müller and A. Bultheel, *Report TW253*, Katholieke Univ. Leuven, Leuven, 1997.

[43] P. Nevai, *Orthogonal Polynomials*, Memoirs Amer. Math. Soc., Vol. 213, Amer. Mat. Soc., Providence, R.I., 1979.
[44] P. Nevai, *Géza Freud, orthogonal polynomials and Christoffel functions, A case study*, J. Approx. Theory **48** (1986), 3–167.
[45] X. Nie, D. Raghuramirddy, and R. Unbehauen, *Orthonormal expansion of stable rational transfer functions*, Electron. Lett. **27** (1991), 1492–1494.
[46] K. Pan, *Strong and weak convergence of orthonormal systems of rational functions on the unit circle*, J. Comput. Appl. Math. **46** (1993), 427–436.
[47] K. Pan, *Extensions of Szegő's theory of rational functions orthogonal on the unit circle*, J. Comput. Appl. Math. **62** (1995), 321–331.
[48] K. Pan, *Convergence of orthogonal rational functions outside of the unit circle*, J. Comput. Appl. Math. **69** (1996), 345–351.
[49] K. Pan, *On the convergence of rational functions orthogonal on the unit circle*, J. Comput. Appl. Math. **76** (1996), 315–324.
[50] P.K. Suetin, *Classical Orthogonal Polynomials*, Nauka, Moscow, 1976. (Russian)
[51] G. Szegő, *Beiträge zur Theorie der Toeplitzschen Formen*, Math. Z. **6** (1920), 167–202.
[52] G. Szegő, *Beiträge zur Theorie der Toeplitzschen Formen. II*, Math. Z. **9** (1921), 167–190.
[53] G. Szegő, *Orthogonal Polynomials*, Amer. Math. Soc. Colloq. Publ., vol. 23, 4th ed., Amer. Math. Soc., Providence, R. I., 1975.
[54] A.K. Taslakyan, *Some properties of Legendre quasi-polynomials with respect to a Müntz system*, Mathematics, Èrevan University, Èrevan **2** (1984), 179–189. (Russian, Armenian summary)
[55] J.L. Walsh, *Interpolation and Approximation in the Complex Plane*, 5th ed., Amer. Math. Soc., Providence, R.I., 1969.

Bratislav Danković
University of Niš
Faculty of Electronic Engineering
Department of Automatics, P.O. Box 73
18000 Niš, Serbia – Yugoslavia

Gradimir V. Milovanović
University of Niš
Faculty of Electronic Engineering
Department of Mathematics, P.O. Box 73
18000 Niš, Serbia – Yugoslavia

E-mail: grade@efnis.elfak.ni.ac.yu

Snežana Lj. Rančić
University of Niš
Faculty of Electronic Engineering
Department of Automatics, P.O. Box 73
18000 Niš, Serbia – Yugoslavia

J-L GARCIA-ROIG

On the group of isometries and a system of functional equations of a real normed plane

1. The group of isometries of a real normed plane.

The problem considered in this section is an interesting case of the general geometric problem of describing the group of isometries of a given metric space and, indeed, may be considered as a generalization of [9].

To begin with, if $\|\cdot\|$ stands for an arbitrary norm of the real plane $\mathbb{R}^2$, observe that translations are obviously (bijective) isometries (since the distance induced by a norm is translation-invariant) and therefore, if f is any isometry from $\mathbb{R}^2$ onto itself, then $x \mapsto f(x) - f(0)$ is also an isometry leaving the origin fixed. Furthermore, by [2] Th. 11, p. 144, all bijective isometries are necessarily affine maps (so that those fixing the origin are automatically linear automorphisms). These considerations force the whole group of isometries to be equal to the semidirect product of the group of all translations (isomorphic with $\mathbb{R}^2$) by the group of linear isometries, in the same way as it occurs in the case of the plane with its usual metric (cf. [9]) and allow us in what follows, to concentrate only on the isometries of the real plane onto itself which leave the origin fixed, i.e., on the linear isometries. (These arguments also hold for $\mathbb{R}^n$ but we refer only to $\mathbb{R}^2$ for the sake of simplicity).

Now, recall that the unit sphere $S = \{x \in \mathbb{R}^2 : \|x\| = 1\}$ of an arbitrary norm of $\mathbb{R}^2$ consists of a closed continuous curve (remember that we are working in the finite-dimensional case) around the origin which together with its "inner part" (i.e., the unit ball $B = \{x \in \mathbb{R}^2 : \|x\| \leq 1\}$) constitutes a convex set having finite (Lebesgue) measure (cf. [8] Ch. 2). If f is a linear automorphism (i.e., $f \in GL_2(\mathbb{R})$) leaving S invariant (this is the case if f is an isometry), then, by linearity, f also leaves the unit ball B invariant and, consequently (see [10], Ch. V, Lemma 5.1.d.) its determinant must be ± 1. In other words, a linear isometry either belongs to the special linear group $SL_2(\mathbb{R})$ or it has determinant -1. Observe too that in our situation, if f has a real eigenvalue, it has to be necessarily ± 1 for, otherwise, S would not be preserved by f.

In any case, for such a map f (which will be thought of as a 2×2 real matrix), its characteristic polynomial is $X^2 - \tau X \pm 1$, where $\tau=$ trace of f, and $\pm 1=$ determinant of f. This polynomial has discriminant $\Delta = \tau^2 \mp 4$. So, in case det $f = -1$, Δ is always positive and, therefore, f has real eigenvalues. As these have to be ± 1, as remarked above, we conclude that $\tau = 0$ and the eigenvalues are $+1$ and -1 (and, consequently, f has order 2). And in case det $f = +1$ (i.e., $f \in SL_2(\mathbb{R})$), f has real eigenvalues if and only if $|\tau| \geq 2$ but, as these have to be ± 1, we conclude that this is the case if and only if $|\tau| = 2$ and that the eigenvalues have to be both $+1$ or both -1 (for $X^2 \mp 2X + 1) = (X \mp 1)^2$). Now, if $f \neq \pm$ identity, then f is conjugate to either

$$\begin{pmatrix} \pm 1 & \beta \\ 0 & \pm 1 \end{pmatrix} \quad \text{or} \quad \begin{pmatrix} \pm 1 & 0 \\ \gamma & \pm 1 \end{pmatrix},$$

where β and γ are (real and) nonzero. As

$$\begin{pmatrix} 1 & \beta \\ 0 & 1 \end{pmatrix}^n = \begin{pmatrix} 1 & n\beta \\ 0 & 1 \end{pmatrix},$$

and similar formulae hold for the other matrices, we gather that a bounded convex set aroung the origin cannot be invariant under such maps which, therefore, have to be excluded from our considerations.

Thus we are finally left with elements $f \in SL_2(\mathbb{R})$ having no real eigenvalues, i.e., whose trace τ verifies $|\tau| < 2$, and for these we have a curious geometric property as the next lemma shows.

Lemma 1. *An element f in $SL_2(\mathbb{R})$ leaves invariant a centered ellipse (and consequently its homothetic ones with respect to the origin) if and only if either* $|$ trace $(f)| < 2$ *or* $f = \pm$ identity.

Proof. A centered ellipse corresponds algebraically to an equation of type

$$Ax^2 + 2Bxy + Cy^2 = 1, \tag{1}$$

where the symmetric (real) matrix

$$\begin{pmatrix} A & B \\ B & C \end{pmatrix} \tag{2}$$

represents a positive definite quadratic form.

If we apply a transformation of type

$$\begin{aligned} x &= \alpha x' + \beta y' \\ y &= \gamma x' + \delta y' \end{aligned} \quad \text{with} \quad \begin{pmatrix} \alpha & \beta \\ \gamma & \delta \end{pmatrix} \in SL_2(\mathbb{R}),$$

which leaves invariant the ellipse (1), we obtain the conditions

$$\begin{aligned} \alpha^2 A + 2\alpha\gamma B + \gamma^2 C &= A, \\ \alpha\beta A + (\alpha\delta + \beta\gamma)b + \gamma\delta C &= B, \\ \beta^2 A + 2\beta\delta B + \delta^2 C &= C, \end{aligned} \tag{3}$$

But a simple computation, bearing in mind $\alpha\delta - \beta\gamma = 1$, shows that

$$\begin{vmatrix} \alpha^2 - 1 & 2\alpha\gamma & \gamma^2 \\ \alpha\beta & \alpha\delta + \beta\gamma - 1 & \gamma\delta \\ \beta^2 & 2\beta\delta & \delta^2 - 1 \end{vmatrix} = 0,$$

which implies the existence of nontrivial solutions of the system (3). These solutions are readily seen to be of the form

$$\frac{A}{2\gamma} = \frac{B}{\delta - \alpha} = \frac{C}{-2\beta}.$$

If it happened that $\beta = \gamma = 0$ then $\begin{pmatrix} \alpha & \beta \\ \gamma & \delta \end{pmatrix}$ would be diagonal and in our circumstances $\alpha = \delta = \pm 1$, so that $\begin{pmatrix} \alpha & \beta \\ \gamma & \delta \end{pmatrix} = \pm$ identity, which is a trivial situation. Were this not the case, then we can take either A or C positive and as (for some nonzero real λ)

$$\begin{vmatrix} A & B \\ B & C \end{vmatrix} = \lambda^2 \begin{vmatrix} 2\gamma & \delta - \alpha \\ \delta - \alpha & -2\beta \end{vmatrix} = \lambda^2[4 - (\alpha + \delta)^2],$$

we infer that (2) is positive definite if and only if $|\alpha + \delta| < 2$. Reading now in the reversed order we conclude the proof of the lemma.

As a matter of fact, for a centered ellipse it is very easy to obtain all linear automorphisms of the real plane leaving it invariant and to interpret the result geometrically. As the interesting case is that of elements of $SL_2(\mathbb{R})$, we state the following.

Lemma 2. *An element of $SL_2(\mathbb{R})$ leaves the ellipse*

$$\frac{x^2}{a^2} + \frac{y^2}{b^2} = 1$$

invariant if and only if it is of the form

$$\begin{pmatrix} \cos\theta & -k\sin\theta \\ \frac{1}{k}\sin\theta & \cos\theta \end{pmatrix}$$

for some $\theta \in \mathbb{R}$, where $k = a/b$.

Proof. If, as before,

$$\begin{aligned} x &= \alpha x' + \beta y' \\ y &= \gamma x' + \delta y' \end{aligned} \qquad \text{with} \qquad \begin{pmatrix} \alpha & \beta \\ \gamma & \delta \end{pmatrix} \in SL_2(\mathbb{R}),$$

the condition that this transformation leaves the above ellipse invariant is readily seen to be equivalent to the 3 conditions

$$\frac{\alpha^2}{a^2} + \frac{\gamma^2}{b^2} = \frac{1}{a^2}, \qquad \frac{\beta^2}{a^2} + \frac{\delta^2}{b^2} = \frac{1}{b^2}, \qquad \text{and} \qquad \frac{\alpha\beta}{a^2} + \frac{\gamma\delta}{b^2} = 0,$$

or, in other words, that the couple of vectors $(\alpha, \frac{a}{b}\gamma)$ and $(\frac{b}{a}\beta, \delta)$ constitute an orthonormal basis of $\mathbf{R}^2$ with respect to the usual dot product. Consequently, if we set $k = \frac{a}{b}$, there exists θ in $\mathbf{R}$ such that

$$\begin{pmatrix} \alpha & \beta \\ \gamma & \delta \end{pmatrix} = \begin{pmatrix} \cos\theta & -k\sin\theta \\ \frac{1}{k}\sin\theta & \cos\theta \end{pmatrix}. \qquad \text{Q.E.D.}$$

Remark. If $M_k(\theta)$ stands for this latter matrix, it is convenient to think of $M_k(\theta)$ as projecting "vertically" the points of the ellipse considered upon the circle of radius a, centered at the origin, rotating then an angle θ along this circle and finally projecting back again "vertically" from this circle onto the ellipse.

In this sense, it is convenient to associate an "angle" to each point P of an ellipse (see figure 1): the angle Φ of P is the usual angle of its projection P' on the outer tangent circle. This makes us think of $M_k(\theta)$ as a usual rotation but acting on points of an ellipse.

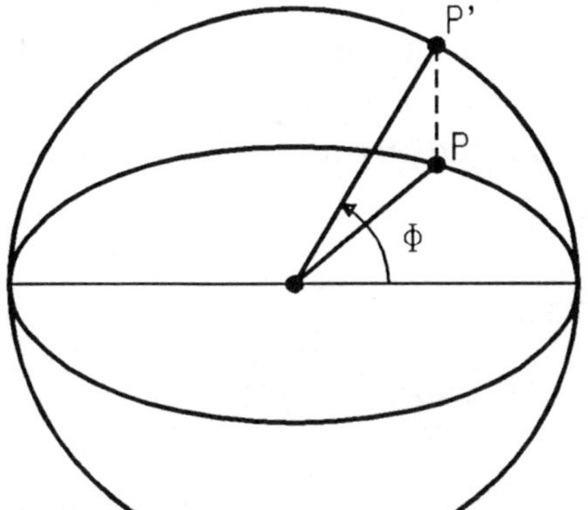

fig. 1

Moreover, this makes clear, for instance, the formula $M_k(\theta)M_k(\theta') = M_k(\theta + \theta')$ which, of course, can be checked by direct computation. (In fact, as θ varies, the map $\varphi : M_k(\theta) \mapsto \frac{\theta}{2\pi} + \mathbf{Z}$ is a group-isomorphism from the multiplicative group of matrices $M_k(\theta)$, for all θ, onto the additive group $\mathbf{R}/\mathbf{Z}$).

Furthermore, $M_k(\theta)$ has finite order if and only if $\theta \in 2\pi\mathbf{Q}$. In fact, it is quite easy to see that the finite subgroups of $\mathbf{R}/\mathbf{Z}$ coincide with the finite subgroups of $\mathbf{Q}/\mathbf{Z}$, which are all cyclic (see [5] Ex. 13, p. 55), and that these subgroups exhaust all those which are not dense in $\mathbb{R}/\mathbb{Z}$. In particular, if θ and 2π (or if we prefer, $\frac{\theta}{2\pi}$ and 1) are linearly independent over $\mathbb{Q}$, i.e., if $M_k(\theta)$ has infinite order, then $\frac{\theta}{2\pi}$ generates a dense subgroup of $\mathbb{R}/\mathbb{Z}$: geometrically speaking this means that the orbit $\{M_k(\theta)^n(P) \mid n \in \mathbb{Z}\}$ of a point P on the ellipse $\frac{x^2}{a^2} + \frac{y^2}{b^2} = 1$ invariant under $M_k(\theta)$ –which has infinite order– is a dense subset of this ellipse.

As any other centered ellipse can be obtained from $\frac{x^2}{a^2} + \frac{y^2}{b^2} = 1$ by a suitable rotation ρ, it is clear that the group of elements of $SL_2(\mathbb{R})$ leaving invariant such an ellipse is the conjugate $\rho\{M_k(\theta)\}\rho^{-1}$ of the group of matrices $\{M_k(\theta)\}$ as θ varies and, obviously, the assertion concerning the dense subset of the ellipse still holds if $\rho M_k(\theta)\rho^{-1}$ (or, equivalently, $M_k(\theta)$) has infinite period.

Concerning different values of k, we have

Lemma 3. *If A and B are conjugates (under suitable rotations) to $M_k(\theta)$ and $M_{k'}(\theta')$ and both other than* $\pm$ identity*) belong to a group G, with $k \neq k'$, then G is infinite and cannot be contained in the group of linear isometries of a norm $\|\cdot\|$ in $\mathbb{R}^2$.*

Proof. The interesting thing is to see what happens if both $M_k(\theta)$ and $M_{k'}(\theta')$ have finite order (necessarily ≥ 3). But first, by rescaling (if necessary) the second axis of coordinates, observe that we can reduce our problem to the case when the lowest of the values k and k' is 1, so that, geometrically speaking, if we think of a couple of (centered) ellipses invariant respectively under A and B, i.e., ellipses obtained by rotating, if necessary, around the origin those of equations

$$\frac{x^2}{a^2} + \frac{y^2}{b^2} = 1,$$

for $\frac{a}{b} = k$ or k', we can assume that one of them is a circle and then, by substituting the other by a homothetic one (with respect to the origin), we can assume the second ellipse E to be tangent and contained inside the circle. Furthermore, in order to simplify our arguments, the tangent points may be thought of to lie on the horizontal axis.

We are now going to prove that the orbit of one of these two points, say P (see figure 2), under suitable transforms of the group generated by our two elements which, under our present circumstances, may be assumed to be $A = M_1(\theta)$ and $B = M_k(\theta')$, with $k > 1$, is infinite.

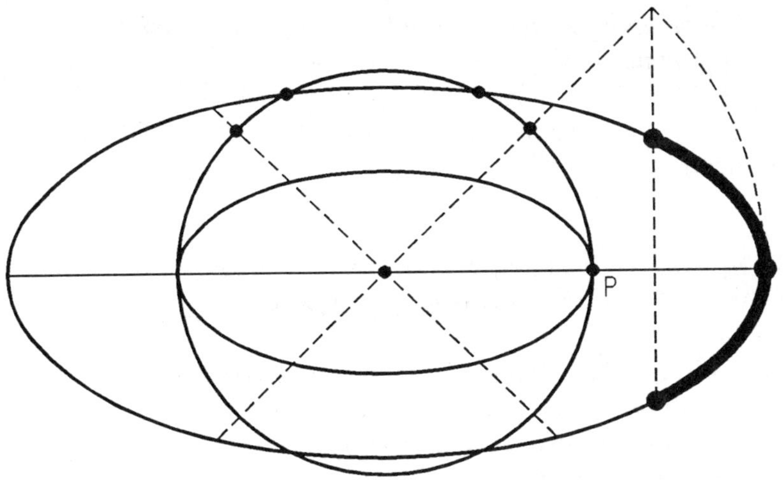

fig. 2

To this end, apply a suitable power of A to P so that the transform P_1 of P lies between 45° and 135° (this can always be done, for, we have $\theta = k \cdot \frac{360^o}{n}$, with k, n coprime and $n \geq 3$). Then apply (the inverse of) a suitable power of B to P_1 so that the new transform P_2 of P_1 (which has to lie on the homothetic ellipse to E throught P_1) has an "angle" (see the remark after Lemma 2) lying between -45° and 45° (the thick line on figure 2) or between 135° and 225° (this can be done by a reason similar to the previous one, but the reader is advised to perform some drawings on his own in order to realise all the details). Then, again, a suitable power of A will transform P_2 into P_3 having an angle either between 45° and 135° or between 225° and -45°, then a suitable power of B, etc. It is easily seen that this process can be continued ad infinitum and that the radii of the circles through P, P_1, through P_2, P_3, etc. tend to infinity (use a variant of a geometric progression argument). From these facts our assertion follows immediately.

Now we can establish the following

Theorem 1. *If $G \subset SL_2(\mathbb{R})$ is an infinite group of linear isometries of a norm $\|\cdot\|$ in $\mathbb{R}^2$, then $\|\cdot\|$ is induced by an inner product.*

Proof. By lemma 3, G can be assumed to be contained in a group $\{M_k(\theta) \mid \theta \in \mathbb{R}\}$ for a fixed k. In this case the unit sphere of the norm $\|\cdot\|$ is forced to be (geometrically) an ellipse, and this case is easily seen to correspond to that where the norm comes from an inner product. Q.E.D.

Remark. In the above theorem it is not required that G contains elements of infinite order. For instance, with our previous notations, $\varphi(G)$ can be an infinite subgroup of $\mathbb{Q}/\mathbb{Z}$, the elements of which have all finite order.

Corollary. *The group of all linear isometries (of determinant* ± 1*) of* $(\mathbb{R}^2, \|\cdot\|)$ *is finite if and only if* $\|\cdot\|$ *is not induced by an inner product.*

Proof. Trivial, from theorem 1, if G is contained in $SL_2(\mathbb{R})$. Otherwise, consider the map "determinant":

$$\det : G \to \{\pm 1\},$$

which is clearly a surjective group-homomorphism whose kernel G' has index 2 in G, so that, G' is finite if and only if G is so. Q.E.D.

Examples. 1. (cf. [9]). The group of all linear isometries of the taxicab norm (whose unit sphere is the square with vertices at $(\pm 1, 0)$, $(0, \pm 1)$ is the dihedral group D_4 (which is the symmetry group of a square). It consists of four rotations (= the elements of $D_4 \cap SL_2(\mathbb{R})$) and four symmetries (=(in notation of corollary) the four elements of $G \backslash G'$). This norm is not derivable from an inner product.

2. Consider the convex octagon with consecutive vertices: $(1,0)$, $(\frac{1}{2}, 1)$, $(0,1)$, $(-1, \frac{1}{2})$, $(-1, 0)$, $(-\frac{1}{2}, -1)$, $(0, -1)$ and $(1, -\frac{1}{2})$. It makes sense to take it as a unit ball for a norm in $\mathbb{R}^2$ (recall Minkowski's functionals: see for instance [7], p. 24). Then the group of all linear isometries admits no elements of determinant -1 (an easy exercise) and therefore consists of just 4 rotations.

These two examples essentially show that in the case of finite groups of linear isometries, both cases actually occur: those which admit elements of determinant -1 and those which do not. This clearly, cannot be the case if G is infinite, for then, the unit sphere has to be an ellipse, which always admits automorphisms of determinant -1 leaving it invariant.

2. On some functional equations involving linear mappings and norms of the real plane.

If E and F are (real) inner product spaces then it is easily seen ([3]) that any (continuous) solution $f : E \to F$ of the system of functional equations (for these see [1], [2]):

$$(4) \begin{cases} \text{(i)} & f(tu) = tf(u), \quad \text{for all } t > 0 \text{ and } u \text{ in } E, \\ \text{(ii)} & f(u+v) = f(u) + f(v), \quad \text{whenever } \|u\| = \|v\|, \text{ and} \\ \text{(iii)} & \|f(u)\| = K\|u\|, \quad \text{for some positive constant } K, \end{cases}$$

is a linear transformation (cf. [4] too).

There immediately arises the question of whether a similar result can be obtained if E, F are no longer inner product spaces, but just (real) linear normed spaces.

The aim of this section is to deal with this problem in its simplest case, i.e., when $E = F = \mathbb{R}^2$ with an arbitrary norm $\|\cdot\|$. Thus, we will assume in the sequel that f is a map from $\mathbb{R}^2$ into itself and, moreover, K will be supposed to be equal to 1 (this is indeed the case if f is replaced by $\frac{1}{K}f$).

As a matter of fact we have:

Proposition 1. *Any solution f of* (4) *is continuous.*

Proof. From (ii) it easily follows $f(0) = 0$ and $f(-x) = -f(x)$. This, together with (iii), entails that f is continuous at zero. Now let $x_n \to x \neq 0$. Then, as

$$\left\| \frac{\|x_n\|}{\|x\|} x \right\| = \|x_n\| = \| - x_n\|,$$

we get

$$f\left(\frac{\|x_n\|}{\|x\|}x - x_n\right) = f\left(\frac{\|x_n\|}{\|x\|}x\right) - f(x_n) = \frac{\|x_n\|}{\|x\|}f(x) - f(x_n),$$

so that

$$f(x_n) = \frac{\|x_n\|}{\|x\|}f(x) - f\left(\frac{\|x_n\|}{\|x\|}x - x_n\right) \to \frac{\|x\|}{\|x\|}f(x) - 0 = f(x).$$

Q.E.D.

Proposition 2. *Any solution f of* (4) *is* $\mathbb{R}$*-linear.*

Proof. The proof is most easily conveyed in geometric terms.

Take a couple of $\mathbb{Q}$-linearly independent unit vectors u, v. Then $f(u)$ and $f(v)$ are also linearly independent over $\mathbb{Q}$ (if $f(u) = f(v)$, then, as $f(-u) = -f(u)$, we get $0 = f(-u+v)$, a contradiction since $-u+v \neq 0$, and similarly if $f(u) = -f(v)$).

After this little remark, forced by (ii), we construct $f(u+v) := f(u)+f(v)$. If $\alpha(u+v)$ is the unique unit vector on the halfline spanned by $u+v$ (for some $\alpha \geq \frac{1}{2}$), then, by (i) and (ii), we get $f(\alpha u + \alpha v) = \alpha f(u) + \alpha f(v)$, which is, by (iii), the unit vector on the halfline spanned by $f(u) + f(v)$.

We now proceed in this way with the couples of vectors u and $\alpha u + \alpha v$ on the one hand and $\alpha u + \alpha v$ and v on the other, i.e. we set for the first couple

$$f(u + (\alpha u + \alpha v)) := f(u) + f(\alpha u + \alpha v),$$

and similarly for the second. Observe that this relation may be rewritten like this:

$$f((1+\alpha)u + \alpha v) = (1+\alpha)f(u) + \alpha f(v),$$

and then pay attention to the unique unit vector $\beta((1+\alpha)u + \alpha v)$, for some $\beta \geq \frac{1}{2}$, on the halfline spanned by $u + \alpha(u+v)$.

We continue this process indefinitely: suppose that after k steps we have a finite collection of (ordered) unit vectors starting with u and finishing with v, for which linearity of f applied to these vectors holds. The next step consists of inserting in between each couple of consecutive vectors the (unique) unit vector on the halfline spanned by the sum of these consecutive vectors. (Though not exactly, this reminds us of the construction of Farey sequences in Number Theory (see [6]).) Now let $\lambda u+\mu v$ and $\lambda' u+\mu' v$ be two consecutive unit vectors of the kth step. Recall that (induction hypothesis) for these hold

$$f(\lambda u+\mu v)=\lambda f(u)+\mu f(v) \quad \text{and} \quad f(\lambda' u+\mu' v)=\lambda' f(u)+\mu' f(v).$$

Now consider the unit vector

$$w=\rho[(\lambda+\lambda')u+(\mu+\mu')v],$$

for a (unique) suitable positive real ρ. Then, by (i), (ii) and the relations three lines above, we have

$$\begin{aligned} f(w)&=\rho f[(\lambda+\lambda')u+(\mu+\mu')v]=\rho f[(\lambda u+\mu v)+(\lambda' u+\mu' v)]\\ &=\rho[f(\lambda u+\mu v)+f(\lambda' u+\mu' v)]=\rho[\lambda f(u)+\mu f(v)+\lambda' f(u)+\mu' f(v)]\\ &=\rho[(\lambda+\lambda')f(u)+(\mu+\mu')f(v)]. \end{aligned}$$

If the ratios μ/λ of the linear combinations $\lambda u+\mu v$ occurring in the steps made above are dense in $[0,\infty)$, then (proceeding in the same way with $\pm u$ and $\pm v$) by the continuity proved in the previous proposition, we infer that there exists a unique map $f:E\to F$ satisfying now, for all real λ,μ,

$$f(\lambda u+\mu v)=\lambda f(u)+\mu f(v),$$

i.e., that f is linear and we will be done.

In order to prove this density property, we are going to proceed in as simple geometric terms as we can:

Observe that as $\mathbb{R}^2$ is finite-dimensional, if we also consider the usual euclidean norm $\|\cdot\|_{\text{eucl}}$, then there exist two positive constants $L\geq l$ such that

$$l\leq \|x\|_{\text{eucl.}}\leq L,$$

for all x such that $\|x\|=1$.

Now draw the triangle (see figure 3) on two vectors a, b forming an angle A of not more than 90° and whose lengths are L and l respectively. We want to find out some relations between angles B and A.

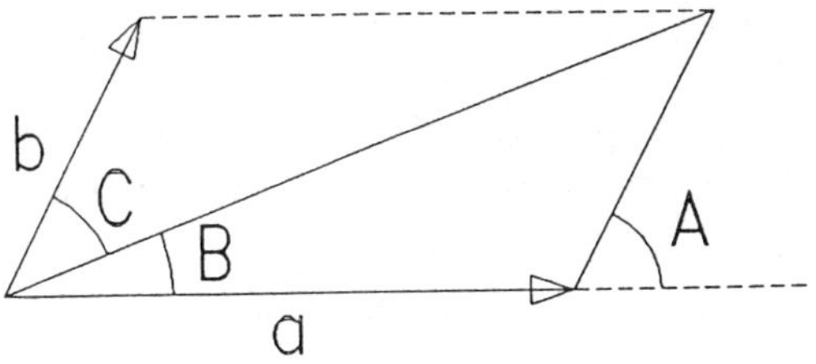

fig. 3

Prolong (see figure 4) the shortest side n times (archimedean property) so that $nl \geq L$ and consider the isosceles triangle of fig. 4

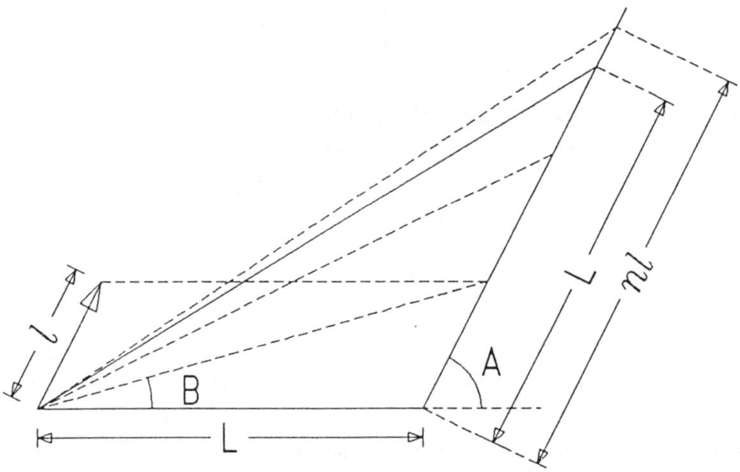

fig. 4

to realise (by very simple geometric arguments) that $nB \geq \frac{A}{2}$ or written otherwise, $B \geq \frac{A}{2n}$.

[If $A > 90^o$ then by splitting, i.e. by considering the question for a and $a+b$, and for $a+b$ and b, we arrive at similar results, unless some angle still remains obtuse, in which case we iterate the necessary number of times]

This argument trivially entails that the inequality $B \geq \frac{A}{2n}$ (which clearly entails $A - B \leq A\left(1 - \frac{1}{2n}\right)$) also holds if the euclidean norms of a, b just lie on the interval $[l, L]$, and is the key point to prove the density property, for, the sequences

$$\left(\frac{1}{2n}\right)^k \qquad \text{and} \qquad \left(1 - \frac{1}{2n}\right)^k$$

both tend to zero when $k \to \infty$. Q.E.D.

Remark. Obviously, a simple inspection of the proofs shows that the two preceding propositions hold for arbitrary E, not only for $E = \mathbb{R}^2$ (if necessary, work with nets instead of sequences in the proof of proposition 1).

Now that we have proved that all solutions f of (4) are linear, we can apply the results of Section 1.

So we have:

Theorem 2. *Let $\|\cdot\|$ be a norm of $\mathbb{R}^2$. Then all solutions f of the system (4) of functional equations for $f : \mathbb{R}^2 \to \mathbb{R}^2$ are necessarily linear. Obviously $f = \pm K\cdot$ identity is always a solution. If there is no other solution, then $\|\cdot\|$ can be quite arbitrary. Otherwise, i.e. if there is a solution of (4) other than $\pm K\cdot$ identity, then either $\|\cdot\|$ is induced by an inner product or $\|\cdot\|$ is of a very special type. More precisely, if (4) admits an infinite number of solutions, then $\|\cdot\|$ is induced by an inner product. In the remaining cases we distinguish two possibilities:*

i) *Apart from $K\cdot(\pm$ identity$)$, there are no other solutions having positive determinant. In this case the whole group of solutions of (4) is $\{\pm f, \pm K\cdot$ identity $\}$, with f conjugate to $\begin{pmatrix} 1 & 0 \\ 0 & -1 \end{pmatrix}$.*

(An example: take as unit sphere the hexagon on vertices (1,0), (1,1), (0,1), (−1,0), (−1,−1) and (0,−1)).

ii) *If there are other solutions of positive determinant, then the unit sphere of $\|\cdot\|$ is obtained from a suitable curve under the action of a linear map (of positive determinant) having finite order (such that the orbit of the curve –together with its inner part– yields a closed convex set symmetric with respect to the origin, which has to be an interior point).*

In fact, such a linear map can be chosen to be conjugate (under a rotation) to

$$K\begin{pmatrix} \cos\theta & -k\sin\theta \\ \frac{1}{k}\sin\theta & \cos\theta \end{pmatrix},$$

with $\theta = \dfrac{2\pi}{n}$, for some positive integer $n \geq 3$.

(An easy example of this last possibility is provided by the unit sphere consisting of a (centered) regular convex polygon with an even number of sides: of course the corresponding norm does not come from an inner product! (cf. [9] for the case of a square)).

Proof. Immediate from Proposition 2 and Section 1.

Remarks.

1. It is not difficult to realise that if (4) has an infinite number of solutions, then there always exist solutions having infinite orders (modulo homotheties, if $K \neq 1$).
2. At the beginning of this section we have assumed that both $E = F = \mathbb{R}^2$ and that $\|\cdot\|_E = \|\cdot\|_F$ and have solved (4) under this condition. Let S be the set of its solutions (actually a group under composition). Assume now that $\|\cdot\|_E \neq \|\cdot\|_F$. Well, if there is a solution $g_0 : E \to F$ of (4) in these circumstances, then obviously all the solutions $g : E \to F$ of (4) are given by: $g = g_0 \circ f$, for $f \in S$. This can be expressed simply by saying that in the case of different norms on E and F, either (4) has no solutions or it has, in which case the norm on F is obtained from that of E by applying a suitable linear automorphism of $\mathbb{R}^2$.

Acknowledgement.

The author is deeply indebted to Prof. Claudi Alsina (Barcelona) who suggested the system of functional equations dealt with in Section 2 and for his help and constant encouragement during the preparation of this paper.

References.

[1] Aczél, J., *Lectures on Functional Equations and Their Applications.* Academic Press, New York, 1966.

[2] Aczél, J. and Dhombres, J., *Functional Equations in Several Variables.* Cambridge Univ. Press, 1989.

[3] Alsina, C., On the functional equation of the bisectrix transform, to appear.

[4] Alsina, C. and Garcia-Roig, J-L., On a conditional Cauchy equation on rhombuses, *Functional Analysis, Approximation Theory and Numerical Analysis.* Edited by J.M. Rassias. World Scientific Pub. Co. (1994), pp. 5-7.

[5] Lang, S.: *Algebra*, 2nd edition, Addison-Wesley Pub. Co., 1984.

[6] Niven, I. and Zuckerman, H.S., *An introduction to the Theory of Numbers*, John Wiley & Sons, Inc., New York, 1966.

[7] Rudin, W.: *Functional Analysis.* Tata McGraw-Hill Pub. Co., 1974.

[8] Rudin, W.: *Real and Complex Analysis*, 3rd edition. McGraw-Hill Book Co., 1987.

[9] Schattschneider, D.J.: The Taxicab Group, Amer. Math. Monthly, 91 No. 7 (1984), 423-428.

[10] Williamson, J.H.: *Lebesgue Integration.* Holt, Rinehart & Winston, Inc. 1962.

Secció de Matemàtiques i Informàtica. E.T.S.A.B.
Univ. Politècnica de Catalunya
Diagonal 649, 08028 Barcelona (Spain)

SAMUEL S HOLLAND, JR

The eigenvalues of the sum of two projections

Noncommutativity and infinite dimensionality seem to lie at the source of the mysteries of Hilbert space. Consider a theorem of Peter Fillmore [1]: given any bounded selfadjoint operator T on a separable infinite-dimensional complex Hilbert space, there exists a positive real number α such that $\alpha T + 4I$ equals the sum of eight or fewer projections. Not a linear combination: simply a sum. Put another way, Fillmore's theorem says that **after scaling, any bounded selfadjoint operator equals the sum of eight or fewer projections.**

In this expository paper we consider the simplest case: the sum of two noncommuting projections. We aim to explain how we may construct projections E and F so that their sum $E + F$ has for eigenvalues any prescribed sequence $1 \pm \lambda_i$, $0 < \lambda_i < 1$, each with preassigned multiplicity.

Consider first two one-dimensional subspaces M and N which intersect at angle θ (see Figure 1). The projection operator E belonging to the subspace M has the formula $Ex = <x, m> m$ for a fixed unit length vector m in our subspace M. Here the symbol $<\cdot,\cdot>$ stands for the inner or dot product defining our complex Hilbert space. This inner product (positive definite Hermitian form) defines a length $\| x \| = <x, x>^{1/2}$ and a corresponding metric $d(x, y) = \| x - y \|$ that provides the basis for topological concepts, such as "closed", etc. When vectors m and n meet at angle θ we have

$$\cos\theta = \frac{<m, n>}{\| m \| \, \| n \|}$$

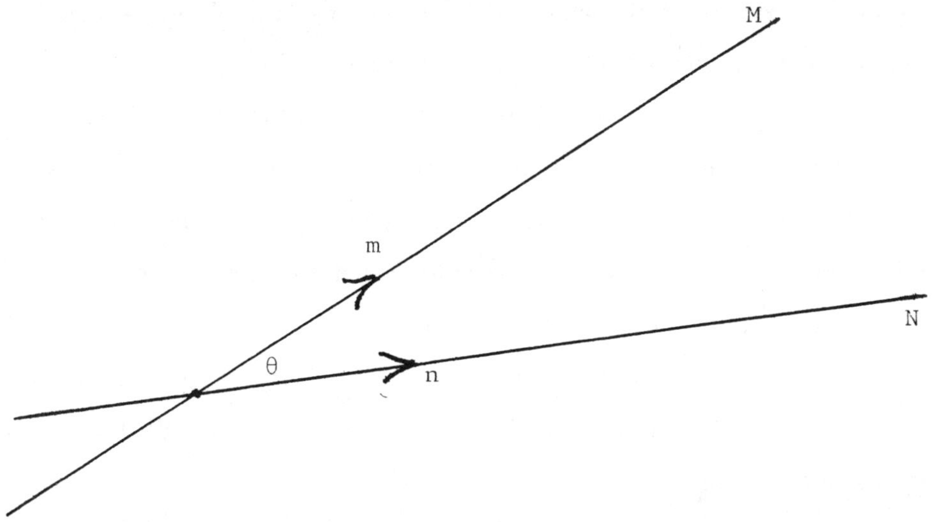

FIGURE 1

In our simple two-dimensional configuration, the operator $E+F$, E the projection on M, F the projection on N, has the eigenvalues $1+\lambda$ and $1-\lambda$, where $\lambda = \cos\theta$. These eigenvalues have for their corresponding eigenspaces the lines that bisect the angles between M and N. Because if m and n are vectors of unit length in their respective subspaces M and N, then

$$\begin{aligned}(E+F)(m+n) &= Em + En + Fm + Fn \\ &= m + \langle n, m\rangle m + \langle m, n\rangle n + n \\ &= (1+\lambda)m + (1+\lambda)n \\ &= (1+\lambda)(m+n).\end{aligned}$$

Remembering that we call the complex number μ an **eigenvalue** of the operator T when $Tx = \mu x$ for some nonzero vector x, we see from this calculation that $1+\lambda$ qualifies as an eigenvalue for $E+F$. And using the vector $m-n$ in place of $m+n$, we see that $1-\lambda$ is also an eigenvalue of $E+F$.

The nonzero vectors x for which $Tx = \mu x$ we call the **eigenvectors** for μ. The totality of eigenvectors belonging to a fixed eigenvalue μ, together with the zero vector, make up a closed subspace called the **eigenspace** for μ. In the example under consideration, the line determined by the vector $m+n$, which bisects one angle between M and N, qualifies as the eigenspace for $1+\lambda$; the line determined by $m-n$ which bisects the other angle, is the eigenspace for $1-\lambda$.

By varying the angle between our lines M and N, we thus secure projections E and F such that $E+F$ has eigenvalues $1 \pm \lambda$ for any given λ satisfying $0 < \lambda < 1$. Each of these eigenvalues has multiplicity one (the multiplicity equals the dimension of the eigenspace).

There is a higher dimensional generalization of the configuration of Figure 1. We can construct two equal-dimensional closed subspaces M and N, either finite - or infinite-dimensional, such that for their corresponding projections E and F, the operator $E+F$ has only the eigenvalues $1 \pm \lambda$, where λ equals the "cosine of the angle" between M and N. These eigenvalues each have multiplicity equal to the common dimension of M and N.

How can we speak sensibly of an "angle" between subspaces M and N? We shall put our sought-after property this way. Given any nonzero vector m in our subspace M, consider the angle θ between m and its projection Fm on N:

$$\cos\theta = \frac{<m, Fm>}{\|m\| \, \|Fm\|}.$$

Can we so arrange the subspaces M and N so that this $\cos\theta$ stands constant, independent of the choice of the vector m in M? Yes we can. Of course we want also to demand that given any n in N, the cosine of the angle between n and En does not depend on the choice of n. Moreover we want to get the same (constant) $\cos\theta$ as we got in the first case.

1, Theorem. *Let M and N represent closed subspaces of our Hilbert space, E and F their corresponding projections, and let λ stand for a fixed real number* *satisfying* $0 < \lambda < 1$. *Then the following two statements are equivalent:*

(1) *For every nonzero* m *in* M *and nonzero* n *in* N,
$$\lambda = \frac{<m, Fm>}{\| m \| \| Fm \|} = \frac{<En, n>}{\| En \| \| n \|},$$

(2) $EFE = \lambda^2 E$ *and* $FEF = \lambda^2 F$.

Either statement implies that:

(3) $M \cap N = M^{\perp} \cap N = M \cap N^{\perp} = 0$, *and the spectrum of* $E + F$ *consists solely of the eigenvalues* $1 \pm \lambda$, *and possibly the eigenvalue* 0.

Before getting into the proof, I shall explain some of the notation introduced. The notation $M^{\perp}$ stands for the closed subspace consisting of all vectors orthogonal to every vector in $m : x \in M^{\perp} \Longleftrightarrow <x, m> = 0$ for every $m \in M$. Given any closed subspace M of our Hilbert space H, we have $H = M + M^{\perp}$ in the sense that every $x \in H$ has a unique representation $x = m + k$, $m \in M$, $k \in M^{\perp}$. The operator E on H defined by setting $Ex = m$ we call the **projection on** M. In the case of a one-dimensional M, we get back the formula for E introduced earlier. A projection E has the properties $E^2 = E^* = E$, where E^* is the Hilbert-space adjoint of E defined by the equation $<Ex, y> = <x, E^* y>$. These algebraic properties characterize projections: If a bounded operator T satisfies $T^2 = T^* = T$, then T is the projection on the closed subspace M determined by $x \in M \Longleftrightarrow Tx = x$.

Consider now the implication (1) $\Longrightarrow$ (2). Using the algebraic properties of projections, we find $< m, Fm > = < m, F^2 m > = < m, F^* Fm > = < Fm, Fm > = \| Fm \|^2$, so (1) says that $\lambda^2 < m, m > = < Fm, Fm >$ for any $m \in M$. Always $Ex \in M$ for any x, thus $\lambda^2 < Ex, Ex > = < FEx, FEx >$ which leads to $< (EFE - \lambda^2 E)x, x > = 0$ for all x. By polarization, the form $< (EFE - \lambda^2 E)x, y >$ must vanish identically, so $EFE - \lambda^2 E = 0$. Similarly, the other half of (1) implies $FEF = \lambda^2 F$. Thus (1) $\Longrightarrow$ (2). To prove the implication (2) $\Longrightarrow$ (1), just reverse the steps. Thus conditions (1) and (2) are equivalent.

To signify that the subspaces M and N (or the corresponding projections E and F) satisfy the equivalent conditions (1) and (2), we shall use the well-justified descriptive phrase M **and** N **(or** E **and** F **) make a uniform angle** θ **with each other**, $\lambda = \cos\theta$. Clearly, for such pairs of subspaces we can sensibly speak of the angle between them.

We may easily construct such pairs. Select an orthonormal set $\{m_1, m_2, ..., m_p\}$ in our space, $p = \infty$ also allowed. Select another orthonormal set $\{k_1, k_2, ..., k_p\}$ orthogonal to the first one, and define $n_i = \lambda m_i - (1-\lambda^2)^{1/2} k_i$. Then the closed subspace spanned by the m_i makes a uniform angle $\lambda = \cos\theta$ with the closed subspace N spanned by the n_i .

We can construct uniform angle pairs in another way. We call a bounded operator W on our Hilbert space H **partially isometric** if there exists a closed subspace M of H so that, restricted to M, W preserves the inner product, $<Wm, Wm>=<m, m>$ for all $m \in M$, and $W \equiv 0$ on $M^{\perp}$. A partial isometry W defines an inner-product-preserving map of the domain subspace M onto another closed subspace K. The adjoint operator W^*, also a partial isometry, gives the inverse mapping to W. It maps the range space K back onto M. Hence $W^*W = E$, the projection on M, and $WW^* = B$, the projection on K.

2, Lemma. *Given a projection* E *and a partial isometry* W *such that* $WW^* = E$ *and such that* $B = W^*W$ *is orthogonal to* E, *then for any fixed real* λ *satisfying* $0 < \lambda < 1$, *the formula*

$$F = \lambda^2 E + (1-\lambda^2)B - \lambda(1-\lambda^2)^{1/2}(W + W^*)$$

defines a projection F *that makes a uniform angle* θ *with* E, $\lambda = \cos\theta$.

The proof consists in verifying the equations $F^* = F^2 = F$, which check that F is a projection, and in verifying the equations $EFE = \lambda^2 E$, $FEF = \lambda^2 F$, which check that E and F make a uniform angle with each other. To verify, just do the straightforward algebra and use the equations $EB = 0$, $EW = W = WB$.

Lemma 2 shows that we can find a subspace N making a uniform angle with a given subspace M provided only that $\dim(M^\perp) \geq \dim(M)$, because when the dimension of $M^\perp$ exceeds the dimension of M then we can always map M isometrically onto $K \subseteq M^\perp$. Thus if $\dim(M^\perp) = \infty$, or (in the finite-dimensional case) if $\dim(M) \leq \frac{1}{2}\dim(H)$, we can always find a companion subspace to M that makes a uniform angle with it. Thus uniform angle pairs exist in abundance.

And, exactly as in the two-dimensional case depicted in Figure 1, the general uniform angle pair admits two angle bisectors:

3, Lemma. *If the projections E and F have a uniform angle $\lambda = \cos\theta$ between them, $0 < \lambda < 1$, then each of the projections*

$$G^{\pm} = \frac{1}{2(1 \pm \lambda)}(E + F \pm \frac{1}{\lambda}(EF + FE))$$

makes a uniform angle $\theta/2$ with both E and F. Both $G^{\pm}$ have dimension equal to the common dimension of E and F.

The proof consists in verifying the equations $G^* = G^2 = G$, which check that G is a projection, and in verifying the equations $GEG = \mu^2 G$, $EGE = \mu^2 E$, $GFG = \mu^2 G$, $FGF = \mu^2 F$ which check that G makes an equal uniform angle $\theta/2$ with both E and F, $\mu = \cos(\theta/2)$, $\mu^2 = \frac{1}{2}(1 + \lambda)$. Just do the straightforward algebra, and use the equations $EFE = \lambda^2 E$, $FEF = \lambda^2 F$.

Similarly, one can check the equation $SES = F$ for the operator $S = 2G - I$. The operator S is selfadjoint $(S^* = S)$ and unitary $(SS^* = S^*S = I)$, thus defines a special kind of isometric map of our Hilbert Space onto itself that we call a **symmetry**. The operator S leaves the vectors in G pointwise fixed, and reflects other vectors through G. The equation $SES = F$ then has this interpretation: the reflection through the angle bisector G takes E onto F. Exactly what we would expect from our protype example. Thus the projections in a uniform angle pair have the same dimension; in particular $E, F,$ and G all have the same dimension. That completes the proof of Lemma 3.

At this point we should remind ourselves that we have not yet proved that statement (3) of Theorem 1 follows from statement (1) and (2); we have not yet established the connection between uniform angle pairs and eigenvalues. Before we do this, let us consider one more nice simple geometric property of the two-line configuration in Figure 1. In that figure, draw a circle of unit radius, center at the intersection of M and N. From any point p on that unit circle, drop perpendiculars to M and N, and label the points of intersection x and y respectively. Then the distance $d(x, y)$ between x and y does not depend on the choice of the original point p on the unit circle. In fact $d(x, y) = (1 - \lambda^2)^{1/2}$ where $\lambda = \cos\theta$.

This geometrical fact carries over verbatim to the general case: Consider closed subspaces M and N with corresponding projections E and F. **If E makes a uniform angle θ with F, then**

$$\| Ex - Fx \| = (1 - \lambda^2)^{1/2} \| x \|$$

for all $x \in M \vee N$, where $\lambda = \cos\theta$. (By $M \vee N$ we mean the closure of the subspace spanned by the $m + n$, $m \in M$, $n \in N$)

Let us now get on with the remaining arguments to establish Theorem 1. Refer to statement (3) there. We need to show that if closed subspaces M and N make a uniform angle θ with each other then $M \cap N = M^\perp \cap N = M \cap N^\perp = 0$ and the spectrum of $E + F$ consists exactly of the eigenvalues $1 \pm \lambda$, $\lambda = \cos\theta$, together with (possibly) the eigenvalue 0. (E is the projection on M, F the projection on N.)

Suppose that M and N are a uniform angle pair. A vector x that lies in the intersection $M \cap N$ satisfies both $Ex = x$ and $Fx = x$. Using $EFE = \lambda^2 E$ we get $x = \lambda^2 x$ which forces $x = 0$. Thus $M \cap N = 0$. If $x \in M^\perp \cap N$, then $Ex = 0$ and $Fx = x$ which leads to $0 = \lambda^2 x$, thus $x = 0$ again. Hence $M^\perp \cap N = 0$, and similarly $M \cap N^\perp = 0$.

Closed subspaces that satisfy the intersection conditions $M \cap N = M^\perp \cap N = M \cap N^\perp = 0$ we sometimes refer to as in "general position". The subspace $M \cap N$ is the eigenspace of $E+F$ that belongs to the eigenvalue 2, and $(M^\perp \cap N) \oplus (M \cap N^\perp)$ the eigenspace for the eigenvalue 1. Thus the just-verified conditions tell us that $\lambda = 1$ and $\lambda = 2$ are not eigenvalues of $E + F$. The possible eigenvalue $\lambda = 0$ has for eigenspace $M^\perp \cap N^\perp$. Thus $\lambda = 0$ will be an eigenvalue of $E + F$ exactly when $M^\perp \cap N^\perp \neq 0$.

The **spectrum** of operator T consists of those complex numbers λ for which $T - \lambda I$ fails to have a two-sided inverse in the algebra of all bounded operators on our Hilbert space. The **resolvent set** consists of the remaining points. Statement (3) of Theorem 1 asserts that $1 + \mu$ always lies in the resolvent set of $E + F$ if $0 < |\mu| < 1$, $|\mu| \neq \lambda$. And one may check this by verifying directly that the operator R defined by $(\lambda^2 - \mu^2)((1 + \mu)R + I) = EF + FE + \mu(E + F)$ is a two-sided inverse of $E + F - (1 + \mu)I$ when $0 < |\mu| < 1$, $|\mu| \neq \lambda$. Just use the equations $EFE = \lambda^2 E$ and $FEF = \lambda^2 F$.

To complete the proof that $(2) \implies (3)$ we need finally to verify that $1 \pm \lambda$ are eigenvalues of $E + F$. But we already have a formula for the angle bisectors of E and F given in Lemma 3. A direct calculation then shows that indeed these angle bisectors are the eigenspaces for the eigenvalues $1 \pm \lambda$. We have completed the proof that $(2) \implies (3)$ (in Theorem 1).

So the simple configuration depicted in Figure 1, which consists of two lines M and N intersecting at angle θ, generalizes to any dimension. In the generalized configuration, the closed subspaces M and N intersect at uniform angle θ, they possess two closed angle-bisecting subspaces $G^\pm$ that make uniform angle $\theta/2$ with each, and for their corresponding projections E and F, the operator $E + F$ has $1 \pm \lambda$ as its eigenvalues with corresponding eigenspaces $G^\pm$ ($\lambda = \cos\theta$).

So much for the case of one pair of eigenvalues $\{1-\lambda,\ 1+\lambda\}$. Given a finite or countably infinite or countably infinite sequence of distinct real $\lambda_i,\ 0<\lambda_i<1,\ i=1,2,...,$ we seek now two projections E and F such that $E+F$ has precisely the eigenvalues $1\pm\lambda_i$ each with a preassigned multiplicity n_i, n_i a positive integer or ∞.

Select in our Hilbert space an orthogonal family of closed subspaces of dimension $2n_i, i=1,2,....$ Then, within each L_i, construct two closed subspaces M_i and N_i each of dimension n_i that make a uniform angle θ_i, $\lambda_i=\cos\theta_i$. Set $M=\oplus M_i$, $N=\oplus N_i$, E the projection on M, F the projection on N. Then $E+F$ has the desired eigenvalue spectrum, namely the $1\pm\lambda_i$ are exactly the eigenvalues of $E+F$, each with preassigned multiplicity n_i. the limit points of the $1\pm\lambda_i$, while not eigenvalues, also belong to the spectrum of $E+F$, the "continuous spectrum". All other numbers belong to its resolvent set (except possibly 0.)

For example, we may take for the λ_i all rationals between 0 and 1, and for each n_i take $n_i=\infty$. Then all rationals in $(0,1)\cup(1,2)$ are eigenvalues of $E+F$, each of infinite multiplicity, and all other points in $[0,2]$ are limit points of eigenvalues.

A strong uniqueness goes along with this construction principle.

4, Theorem. *Suppose that the nonzero projections* E,F *satisfy* $E\cap F=E^{\perp}\cap F=E\cap F^{\perp}=0$. *Let* I *represent a set of integers, designate by* $\{1\pm\lambda_i\ :\ i\in I\}$ *the distinct nonzero eigenvalues of* $E+F$, *and for each* $i\in I$ *let* n_i *stand for the multiplicity of* $1+\lambda_i$, $1\le n_i\le\infty$. *Then we can write* E *and* F *as the sum of orthogonal subprojections.*

$$E=\bigoplus_{i\in I}E_i\oplus E_\infty \qquad F=\bigoplus_{i\in I}F_i\oplus F_\infty$$

such that

(1) $(E_i\vee F_i)\perp(E_j\vee F_j)$ *when* $i\neq j$, $i=\infty$, $j=\infty$ *included,*

(2) E_i *and* F_i *make a uniform angle* θ_i *with each other,* $\cos\theta_i = \lambda_i$, *and*

(3) $E_\infty + F_\infty$ *has no eigenvalues except possibly* 0 *. These conditions determine the* E_i *and* F_i *uniquely. We have* $\dim(E_i) = \dim(F_i) = n_i$.

My somewhat complicated proof of this makes perhaps not too interesting reading. I have copies of a fully detailed treatment available for those who might wish to see it.

The projections E_∞ and F_∞ in Theorem 4 may actually occur. For example, let $\{e_1, e_2, ..., g_1, g_2, ...\}$ be an orthonormal basis for our Hilbert space, let E represent the projection on the closed subspace spanned by $\{e_1, e_2, ...\}$, and F the projection on the closed subspace spanned by the orthonormal set $\{f_1, f_2, ...\}$ where $f_i = 1/2(e_i + e_{i+1} + g_1 - g_{i+1})$. Then $E + F$ has no eigenvalues. I thank Mark Leeper for his help with this example.

A uniform angle pair of projections forms perhaps the simplest noncommuting pair of elements in the $*$-algebra of operators on our Hilbert space. Such pairs have pure algebraic characterizations that have proved useful in orthomodular lattice theory [2], and may eventually prove useful in the theory of Hermitian forms over $*$-fields and in the theory of $*$-rings. Indeed, the orthomodular law, $M \leq N \Longrightarrow N = M \oplus (N \wedge M^\perp)$, for closed subspaces M, N of our Hilbert space H, (combined with the noncommutativity and infinite dimensionality referred to in the opening sentence of this paper) adapted to a purely algebraic context, can be used to characterize real, complex, or quaternionic Hilbert space [3].

References

[1] P. A. Fillmore, *On Sums of Projections*, J. Functional Analysis **4** (1969), 146-152.

[2] S. S. Holland, Jr., *Isomorphisms Between Interval Sublattices of an Orthomodular Lattice*, Hiroshima Math. J. **3** (1973), 227-241.

[3] S. S. Holland, Jr., *Orthomodularity in Infinite Dimensions; a Theorem of M. Solèr*, Bull. Amer. Math. Soc. **32** (1995), 205-234.

Department of Mathematics and Statistics
University of Massachusetts Amherst, Box 34515
Amherst, Massachusetts 01003-4515
USA
holland@math.umass.edu

PL KANNAPPAN

On inner product spaces IV

Introduction

Let E be a normed linear space (n.l.s.) over the reals $\mathbb{R}$. A mapping $A : E \to \mathbb{R}$ is said to be *additive* if $A(x+y) = A(x)+A(y)$ for all $x, y \in E$. A map $J : E \to \mathbb{R}$ is called *Jensen* if $J(x+y) + J(x-y) = 2J(x)$, for $x, y \in E$. If J is Jensen, there exists an additive A such that $J(x) = A(x) + c$, where c is a constant. A map $B : E \times E \to F$ is called *biadditive* provided B is additive in each variable. A mapping $q : E \to \mathbb{R}$ is a *quadratic functional* if q satisfies the *quadratic functional equation*

$$q(x+y) + q(x-y) = 2q(x) + 2q(y), \quad \text{for } x, y \in E. \tag{QE}$$

If the norm satisfies either the Jordan-von Neuman identity [3], that is the parallelogram identity

$$||x+y||^2 + ||x-y||^2 = 2||x||^2 + 2||y||^2 \tag{*}$$

or

$$\sum_{i=1}^{n} ||x_i - \bar{x}||^2 = \sum_{i=1}^{n} ||x_i||^2 - n||\bar{x}||^2 \tag{**}$$

for some $n \geq 2, x_1, \ldots, x_n \in E$ and $\bar{x} = \frac{1}{n}\sum_{i=1}^{n} x_i$, [6, 7], then the n.l.s. is an i.p.s. Here we treat functional equations which are generalizations of (**) in the next section. Our main equation for consideration is

$$\sum_{i=1}^{n} \phi(x_i - \bar{x}) = \sum_{i=1}^{n} \phi(x_i) - n\psi(\bar{x}). \tag{ME}$$

Using (ME), we solve

$$\sum_{i=1}^{n} f(||x_i - \bar{x}||) = \sum_{i=1}^{n} f(||x_i|| - nf(||\bar{x}||), \tag{10}$$

and

$$\sum_{i=1}^{n} g(||x_i - \bar{x}||^2) = \sum_{i=1}^{n} g(||x_i||^2) - ng(||\bar{x}||^2). \tag{11}$$

Solution of the equations, (ME), (10) *and* (11).

We make use of the following result:

Result 1 [1]: The functional $q : E \to \mathbb{R}$ satisfies the quadratic equation (QE) if and only if, there exists a symmetric biadditive functional $B : E \times E \to \mathbb{R}$ such that $q(x) = B(x, x)$. This B is unique.

First we treat the equation (ME) and prove the following theorem.

Theorem 1. *Let $\phi, \psi : E \to \mathbb{R}$ satisfy the functional equation* (ME) *for $x_1, \ldots, x_n$ in E and $\bar{x} = \frac{1}{n}\sum_{i=1}^{n} x_i$ for a fixed $n \geq 2$. Then*

$$\begin{aligned} \phi(x) &= q(x) + A(x) + c, = B(x,x) + A(x) + c \\ \psi(x) &= q(x) + A(x) = B(x,x) + A(x). \end{aligned} \tag{1}$$

for $x \in E$, where q quadratic satisfies (QE), A is additive, B is biadditive, and c is a constant.

Proof. Set $x_1 = x_2 = \ldots = x_n = x$ in (ME) to get

$$\psi(x) = \phi(x) - \phi(0), \quad \text{for } x \in E. \tag{2}$$

Now (ME) can be written as

$$\sum_{i=1}^{n} \phi(x_i - \bar{x}) = \sum_{i=1}^{n} \phi(x_i) - n\phi(\bar{x}) + n\phi(0). \tag{3}$$

First let n be even say $2k$. Put $x_1 = \ldots = x_k = x, x_{k+1} = \ldots = x_{2k} = y$ in (3) to have

$$\phi\left(\frac{x-y}{2}\right) + \phi\left(\frac{y-x}{2}\right) = \phi(x) + \phi(y) - 2\phi\left(\frac{x+y}{2}\right) + 2\phi(0). \tag{4}$$

Next, let n be odd say $2k+1$. Set $x_1 = \ldots = x_k = x, x_{k+1} = \ldots = x_{2k} = y, x_{2k+1} = \frac{x+y}{2}$ in (3) to obtain

$$
\begin{aligned}
k\phi\left(\frac{x-y}{2}\right)+k\phi\left(\frac{y-x}{x}\right)+\phi(0) &= k\phi(x)+k\phi(y)+\phi\left(\frac{x+y}{2}\right) \\
&- (2k+1)\phi\left(\frac{x+y}{2}\right)+(2k+1)\phi(0),
\end{aligned}
$$

which is (4). Now we solve (4).

Replacing x by $2x$ and y by $2y$ in (4), we have

$$2\phi(x+y)+\phi(x-y)+\phi(y-x)=\phi(2x)+\phi(2y)+2\phi(0). \tag{5}$$

Define

$$2q(x)=\phi(x)+\phi(-x)-2\phi(0), \quad \text{for } \ x\in E, \tag{6}$$

so that (5) can be rewritten as

$$\phi(2x)+\phi(2y)=2q(x-y)+2\phi(x+y),$$

that is,

$$\phi(u+v)+\phi(u-v)=2\phi(u)+2q(v), \quad \text{for } \ u,v\in E. \tag{7}$$

Note that q is even and $q(0)=0$. Change u to $-u$ in (7) to yield

$$\phi(-u+v)+\phi(-u-v)=2\phi(-u)+2q(v). \tag{8}$$

Adding (7) and (8) and utilizing (6), we obtain

$$q(u+v)+q(u-v)=2q(u)+2q(v), \tag{QE}$$

that is, q is quadratic so that $q(x)=B(x,x)$, for a biadditive B, by Result 1. Subtracting (8) from (7), we get

$$A(u+v)+A(u-v)=2A(u)$$

which is Jensen, where

$$A(u)=\phi(u)-\phi(-u), \quad \text{for } \ u\in E. \tag{9}$$

Note that $A(0)=0$ and A is odd. Thus A is additive. Adding (6) and (9) and from (2), we obtain the sought for result (1). This proves the theorem.

Remark 1. In theorem 1, ϕ and ψ are obtained without assuming any regularity condition.

Corollary 1. *Suppose the norm of E satisfies the condition (**).*

Define $\phi(x) = ||x||^2$. Then $\phi(0) = 0$, ϕ is even and (SR) becomes (ME) with $\psi(x) = \phi(x)$. Then by theorem 1, $A = 0, q(x) = \phi(x) = ||x||^2$ satisfies the parallelogram identity (*). Hence E is an i.p.s., a result obtained in [6] and [7].

Theorem 2. *Let $f : \mathbb{R}^+ \to \mathbb{R}$ be continuous and satisfy the functional equation*

$$\sum_{i=1}^{n} f(||x_i - \bar{x}||) = \sum_{i=1}^{n} f(||x_i||) - nf(||\bar{x}||), \tag{10}$$

for $n \geq 2, x_1, \ldots, x_n \in E, \bar{x} = \frac{1}{n}\sum_{i=1}^{n} x_i$ where $\mathbb{R}^+$ is the set of non-negative reals. Then E is an i.p.s.

Proof. Define $\phi(x) = f(||x||)$, for $x \in E$. Note that ϕ is even, $\phi(0) = f(0) = 0$ (take all $x_i = 0$ in (10)) and ϕ satisfies (ME) with $\psi = \phi$. Then by theorem 1,

$$f(||x||) = \phi(x) = q(x)$$

satisfies $f(||x+y||) + f(||x-y||) = 2f(||x||) + 2f(||y||)$ whose solution is $f(t) = ct^2$ for $t \in \mathbb{R}^+$ (see [2] in (1.10)) and some constant c. Then the norm satisfies (*) so that E is an i.p.s.

Theorem 3. *Let $g : \mathbb{R}^+ \to \mathbb{R}$ be continuous and satisfy the functional equation*

$$\sum_{i=1}^{n} g(||x_i - \bar{x}||^2) = \sum_{i=1}^{n} g(||x_i||^2) - ng(||\bar{x}||^2), \tag{11}$$

for $n \geq 2, x_1, x_2, \ldots, x_n \in E, \bar{x} = \frac{1}{n}\sum_{i=1}^{n} x_i$. Then E is an i.p.s.

Proof. Define $\phi(x) = g(||x||^2)$, for $x \in E$. As before note that ϕ is even, $\phi(0) = 0$ and ϕ satisfies (ME) with $\psi = \phi$. Then by theorem 1,

$$g(||x||^2) = \phi(x) = q(x)$$

satisfies

$$g(||x + y||^2) + g(||x - y||^2) = 2g(||x||^2) + 2g(||y||^2). \tag{12}$$

By induction, $g(||nx||^2) = n^2 g(||x||^2)$ and that $g(||x||^2) = g(||n\frac{x}{n}||^2) = n^2 g(||\frac{x}{n}||^2)$. Hence for any rational r and real $t \geq 0, g(r^2t^2) = r^2 g(t^2)$. By continuity, $g(p^2t^2) = p^2 g(t^2)$ for any $p \in \mathbb{R}^+, t \in \mathbb{R}^+$. Hence $g(p^2) = cp^2$, for any $p \in \mathbb{R}^+$. Thus (12) implies that the norm satisfies (*) or (11) that is, g satisfies (**) so that E is an i.p.s.

References

1. Aczél, J. and Dhombres, J., *Functional Equations in several variables*, Cambridge University Press, Cambridge, 1989.

2. Amir, D., *Characterizations of inner product spaces*, Birkhauser Verlag, Basel, 1986.

3. Jordan, P. and von Neumann, J., *On inner products in linear metric spaces*, Ann. Maths., 36 (1935), 719-723.

4. Kannappan, PL, *On inner product spaces -II*, MathWare and Soft. Comp., vol 11 (1995), 61-70.

5. Kannappan, PL, *Quadratic functional equation and inner product spaces*, Results in Math., vol 27 (1995), 368-372.

6. Rassias, T.M., *New Characterizations of inner product spaces*, Bull. Sci. Math. (2) 108, (1984), 95-99.

7. Sundaresan, K., *Characterization of inner product spaces*, Math. Student (1961), 41-45.

Department of Pure Mathematics
University of Waterloo
Waterloo, ON, Canada N2L 3G1

MANFRED KRACHT AND ERWIN KREYSZIG

On Hilbert space theory and kernel functions

Dedicated to Professor Boro Döring on the occasion of his 65th birthday

1. Inner Products Until 1900

Hilbert spaces first appeared in Hilbert's spectral theory of Fredholm integral equations. This theory had many classical roots reaching far back into the 19th century and even into Euler's times. The most important of these early ideas and their evolution will be discussed in this chapter, beginning with Fourier series (Sec. 1.1) and proceeding to Sturm-Liouville theory (Sec. 1.2) and integral equations (Sec. 1.3), and culminating in Fredholm's famous work of 1900 (Sec. 1.4).

1.1 Inner Products in the Theory of Trigonometric Series

Inner products in Hilbert spaces were motivated and foreshadowed in several parts of classical analysis, notably in three- and n-dimensional Euclidean geometry, Fourier series, and more general orthogonal expansions. In this section we shall consider the main ideas and landmarks of the evolution of Fourier series as a chapter in the prehistory of inner product spaces.

If a given real function f can be developed into a *trigonometric series*

$$f(x) = \frac{1}{2}a_0 + \sum_{n=1}^{\infty}(a_n \cos nx + b_n \sin nx), \tag{1.1.1}$$

then because of the orthogonality of the *trigonometric system* $\{1, \cos nx, \sin nx\}$, $n \in \mathbb{N}$, the

Fourier coefficients a_n, b_n of f are

$$a_n = \frac{1}{\pi}\langle f, c_n\rangle, \quad b_n = \frac{1}{\pi}\langle f, s_n\rangle \tag{1.1.2}$$

with $c_n(x) = \cos nx$, $s_n(x) = \sin nx$ and the inner product defined by

$$\langle f, g\rangle = \int_{-\pi}^{\pi} f(x)g(x)\, dx. \tag{1.1.3}$$

Trigonometric series appeared first in astronomy in interpolation to determine the position of planets from those obtained by observation, and almost simultaneously in partial differential equations governing motions of vibrating strings.

This early period involved works by Euler (beginning at age 22), d'Alembert, Clairaut, Lagrange, Johann Bernoulli, and his son Daniel Bernoulli.

Papers of 1753 (*Hist. Acad. Berlin* **9**, 147-172, 173-195) of the latter contained trigonometric series, together with the claim that every possible initial curve of the vibrating string (of length π, say) can be represented by a trigonometric sine series and, thus, the subsequent motion by a series

$$u(x,t) = \sum_{n=1}^{\infty} a_n \sin nx \cos cnt, \tag{1.1.4}$$

where c^2 is the coefficient in the (one-dimensional) *wave equation*

$$u_{tt} = c^2 u_{xx}. \tag{1.1.5}$$

The coefficient formulas (1.1.2) are named after Euler, who discovered them in 1777 (at age 70) by using the *orthogonality of the trigonometric system*, the method we use today. Controversies related to trigonometric series during that early period contributed substantially to the beginning clarification of the concept of function. Cf. [15], 454-460, 502-522; [22], 227-234.

Trigonometric series (1.1.1) with coefficients (1.1.2) are called *Fourier series* (of the corresponding function f), after J.-B. J. Fourier (1768-1830), whose book [9] of 1822, *La théorie analytique de la chaleur*, became a landmark in classical analysis and mathematical physics – A. Sommerfeld, supervisor of the Nobel laureates Bethe, Debye, Heisenberg, and Pauli, called the book the Bible of the mathematical physicist (cf. [25], 1).

Fourier was working on his masterfully written book on heat conduction from 1807 on. He emphasized (on p. 12) the necessity of proceeding from theoretical models all the way to numerical solutions, "a condition necessary in any research, ... without which one arrives only at useless transformations."

His method was purely phenomenological. He was convinced that all phenomena in nature can be modeled mathematically and clarified by mathematics in a complete and satisfactory fashion. This belief was reflected in the elegance of his presentation.

However, Fourier was unable to justify his methods mathematically by (rigorous) proofs; this had to be left to later times. Poisson and Cauchy, historically coordinated (although involved in constant polemics with one another), were not successful either.

G. Lejeune Dirichlet (1805-59), descending from a family of French immigrants to Germany, was in close personal contact with Fourier during his early years in Paris, 1822-27. Two years later, in a basic paper [8], he published sufficient conditions that a trigonometric series (1.1.1) with coefficients (1.1.2) converge and have the sum f. Following an idea by Fourier, he obtained his conditions by evaluating a partial sum and then very carefully investigating the limit as $n \to \infty$. This was the first step into a rigorous theory of Fourier series.

Without going into too many further details, we just mention that the *Habilitationsschrift* of B. Riemann (1826-66) submitted in 1854 (cf. [22], 227-271, not to be confused with Riemann's *Habilitationsvortrag*) also concerned Fourier series. It was most likely inspired by Dirichlet, whose lectures Riemann had attended in Berlin during his absence from Göttingen.

Riemann began his paper as follows, confirming our opinion stated above:

> "Trigonometric series [were] named by Fourier ... [and] play an important role in that part of mathematics which involves completely arbitrary functions; indeed, there is a good reason for claiming that the most essential progress in this for physics so important part of mathematics depended on the clear insight into the nature of these series.
>
> Already in the earliest ... [corresponding] investigations ... there arose the question whether such a completely arbitrary function can be expressed by a series of the above form. This happened in the middle of the last century on the occasion of the investigation of vibrating strings, with which the most famous mathematicians were concerned at that time."

Riemann's paper also contains the definition of the *Riemann integral*. On p. 253 we find the famous Riemann's *localization theorem*, stating that the convergence of a trigonometric series at an x with coefficients (1.1.2) depends only on the behavior of f in an (arbitrarily small) neighborhood of that x.

Riemann's *Habilitationsschrift* was published in 1867 (by Dedekind), a year after Riemann's early death. It was fundamental, one of the sources from which modern analysis was to develop.

As a first instance of this, it inspired G. Cantor (1845-1918) to the creation of his set theory as follows.

In 1870, working on Riemann's paper, Cantor obtained his theorem on the uniqueness of the expansion of a given function into a convergent trigonometric series. Efforts to weaken the assumptions of the theorem (by admitting exceptional sets on which nothing is assumed about the convergence of the series) led Cantor to create his theories of abstract sets (beginning with publication in 1874) and of point sets on $\mathbb{R}$ and in $\mathbb{R}^n$ (published in papers of 1879-84, all in *Mathematische Annalen*).

In that work, Cantor introduced concepts of crucial importance, notably *accumulation point, derived set,* and *everywhere dense set.* These and the resulting notions of *closed set, perfect set, connected set, continuum, domain,* etc. have since become basic in analysis and topology.

1.2 Inner Products in Connection with Eigenfunction Expansions

It is the orthogonality of the trigonometric system that leads to the simple method of determining the Fourier coefficients of a given function f, resulting in the Euler formulas (1.1.2) and a corresponding trigonometric series (1.1.1) for representing f in terms of those orthogonal functions. We have seen that problems in vibrations and heat conduction as well as in astronomy motivated the introduction of that system and those series.

However, those cosine and sine functions constitute just the simplest orthogonal system, and other such systems of Legendre polynomials or Bessel functions appear quite naturally and play an important role in various applications. The extension from trigonometric to generalized Fourier series involving more general orthogonal systems took place sixty years after the discovery of the Euler formulas (1.1.2) and was accomplished by Liouville, using results by Sturm.

To outline that development and the novel ideas involved, we begin by defining the necessary notions and notations.

A system (y_m) of real functions defined on an interval $[a, b]$ is *orthogonal* on $[a, b]$ with respect to a *weight function* p (usually assumed to be positive on $[a, b]$) if

$$\langle y_m, y_n \rangle = \int_a^b p(x) y_m(x) y_n(x)\, dx = 0 \quad (m \neq n), \tag{1.2.1}$$

where $m = 0, 1, \ldots$ and $n = 0, 1, \ldots$. The *norm* $\|y_m\|$ of y_m is defined by

$$\|y_m\| = \langle y_m, y_m \rangle^{1/2}. \tag{1.2.2}$$

The system is *orthonormal* if it is orthogonal and all y_m have norm 1.

Let (y_m) with $y_m \neq 0$ for $m \in \mathbb{N}_0$ be any orthogonal system on $[a, b]$ with respect to a weight function p, and let f be a given function that can be developed into an *orthogonal series* or *generalized Fourier series*

$$f(x) = \sum_{m=0}^{\infty} a_m y_m(x). \tag{1.2.3}$$

Then, as for trigonometric series, the *Fourier constants* a_m of f with respect to that system are obtained by multiplying (1.2.3) by y_n, integrating from a to b, assuming that termwise integration is permissible, and using the orthogonality. This yields a_m in terms of an inner product given by the *generalized Euler formulas*

$$a_m = \langle f, y_m \rangle / \|y_m\|^2. \tag{1.2.4}$$

If (y_m) is orthonormal, we simply have

$$a_m = \langle f, y_m \rangle. \tag{1.2.5}$$

Now the functions of the trigonometric system are solutions of the differential equation

$$y'' + \lambda y = 0 \tag{1.2.6}$$

with suitable parameter λ (depending on the function in the system). Together with boundary conditions at some $x = a$ and $x = b$, as they arise, e.g., in the separation of the wave equation (1.1.5) or the heat equation (u = temperature)

$$u_t = c^2 u_{xx} \tag{1.2.7}$$

or some other partial differential equation of physics, this constitutes an *eigenvalue problem*, with eigenvalues λ to be determined such that the problem has a nontrivial solution (an *'eigenfunction'*).

The earliest determination of individual eigenvalues seems to have occurred in works by Taylor (1713, for the vibrating string), Johann Bernoulli (1728, for the hanging cable or chain fixed at the upper end), and his son Daniel Bernoulli (1732-33, also for the cable or chain).

Work on substantially more general eigenvalue problems began only a few years later, in 1836 and 1837, by J. Liouville (1809-82) and C. F. Sturm (1803-55) (*J. de Math.* **1,** 253-265, **2,** 16-35, 418-436, and **1,** 106-186, respectively), and resulted in generalized Fourier series involving eigenfunctions, so-called *eigenfunction expansions.*

The *Sturm-Liouville equations* considered were self-adjoint and of the form

$$(r(x)y')' + (q(x) + \lambda p(x))y = 0 \tag{1.2.8}$$

together with the *Sturm-Liouville boundary conditions*

$$\begin{aligned} k_1 y(a) + k_2 y'(a) &= 0 \\ \ell_1 y(b) + \ell_2 y'(b) &= 0. \end{aligned} \tag{1.2.9}$$

Theorem. If p, q, r, r' are continuous and $p(x) > 0$, $r(x) > 0$ on $[a, b]$, then the Sturm-Liouville problem (1.2.8), (1.2.9) has infinitely many eigenvalues, and eigenfunctions corresponding to numerically different eigenvalues are orthogonal on $[a, b]$ with respect to the weight function p.

Various second-order differential equations can easily be written in the form (1.2.8); best known among them are Legendre's and Bessel's equation. The theorem then allows representing given functions by eigenfunction expansions (1.2.3), for instance, in terms of Legendre polynomials or Bessel functions (cf. [19], 241-255).

Sturm and Liouville wanted to include all problems considered by Fourier and later by Poisson. Their novel idea was the investigation of general properties of solutions directly from the equation (1.2.8), taking into account (1.2.9), without assuming the possibility of actually solving the problem – which would not be possible in general, simply because such a problem will in most cases define new functions.

Hence this was an absolutely novel theory obtained by a shift of interest from finding solutions to proving existence, and a beginning in qualitative theory of differential equations. Of equal importance is the fact that the above theorem seems to be the earliest on (nonalgebraic) eigenvalue problems.

Sturm was a protégé of Fourier. His main contribution to the theory was the proof of the existence of infinitely many eigenvalues $\lambda_1 < \lambda_2 < \ldots < \lambda_n < \ldots$ tending to infinity as $n \to \infty$. Using this and other results of Sturm, Liouville proved convergence of the eigenfunction expansion by using approximate representations of the eigenfunctions and approximations of the eigenvalues.

This result was rigorously established only much later when more about different kinds of convergence was known and the totality (completeness; cf. [17], 168) of the set of eigenfunctions could be proved. Based on important later work by Poincaré, a complete proof of the expansion theorem for f piecewise C^2 was finally given by A. Kneser in 1904 (*Math. Annalen* **58,** 81-147), the year in which Hilbert published his first paper on integral equations!

To avoid misunderstandings: *algebraic* eigenvalue problems occurred long before Sturm-Liouville's work. In 1762, Lagrange considered systems

$$y_j'' = \sum_{k=1}^{n} a_{jk} y_k \qquad (j = 1, \ldots, n)$$

with constant coefficients a_{jk} and y_j depending on t. Assuming solutions $y_j = x_j \exp \lambda t$, he obtained the linear algebraic system

$$\lambda^2 x_j = \sum_{k=1}^{n} a_{jk} x_k,$$

which shows that (in modern terms) λ^2 must be an eigenvalue of the coefficient matrix $[a_{jk}]$.

1.3 Role of Integral Equations on the Way to Hilbert Space

Hilbert spaces appeared in connection with Hilbert's path-breaking spectral theory. A first step into spectral theory was made by Sturm and Liouville, as we have just seen, but progress beyond that theory needed a much more general frame in which the corresponding problems could be formulated. Such a wider environment was provided by *integral equations* arising from boundary value and eigenvalue problems for the partial differential equations of physics, one of the two great fields of research in analysis during the second half of the 19th century (the other being complex analysis).

In the first place, this concerned the *Dirichlet problem* for the Laplace equation

(1.3.1) $$\Delta u = 0 \quad \text{in } \Omega, \qquad u|_{\partial\Omega} = f$$

where Ω is a domain in the plane or in three-space, $\partial\Omega$ is its boundary, and f is a given function whose values the solution u must approach on the boundary.

Dirichlet (whose work on Fourier series we have mentioned in Sec. 1.1) considered the corresponding *Dirichlet integral*

(1.3.2) $$D(u) = \int_\Omega |\text{grad } u|^2 \, dV$$

and claimed that there exists a function u that minimizes the integral (1.3.2) in the family of continuously differentiable functions which equal f on $\partial\Omega$, and this u is a solution of the Dirichlet problem (1.3.1) (for continuous f in a bounded domain with smooth boundary). Riemann, having studied under Dirichlet in Berlin, called this conclusion *Dirichlet's principle* (although the principle had been used earlier by Gauss in 1839 and by Lord Kelvin in 1847). Weierstrass pointed out, first privately and in 1870 in an address delivered to the Berlin Academy (cf. [29], II, 49-54) that the principle was faulty, the conceptual error being caused by not distinguishing between greatest lower bound (the integral (1.3.2) is certainly bounded below by 0) and minimum.

This breakdown of Dirichlet's principle initiated a search for valid existence proofs for the Dirichlet problem (1.3.1) and resulted in the creation of three ingenious new methods, the *alternating method* by H. A. Schwarz (1870), the *sweeping-out process* by Poincaré (1887), and an *integral equation method* by C. Neumann (1870), which he called the 'method of the arithmetic mean' and published in the *Berichte und Verhandlungen der Sächsischen Akademie der Wissenschaften, Leipzig.*

Special integral equations had occurred before, notably Abel's integral equation in 1823 in the study of the brachystochrone (the curve of fastest descent under gravity from one point to another in a vertical plane) and several integral equations considered by Liouville, some in connection with Sturm-Liouville problems. However, these applications did not arouse much interest in integral equations, in contrast to Neumann's work, which attracted wide attention.

Neumann's idea for an existence proof of the Dirichlet problem (1.3.1) consisted in assuming the solution u to be the potential of a layer of dipoles on $\partial\Omega$ of unknown density σ, say,

$$u(P) = \frac{1}{2\pi}\int_{\partial\Omega} \sigma(Q)\frac{\partial}{\partial\nu}\left(\frac{1}{r}\right)\, dS, \tag{1.3.3}$$

where $P \in \Omega$, $Q \in \partial\Omega$, $r = d(P,Q)$ is the distance from P to Q, and ν is the outer normal of Ω. Substitution into (1.3.1) then gives the integral equation

$$\sigma(Q) + \frac{1}{2\pi}\int_{\partial\Omega} \sigma(Q^*)\frac{\partial}{\partial\nu}\left(\frac{1}{r^*}\right)\, dS = f(Q), \tag{1.3.4}$$

where $Q^* \in \partial\Omega$ and $r^* = d(Q,Q^*)$. Neumann solved this equation iteratively by successive approximation, setting $\sigma_0 = f$ and

$$\sigma_n = -\frac{1}{2\pi}\int_{\partial\Omega} \sigma_{n-1}\frac{\partial}{\partial\nu}\left(\frac{1}{r}\right)\, dS. \tag{1.3.5}$$

Denoting the second term in (1.3.4) by $K\sigma(Q)$, we obtain (1.3.4) in operator form

$$\sigma + K\sigma = f \tag{1.3.6}$$

and (1.3.5) in the form $\sigma_n = -K\sigma_{n-1} = (-K)^n f$, so that (1.3.6) gives formally

$$\sigma = (I+K)^{-1} f = \sum_{n=0}^{\infty} (-K)^n f, \tag{1.3.7}$$

the famous *Neumann series*. For it, Neumann was able to prove convergence, assuming f to be continuous and Ω to be bounded and convex, with piecewise smooth boundary $\partial\Omega$. (A small gap in the proof went unnoticed for almost seventy years and was finally discovered and closed by Lebesgue (cf. [20], IV, 151-166) in 1937.)

Neumann's accomplishment was a landmark that helped moving integral equations into the center of general mathematical interest.

The term *integral equation* was suggested by the former Weierstrass student du Bois-Reymond in 1888. Du Bois-Reymond expressed his disappointment about the fact that no general method for solving such equations was available. However, he did not have to wait very long. The first such theory for obtaining solutions (in the form of convergent infinite series) as well as existence and uniqueness theorems concerned a general class of integral equations of the form

$$\phi(s) + \lambda \int_a^s k(s,t)\phi(t)\, dt = f(s) \tag{1.3.8}$$

with unknown ϕ and given f and *kernel* k.

Such a theory was given almost simultaneously but independently by J.-M. Le Roux (1863-1949) in 1895 and V. Volterra (1860-1940) in 1896. These equations (1.3.8) became known as *Volterra equations* (of the second kind), mainly because Volterra's papers (cf. several notes in [26], II) included an explicit formula for a solution in terms of a convergent series involving iterated kernels and emphasized more clearly the analogy with the solution of a linear system of algebraic equations.

The idea as such was far from new. Indeed, beginning one hundred and sixty years before Le Roux and Volterra with Bernoulli's treatment of the vibrating string as analog of finitely many equidistant vibrating mass points on a wire (with $n = 2, 3, 4$ in Bernoulli's papers), one had repeatedly attempted using the analogy together with a limit process for establishing rigorous theories (of differential equations, not integral equations), without success, except in special cases with physically foreseeable solutions (e.g., in 1877 in Lord Rayleigh's *Theory of Sound*).

Unfortunately, the upper limit of integration in (1.3.8) is variable (which makes the proof of uniform convergence quite simple), whereas Neumann's integral equation had fixed limits of integration. (It would not help to integrate from a to b by first defining an extended kernel on a square $a = s$, $t = b$ which is zero in the triangle not needed in (1.3.8) because such a kernel would not be continuous for $s = t$, in general.) A theory that included Neumann's as well as many other equations was created only four years later by Fredholm, to whose celebrated work we turn next.

1.4 Fredholm's Breakthrough, Preparing for Hilbert's Work

Once the key importance of integral equations had been recognized because of Neumann's work, the further development was rapid. In fact, it was I. Fredholm (1866-1927) who published in 1900 (and in detail in 1903 in *Acta Mathematica* **27**, 305-390) a fully satisfactory solution theory that applied to Neumann's equation as well as to many other integral equations, and created a sensation because of its unexpected "simplicity" and its full analogy to the theory of solutions of linear systems of *algebraic* equations.

At that time, Paris was the mathematical center of the world. Most essential for Fredholm was the work by H. Poincaré (1854-1912) on the vibrating membrane governed by the partial differential equation

$$\Delta u + \lambda u = 0 \quad \text{in } \Omega, \qquad u|_{\partial\Omega} = 0. \tag{1.4.1}$$

In 1885, in his famous paper on minimal surfaces (which also contained the Schwarz inequality), H. A. Schwarz had proved the existence of the smallest eigenvalue (corresponding to the fundamental mode of the membrane), and in 1893, É. Picard proved the existence of the first overtone (second eigenvalue). A year later, Poincaré was able to establish the existence of infinitely many eigenvalues, by considering the solution u of the forced vibrations governed by

$$\Delta u + \lambda u = f(x,y) \qquad (f \text{ given}) \tag{1.4.2}$$

as a function of the parameter λ, and he obtained a meromorphic function $u(x,y,\lambda)$, representing a (unique) solution for all λ which are not poles. In the corresponding Mittag-Leffler representation he recognized the poles as eigenvalues and the residues as eigenfunctions of the problem.

Poincaré further proved that for suitable f (not for all f) one obtains all the eigenvalues and eigenfunctions and that for λ not an eigenvalue the nonhomogeneous problem (1.4.2) is uniquely solvable for every f – this now showed the analogy of the transformation to principal axes in analytic geometry.

Poincaré also discovered why A. Beer, a forerunner on Neumann's method, would not always have obtained convergence, had he been concerned with convergence and a rigorous theory at all, which was not the case.

These investigations by Poincaré and the theory by Volterra mentioned before were the only known results which Fredholm could use as a starting point of his own work when he came to Paris in Spring of 1899, accompanied by E. Lindelöf. There he obtained personal contact with Poincaré, Picard, and Hadamard, whose lectures he attended. He stayed in Paris until August of the same year.

Immediately after his return to Stockholm he wrote to Mittag-Leffler on August 8 that he had considered the 'integral equation' (his term)

$$\phi(s) + \lambda \int_0^1 k(s,t)\phi(t)\, dt = f(s) \tag{1.4.3}$$

(not exactly his notation), where ϕ is unknown, f is given, and λ is a parameter. He wrote:

> "One can prove that this equation generally has a solution and that it is a quotient of two entire [power] series in λ ... The denominator ... is ...
>
> $$D(\lambda) = 1 + \sum_{n=1}^{\infty} \frac{(-\lambda)^n}{n!} \int_0^1 \cdots \int_0^1 k\begin{pmatrix} s_1, \ldots, s_n \\ s_1, \ldots, s_n \end{pmatrix} ds_1 \ldots ds_n, \tag{1.4.4}$$
>
> where
>
> $$k\begin{pmatrix} s_1, \ldots, s_n \\ t_1, \ldots, t_n \end{pmatrix} = \det\,[k_{ij}], \quad k_{ij} = k(s_i, t_j), \tag{1.4.5}$$
>
> $i, j = 1, \ldots, n$. Convergence can be proved by a theorem on determinants ..."

He gave a proof himself, but learned later that this was Hadamard's determinant inequality (stating that among all parallelepipeds in $\mathbb{R}^n$ with given lengths of edges the rectangular one has the maximum volume).

He published a short paper about his new theory in 1900 and a full account of it in 1903, as mentioned before. It is interesting that he entitled his paper of 1900 (in French) "On a new method for solving Dirichlet's problem", thus pointing to Neumann's work (which had also been the point of departure for Poincaré's investigations on the vibrating membrane), but his paper of 1903 "On a class of functional equations". This indicates a distinct shift of emphasis.

Written before Frechet's and Hilbert's fundamental works of 1906, Fredholm's two papers belonged to classical analysis. Fredholm's basic idea was the replacement of the integral by Riemann sums, the solution of the resulting linear system of n algebraic equations by Cramer's rule, and the passage to the limit as $n \to \infty$.

He wrote the solution again in the form of an integral equation

$$\phi(s) = f(s) - \lambda \int_0^1 R(s, t, \lambda) f(t)\, dt \tag{1.4.6}$$

with the *resolvent kernel* $R(s, t, \lambda) = D(s, t, \lambda)/D(\lambda)$, the 'Fredholm determinant' as before, and 'Fredholms first minor'

$$D(s, t, \lambda) = k(s, t) + \sum_{n=1}^{\infty} \frac{(-\lambda)^n}{n!} \int_0^1 \cdots \int_0^1 k\begin{pmatrix} s, s_1, \ldots s_n \\ t, s_1, \ldots, s_n \end{pmatrix} ds_1 \ldots ds_n. \tag{1.4.7}$$

In full analogy to the theory of finite linear systems of algebraic equations he proved that either Case (I) or Case (II) holds; this was later called *Fredholm's alternative.*

(I) If λ is not a zero of $D(\lambda)$, then (1.4.3) has exactly one solution, given by (1.4.6).

In this case the homogeneous equation

$$\phi(s) + \lambda \int_0^1 k(s,t)\phi(t)\, dt = 0 \tag{1.4.8}$$

has only the trivial solution $\phi = 0$.

(II) If λ is a zero of $D(\lambda)$ of order m, then the homogeneous equation (1.4.8) has at least one nontrivial solution and at most m linearly independent solutions.

In this case the nonhomogeneous equation (1.4.3) is not solvable for all f, but only for those which satisfy the orthogonality condition

$$\int_0^1 f(s)\psi(s)\, ds = 0 \tag{1.4.9}$$

for every solution ψ of the transposed homogeneous equation

$$\psi(s) + \lambda \int_0^1 k(t,s)\psi(t)\, dt = 0. \tag{1.4.10}$$

2. Hilbert's and Related Works Until 1907

In this chapter we shall investigate the works by Hilbert and others between 1904, the beginning of Hilbert's publications on integral equations, and 1907, the year of the (completely unexpected) amalgamation of Hilbert's and Lebesgue's theories through the Riesz-Fischer theorem. This includes the appearance of Hilbert's most important paper on integral equations and spectral theory in 1906.

2.1 Classical Beginnings: Hilbert's First Paper

The calculus of variations had provided unifying guidelines in at least two main fields of 19th century analysis, complex analysis, where Riemann used the Dirichlet principle, and partial differential equations of physics.

After the breakdown of the Dirichlet principle in consequence of Weierstrass's critique of 1870, there was no more *general* principle for handling various problems of applied mathematics. It seems that, nevertheless, D. Hilbert (1862-1943) before 1900 still set his hope in the calculus of variations.

In Problem 23 of his famous Paris talk of 1900 on unsolved problems Hilbert had drawn attention to Weierstrass's work and to A. Kneser's book of the same year, the first presentation of the "modern" calculus of variations.

Not intimidated by Weierstrass, Hilbert was able to re-establish the Dirichlet principle within proper limits as a valid method of proof. He did this in two papers of 1900 and 1901 (reprinted in 1905). In the first paper he proposed the following more general formulation of the principle (cf. [14], III, 11).

> "Every regular problem of the calculus of variations has a solution as soon as suitable restrictive assumptions with respect to the nature of the given boundary conditions are satisfied and, if necessary, the concept of a solution is suitably generalized. – In what way this principle can serve as a guiding star for finding rigorous and simple existence proofs will be shown by the following examples."

These examples are (a) shortest arcs on a surface, and (b) the Dirichlet problem for the Laplace equation in a plane domain with continuously curved boundary and class C^1 boundary values. (A problem is *regular* if the first coefficient of the Euler-Lagrange equation

$$L_{y'y'}y'' + L_{y'y}y' + L_{y'x} - L_y = 0 \tag{2.1.1}$$

is always positive (or always negative) on the interval of integration; here, $L(x, y, y')$ is the integrand of the integral to be minimized.)

Hilbert's papers initiated the *direct methods* (methods without the use of (2.1.1)), which were to become of basic importance in the existence theory of the calculus of variations.

Hilbert himself made no further attempts in uniformizing analysis by methods of the calculus of variations. A reason could be that he may have recognized integral equations as a more promising instrument for that purpose when he became aware of Fredholm's sensational theory, which happened already in 1901 through a visit and talk of Holmgren in Göttingen. Indeed, in his book [13] he wrote on p. 2:

> "A more detailed study of the subject [of integral equations] led me to the conclusion that the systematic development of a general theory of linear integral equations is of utmost importance to the whole analysis, in particular to the theory of definite integrals and the theory of developments of arbitrary functions into infinite series, furthermore to the theory of linear differential equations and complex analysis as well as to potential theory and the calculus of variations."

Hilbert had lectured on partial differential equations and Neumann's method before the turn of the century. He began systematic work and lectures and seminars on integral equations immediately after Holmgren's visit, drawing intuitive inspiration from the developments that we have discussed in Chap. 1. It took only a year until the first Göttingen Ph.D. thesis on integral equations appeared, written by O. D. Kellogg, one of the 69 Ph.D. theses written under Hilbert's supervision between 1898 and 1910.

In his work, Hilbert concentrated on developing the spectral theory for Fredholm equations of the second kind

$$\phi(s) - \lambda \int_0^1 k(s,t)\phi(t)\, dt = f(s) \tag{2.1.2}$$

with continuous kernel k. He assumed k to be *symmetric*, $k(s,t) = k(t,s)$, which would give him a real spectrum.

Fredholm's theory was a *solution theory*, whose algebraic analog centered around Cramer's rule. Hilbert's spectral theory had as an algebraic analog the *transformation of quadratic forms to principal axes.* Hilbert published his theory in the *Göttinger Nachrichten* of 1904-10 in six papers (*Mitteilungen*) and again as a book [13] in 1912.

In his *First Paper* of 1904, by replacing the integral in (2.1.2) with Riemann sums (just as Volterra and Fredholm had done) he obtained linear systems of algebraic equations of the form

$$\phi_p - \ell \sum_{q=1}^{n} k_{pq}\phi_q = f_p \qquad (p = 1, \ldots, n), \tag{2.1.3}$$

where $\phi_p = \phi(p/n)$, $f_p = f(p/n)$, $k_{pq} = k(p/n, q/n)$. He considered the corresponding form in n variables

$$Q_n(x) = \sum_{p=1}^{n}\sum_{q=1}^{n} k_{pq}x_p x_q \tag{2.1.4}$$

and transformed it orthogonally to principal axes. To take care of the given integral equation, he performed a rigorous limit process as $n \to \infty$. He proved that any continuous symmetric kernel $k \neq 0$ has eigenvalues, even infinitely many if k is *nondegenerate* (not a finite sum of products of a function of s times a function of t). As the analog of the principal axes form he proved

$$\int_0^1\int_0^1 k(s,t)x(s)x(t)\,ds\,dt = \sum_{n=1}^{\infty} \frac{1}{\lambda_n} y_n^2, \tag{2.1.5}$$

where the series on the right is absolutely and uniformly convergent, the 'Fourier coefficients' of x are the inner products

$$y_n = \langle \phi_n, x\rangle = \int_0^1 \phi_n(t)x(t)\,dt \tag{2.1.6}$$

and ϕ_n is the (normalized) eigenfunction corresponding to the eigenvalue λ_n ($1/\lambda_n$ in modern notation).

By establishing (2.1.5) instead of using a 'Fourier series'

$$f(s) = \sum_{n=1}^{\infty} c_n\phi_n(s) \tag{2.1.7}$$

Hilbert avoided the typical difficulties related to convergence and to the contribution of the eigenvalue 0 generally arising in the use of (2.1.7).

2.2 Culmination: Hilbert's Fourth Paper of 1906

Most important of Hilbert's six papers on integral equations is the Fourth Paper (*Vierte Mitteilung*), because of the depth and novelty of its ideas.

Whereas in his First Paper of 1904, Hilbert started from the "concrete" integral equation (2.1.3), in his Fourth Paper he no longer put the integral equation at the beginning, but began with an abstract quadratic form Q defined by

$$Q(x) = \sum_{p=1}^{\infty} \sum_{q=1}^{\infty} k_{pq} x_p x_q \quad (k_{pq} = k_{qp}) \tag{2.2.1}$$

in the terms of an infinite real sequence $x = (x_p)$. He then took those sequences $x = (x_p)$ for which

$$\|x\|^2 = \sum_{p=1}^{\infty} x_p^2 < \infty \tag{2.2.2}$$

(note that $\|x\|$ is the "infinite analog" of the distance in Euclidean n-space). The space whose points $x = (x_p)$ are these sequences is now called the *Hilbert sequence space* and is denoted by ℓ^2. This was the cradle of Hilbert space.

Hilbert called Q *vollstetig* (*completely continuous*; we now say *compact*) if

$$\lim_{n \to \infty} Q(x^{(n)}) = Q(x) \tag{2.2.3}$$

for all $x^{(n)} = (x_p^{(n)})$ with $\|x^{(n)}\| \leq 1$ and $x_p^{(n)} \to x_p$. This componentwise convergence is now called *weak convergence.*

Hilbert's main results on completely continuous forms were

(i) that these forms have a pure point spectrum of (generally) infinitely many eigenvalues λ_j with (in modern terminology) 0 as the only possible accumulation point,

(ii) that Q can be "diagonalized" just like a finite form, to get a *spectral representation*

$$Q(x) = \sum_{\nu} \lambda_\nu y_\nu^2 \tag{2.2.4}$$

by applying an orthogonal transformation

$$y_\nu = \sum_{p} c_{\nu p} x_p \tag{2.2.5}$$

and

(iii) that the spectral theory of these forms includes that of linear integral equations with symmetric continuous kernels.

With respect to *integral equations* these results showed clearly that *essential to the spectral theory is not the representation of the operator by an integral but the compactness of the corresponding form.*

Hilbert also used his *principle of choice* (*Auswahlverfahren*, a special case of the Ascoli-Arzelà theorem) to prove Fredholm's theorems for linear systems of equations for which the corresponding form of the coefficients is compact and the right-hand side is in l^2.

As a second fundamental concept, Hilbert called the form Q in (2.2.1) *stetig* (*continuous*) if $Q(x^{(n)}) \to Q(x)$ for every sequence $(x_p^{(n)})$ in l^2 such that $x^{(n)} \to x$, that is,

$$\sum_{p=1}^{\infty} \left(x_p^{(n)} - x_p \right)^2 \to 0 \quad \text{as} \quad n \to \infty. \tag{2.2.6}$$

This is *strong convergence*, a term coined by E. Schmidt (*Rend. Mat. Palermo* **25** (1908), 53-77), one of Hilbert's most outstanding Ph.D. students, on whose work we shall comment in the next section.

Hilbert showed that for a continuous form the spectrum generally has a discrete part, as before, and a continuous part, which he represented by a *Stieltjes integral*, which Stieltjes had introduced in 1894. Thus the spectral theory of a continuous, not compact form no longer resembled that in the finite-dimensional case.

Hilbert's *results* were path-breaking in functional analysis. His *methods*, however, were those of classical analysis, stretched to the utmost in his very complicated limit processes $n \to \infty$. Hilbert succeded in his truly heroic efforts, although he was hindered in two ways. First, he used his forms instead of vector spaces and linear operators, although Peano and Pincherle had paved the way in that direction more than ten years earlier. Second, topology had not yet been developed sufficiently far, so that he had to invent the necessary tools for himself. A pillar of his theory was his *Auswahlverfahren*, which he used repeatedly with great success. In modern terms this is the compactness of the unit ball for the weak topology.

Nowhere did Hilbert use the word 'space' or the idea of a sequence space or function space – 'Hilbert space' is a term coined by F. Riesz (cf. [23], II, 912) in 1913. Nowhere did Hilbert use any geometrical terms, but it seems clear that in discovering ℓ^2, he must have had a geometric guiding principle.

The further development was of basic importance in the history of Hilbert space. Its roots can be described as follows.

Hilbert had assumed continuity of the kernel. This implied continuity of the eigenfunctions, but only for eigenvalues $\lambda \neq 0$. For $\lambda = 0$, eigenfunctions are defined by

$$\int_0^1 k(s,t)\phi(t)\, dt = 0. \tag{2.2.7}$$

Now one can show that there are continuous kernels which have no continuous eigenfunctions for $\lambda = 0$, but are such that one can still find a continuous f which cannot be represented by

$$f(s) = \int_0^1 k(s,t)x(t)\,dt \tag{2.2.8}$$

with a continuous $x(t)$ (cf. [12], 1365). This *violates the analogy* to the algebraic case, in which the nonsolvability of the equation

$$\sum_{t=1}^{n} k_{st}\phi_t = 0 \tag{2.2.9}$$

implies that $\lambda = 0$ is not an eigenvalue of $[k_{st}]$, so that det $[k_{st}] \neq 0$ and the nonhomogeneous system with arbitrary right-hand side

$$\sum_{t=1}^{n} k_{st}\phi_t = f_s \tag{2.2.10}$$

is solvable.

Hence the creation of a completely analogous theory first required the choice of a suitable function space, a typically functional-analytic idea. In the present case this was the choice of the *Lebesgue space* $L^2([a,b])$, which F. Riesz introduced in the same year 1906 and related to Hilbert space ℓ^2 by the Riesz-Fischer theorem in 1907, as we shall discuss in the next section.

To fully understand the situation historically, we should emphasize the importance of forerunners. In Volterra's and Fredholm's theories it would have been hard to find the forms of the integral equations actually considered without the guidance of problems of physics. In Hilbert's spectral theory, guidance came from classical works, in particular from those discussed in Chap. 1.

In the present case the transition from continuous to square integrable functions began in principle with the Schwarz inequality, discovered in 1859 by Buniakowski, rediscovered and recognized in its importance in 1885 by H. A. Schwarz, and frequently applied as a basic tool by Poincaré. Similarly, Hadamard's determinant inequality, Fredholm's key tool in the convergence proof, can be modified in order to extend Fredholm's theory to the L^2-setting.

2.3 Unification and "Geometrization" of Hilbert Space Theory

The development indicated near the end of the previous section was completed by E. Schmidt (1876-1959) in his Ph.D. thesis of 1905, supervised by Hilbert, on eigenfunction expansions (published in 1907 in *Math. Ann.* **63,** 433-476). In his work, Schmidt developed a new independent approach to the results of Hilbert's First Paper, using not much more than the Schwarz and Bessel inequalities. On p. 435 he wrote:

> "After the proof of some lemmas ... [we give] very simple proofs of Hilbert's theorems [of the First Paper], avoiding the limit process. We first prove the existence of eigenvalues by a method modeled after a famous method of H. A. Schwarz, which in the language of Fredholm's formulas would amount to solving $D(\lambda) = 0$ by Bernoulli's method. From the existence theorem follow the expansion theorems ... Complications [in Hilbert's work] caused by multiple zeros of $D(\lambda)$ do not appear here."

On p. 442 he defined the *Gram-Schmidt orthogonalization process.* (Results similar to some of Schmidt's were obtained independently by W. Stekloff (*Mém. Acad. S.-Petersbourg* (1904), p. 7, etc.).)

Meanwhile, measure theory had developed from work in real analysis by the "older generation" of Italians, Dini, Ascoli, and Arzelà, and the "younger French generation" Borel, Baire, and Lebesgue, culminating in 1902 with the appearance of Lebesgue's famous thesis *Intégrale, longueur, aire.* It was F. Riesz (1880 -1956), elder brother of M. Riesz (1886-1969), who in 1906 (the year in which Fréchet defined metric space) introduced (real) *Lebesgue space* $L^2([a,b])$ (cf. [23], I, 375-377), whose points are (equivalence classes of) square integrable measurable functions, with metric defined by

$$d(f,g) = \left[\int_a^b (f(x) - g(x))^2 dx\right]^{1/2} \tag{2.3.1}$$

and convergence in metric called *mean square* (or *mean*) *convergence.*

Besides to Lebesgue's work, Riesz referred to Fréchet's Ph.D. thesis of 1906, and to E. Schmidt's Ph.D. thesis, in which Schmidt had proved that any total (Schmidt says *complete*; cf. [17], 168) orthonormal set in an inner product space of continuous functions on $[a,b]$ is countable. Riesz proved that $L^2([a,b])$ is *separable* (has a dense subset that is countable), thus generalizing Schmidt's theorem.

The next landmark in the development of Hilbert space was Riesz's discovery of the *Riesz-Fischer theorem* in 1907 cf. [23], I, 378-381). Hilbert made Riesz's result public in his seminar, but just four days earlier, E. Fischer (1875-1959) of Vienna had presented practically the same result in a talk given in Brünn (now Brno)!

In modern language the theorem states that Hilbert sequence space ℓ^2 and Lebesgue space $L^2([a,b])$ are isometrically isomorphic.

This was an unforeseen bridge between Hilbert's theory of integral equations and Lebesgue's theory, for which Riesz was to become second only after Lebesgue himself in showing its power and applicability.

The two short papers of Riesz just discussed contain a confluence of ideas of Lebesgue, Hilbert, Fréchet, Schmidt, and Riesz himself. This indicates that *functional analysis was well on its way of becoming a strong independent discipline of its own.*

Another impulse, essential to the development of inner product and Hilbert space theory, came from a second paper [24] by E. Schmidt published in 1908 (*Rend. Mat.*

Palermo **25**, 53-77), on the solution of systems of linear equations with infinitely many unknowns. We shall have no reason for discussing the actual content of the paper; what interests us here is the change in *notations and terminology* proposed by Schmidt in the first chapter of the paper, a "geometrization" that would help Hilbert space theory to become more comprehensible, more intelligible and "visible" (of course, without loss of rigor!).

Schmidt had moved from Göttingen to Bonn, where his colleague G. Kowalewski (1876-1950) suggested to him the "geometric language" of Hilbert space, as we use it nowadays.

Indeed, referring to Hilbert's Fourth Paper, Fréchet's Ph.D. thesis, and Riesz's papers just discussed, Schmidt gave his first chapter the title "Geometry in a function space." A subtitle was "The Pythagorean theorem" (in connection with orthogonal functions). He denoted sequences in ℓ^2 by A, B, etc., and their *inner product* by

$$(A; B) = (B; A) = \sum_{x=1}^{x=\infty} A(x)B(x). \tag{2.3.2}$$

We would now write $A = (A_n)$, $B = (B_n)$, and

$$(A, B) = \overline{(B, A)} = \sum_{n=1}^{\infty} A_n \overline{B_n}.$$

The term *function space* appears at the beginning, and he says on p. 56:

> "The geometric interpretation of ... the concepts and theorems I owe to Kowalewski. It becomes still clearer if $A(x)$ instead as a function is interpreted as a vector in an infinite dimensional space. The ... length $|A|$ and orthogonality are those introduced by Study (*Math. Ann.* **60,** 372)."

This refers to p. 57, where Schmidt wrote

$$\|A\|^2 = (A; \bar{A}) = \sum_{x=1}^{x=\infty} |A(x)|^2 \tag{2.3.3}$$

($\bar{A}$ the complex conjugate of A). He called $A(x)$ *normalized* (*normiert*) (but said *length*, not *norm*) if $\|A\| = 1$, and $A(x)$ and $B(x)$ *orthogonal* if

$$(A; \bar{B}) = 0, \quad \text{thus} \quad (B; \bar{A}) = 0. \tag{2.3.4}$$

On the next page he stated *Bessel's inequality* in the form

$$\|F\|^2 \geq \sum_{\nu=1}^{\nu=\infty} |(F; \bar{B}_\nu)|^2, \tag{2.3.5}$$

where $(B_\nu(x))$ is an orthonormal system, and the *Schwarz inequality* as

$$\|F\| \, \|G\| \geq |(F;G)|. \tag{2.3.6}$$

He also coined the term *strong convergence*, meaning convergence in the metric, as we use it now. (*Weak convergence* is used on p. 1435 of [12], which was written about 1925.) His *lineares Funktionengebilde* is now called *closed subspace*, and his *Perpendikelfunktion* is the *orthogonal complement* in the projection theorem (p. 64).

2.4 Functionals on Hilbert Space and on Space $C[a, b]$

Continuous linear functionals played a substantial role throughout the evolution of functional analysis. In Hilbert space theory, their role is special. To see this and its ramifications from a historical viewpoint, we discuss and compare the situation with that for space $C[a, b]$, which is not a Hilbert space, but whose investigation has contributed substantial ideas to the development on functionals in general.

Functional analysis, originally the analysis of *functionals* (mappings from any set into the real line $\mathbb{R}$ or the complex plane $\mathbb{C}$) is usually considered to be born in 1887, when Volterra published several notes on functionals defined on a set of continuously differentiable curves (continuously differentiable functions on an interval $[a, b] \subset \mathbb{R}$ with range in $\mathbb{R}$ or in $\mathbb{R}^n$).

This was the earliest study of functionals per se, unrelated to applications in geometry or mechanics, etc. Actually, the term *functional* is of a later date, suggested by Hadamard in 1904 or 1905 (in an undated letter to Fréchet); in Volterra's papers, functionals are called *funzioni dipendenti da linee* (functions depending on curves).

Functional analysis has grown far beyond the task of the analysis of functionals, but the latter task has remained an important part of the field. In particular, this concerns the space of linear functionals on a vector space X, which is usually called the *algebraic dual space* of X, and even much more the *dual space* of a normed space X, which consists of the *continuous* linear functionals defined on X; cf. [17], 106, 120. Investigation of dual spaces is often called *topological duality theory.* It is undertaken mainly as a means of discovering properties of a space from properties of its dual space.

The idea of concentrating on continuous linear functionals and deriving general integral representation formulas for these functionals on practically important spaces seems to have first appeared in work by J. Hadamard (1865-1963), who in 1903 (in *Comptes Rendus (Paris)* **136**, 351-354) represented the continuous linear functionals U on space $C[a, b]$ (the normed space of continuous functions on $[a, b]$ with the maximum norm; cf. [17], 7) by the formula

$$U(f) = \lim_{m \to \infty} \int_a^b f(x) h_m(x) \, dx, \tag{2.4.1}$$

where the h_m are also continuous on $[a, b]$. Hadamard spoke of *opérations fonctionnelles.* He mentioned Volterra, Pincherle, and Bourlet as previous main investigators of functionals in terms of infinite series involving the argument function f and its derivatives. He arrived at this formula by dropping analyticity of f and following "a way traced by Weierstrass and Kirchhoff".

Now the h_m were not uniquely determined by U, and a perfect formula involving an essentially unique auxiliary function $w(x)$ (in general no longer continuous, but always of bounded variation) and an integral (a Riemann-Stieltjes integral) instead of a limit of integrals, was given in 1909 in a famous paper by F. Riesz (cf. [23], I, 400-402), who proved that for a functional U as before,

$$U(f) = \int_a^b f(x)\, dw(x). \tag{2.4.2}$$

This was a deep result, in particular since Riesz could not use the Hahn-Banach theorem (which was discovered only in 1927 and 1929), by which (2.4.2) is now proved (cf. [17], 227-230).

In his short paper Riesz emphasized the importance of a clear definition of the function space on which a functional is defined:

> "For defining what one understands by a linear operation it is first necessary to make precise the *field of functions* [le *champ fonctionnel*]. We consider the totality of real and continuous functions between two fixed numbers, for example, between 0 and 1; for that class we define the *limit function* by the hypothesis of uniform convergence."

This formulation seems remarkable because it shows that Riesz did not regard $C[a,b]$ as a metric space, but just as a totality of functions with a notion of limit defined on it, that is, as a class (L), a *limit space*, in the sense of Fréchet's thesis.

Riesz then defined continuity of his linear functionals U by the now usual requirement that $U(f_n)$ tend to $U(f)$ whenever a sequence $(f_n(x))$ converges with the limit $f(x)$ uniformly on $[a,b]$. And he stated that continuity of a linear functional implies boundedness, "as one easily shows".

Compared to $C[a,b]$, for a Hilbert space $X = L^2(\Omega)$ the problem of representing continuous linear functionals turned out to be much simpler. It was solved two years earlier, in 1907, simultaneously and independently by Fréchet (*Trans. Amer. Math. Soc.* **8**, 433-446) and Riesz ([23], I, 386-388), where in Fréchet's case, $\Omega = [0, 2\pi]$, and in Riesz's paper, Ω was left unspecified, but Riesz most likely also had in mind an interval $[a,b]$. The desired representation was as follows.

Theorem. *Given a continuous linear functional U on a Hilbert space X, there exists a unique $g \in X$ such that for all $f \in X$,*

$$U(f) = \langle f, g\rangle = \int_a^b f(x)g(x)\, dx. \tag{2.4.3}$$

This shows that inner products play a key role in the theory of Hilbert spaces and that this discovery was made early in the development of that theory.

Furthermore, for given U the function g is uniquely determined, as stated, and it can be shown that U and g have equal norm, $\|g\| = \|U\|$. It follows that L^2 is isometrically anti-isomorphic with its dual. This is sometimes briefly expressed by saying that Hilbert space is *self-dual.*

Finally, the Fréchet-Riesz result of 1907 referred to L^2, simply because abstract Hilbert space had not yet been born at that time. But it was not difficult to prove later the representation for an abstract Hilbert space. This was done in 1934 by Riesz.

That self-duality turned out to be a hindrance in the development of duality theory. The latter gained momentum only after Riesz introduced L^p-spaces and their duals and developed the whole corresponding theory in a beautiful paper of only about fifty pages, entitled (in German) *Investigations on systems of integrable functions* (cf. [23], I, 441-489), that was full of novel and exciting ideas and results.

3. From Forms to Bounded and Unbounded Linear Oerators

Hilbert's spectral theory was reformulated in terms of linear operators, as we use it now. It was extended from bounded to unbounded operators in Hilbert space by Weyl (1910), Carleman (1923) and, motivated by quantum physics, by von Neumann (1927-29).

3.1 Continuous Linear Operators on Hilbert Space

The Fourth Paper of 1906 was the culmination point in Hilbert's work on integral equations. Most important in it was the creation of the concepts of completely continuous (i.e., compact) and of continuous quadratic forms and their spectral theories, along with many other basic concepts and methods, notably strong and weak convergence. Formally the paper was still written in the style of classical analysis. Conceptually it gave direction and opened the way to the further development of functional analysis. It is in this sense that we should interpret Dieudonné's statement in [7], p. 110, that Hilbert's Fourth Paper was "a turning point in the history of functional analysis".

At a time when vector spaces and linear operators were already fairly common in circles concerned with "modern" analysis, particularly in Italy (through Peano, Pincherle, etc.), Hilbert's use of quadratic forms (under the influence of Frobenius?) was perhaps not the best idea.

The change of Hilbert's work from quadratic forms to the language of linear operators was accomplished in 1913 in a book (in French) *Systems of Linear Equations in Infinitely Many Unknowns* by F. Riesz (cf. [23], II, 891-1016), written upon invitation by É. Borel as a contribution to his *Collection*, a book series on contemporary mathematics.

In the Preface to that book, Riesz wrote the following.

> "In this volume I want to give a brief report on fundamental ideas, methods, and main results of a theory which one owes almost exclusively to contemporary mathematicians. Personally, I have contributed very little to the theory

> ...[;] that I have undertaken this work is due to my research on certain neighboring subjects, among others orthogonal systems of functions, integral equations, and functional operations. [I shall present] certain parts of the theory under new aspects ... for instance, I can reduce the study of the spectrum of quadratic forms to that of linear functional operations.
>
> Properly speaking, our subject is not a *theory of functions* ... [but it is] a first stage in the *theory of functions of infinitely many variables*, which is just being born but may possibly furnish all the more powerful methods of the whole analysis."

Remarkable in this Preface are on the one hand Riesz's prophetic outlook in the future and on the other hand the characterization of his own role, which seems far too modest; in fact, the reformulation of work by Hilbert and his school in terms of operators should be included into Riesz's formidable list of major functional-analytic accomplishments over a long interval of time of about 35 years, from 1906 to about 1940.

In Chap. I of the book, Riesz gives a historical overview, beginning with Fourier and ending with Poincaré's rigorous introduction of infinite determinants in 1886.

Chapter II is devoted to H. von Koch's work on infinite determinants of 1896, which was a refinement and generalization of Poincaré's results, some of which became the starting point of Fredholm's theory on integral equations in 1900.

The opening of the next chapter reads: "In the present chapter we shall report on the theory of E. Schmidt [of 1908] in a slightly generalized form". That theory concerned infinite systems of equations in Hilbert space, and that "slight generalization" was no less than the transition from Hilbert space l^2 to sequence spaces l^p and l^q, the spaces of sequences (x_k) for which

$$\sum_{k=1}^{\infty} |x_k|^p \quad \text{and} \quad \sum_{k=1}^{\infty} |x_k|^q$$

respectively, are convergent; here, $p > 1$ is fixed and q is such that $1/p + 1/q = 1$.

What Riesz actually created was a solution theory for infinite systems of linear equations (l.c., p. 880; p. 46 of the actual book)

$$\sum_{k=1}^{\infty} a_{ik} x_k = c_i \qquad (i = 1, 2, \ldots) \tag{3.1.1}$$

with coefficients in l^q, for which a solution (x_k) in l^p is wanted. He developed this solution theory parallel to his famous paper of 1910 for systems of integral equations – this extra work was necessary because there is no Riesz-Fischer theorem when $p \neq 2$.

Most important are Chaps. IV and V (comprising almost 2/3 of the book) on the "operator form" of Hilbert's theory in the Fourth Paper. There he coined the name *espace hilbertien* for l^2 (p. 78) and took up the study of *substitutions linéaires*, meaning *continuous* endomorphisms $A : l^2 \to l^2$, whose norm he called *borne* (bound) and denoted by M_A.

Strong convergence of a sequence $(x_k^{(n)})$ in l^2 he defined by

$$\sum_k |x_k - x_k^{(n)}|^2 \to 0 \qquad \text{as } n \to \infty$$

and *convergence à sens ordinaire* by $x_k^{(n)} \to x$ for all k (p. 79).

Riesz established the connection between linear systems of equations and Hilbert's bounded bilinear forms by considering a *tableau* (a_{ik}) and defining a corresponding linear substitution (a continuous linear operator) by (p. 80)

$$x_i' = \sum_{k=1}^{\infty} a_{ik} x_k \qquad (i = 1, 2, \ldots) \tag{3.1.2}$$

where

$$\sum_{i=1} \left| \sum_{k=1} a_{ik} x_k \right|^2 \leq M \sum_{k=1}^{\infty} |x_k|^2, \tag{3.1.3}$$

and, on the other hand, Hilbert's bilinear form by (p. 83)

$$A(x, y) = \sum_{i,k=1}^{\infty} a_{ik} x_k y_i \quad (x, y \in \ell^2). \tag{3.1.4}$$

Next he discussed the transpose and the inverse of an operator. On p. 96 he defined a substitution to be *complètement continue* (i.e., compact) if it maps any convergent sequence onto a strongly convergent one, and showed that this is equivalent to Hilbert's *Vollstetigkeit* of a quadratic form.

On p. 105 he began discussing spectral theory with the resolvent $(E - \lambda A)^{-1}$ (changed to the more usual $(\lambda E - A)^{-1}$ on p. 135, with a remark on the change on p. 123). This was followed by a discussion of sequences (A_n) and series of continuous linear operators, with *uniform convergence* of (A_n) to A defined by

$$M_{A - A_n} \to 0, \tag{3.1.5}$$

the definition of *regular* and *spectral values* on p. 114, a proof of the Hilbert relation on p. 115, and the beginning of residue calculus near the end of Chap. IV.

In Chap. V he considered Hilbert's quadratic forms in infinitely many variables and continued the discussion of the relation to continuous linear operators, culminating with the *spectral representation* (p. 137)

$$f(A) = \int_{-\infty}^{+\infty} f(\xi)\, dA_\xi \tag{3.1.6}$$

where A is self-adjoint and continuous, f is continuous, (A_ξ) is a spectral family of A, and the integration is actually extended over a finite interval only, containing the spectrum of A (cf. [17], 514).

In his book, Riesz created the form of spectral theory that is now commonly used. He also paved the way for his next path-breaking work, an "abstract Fredholm theory" of compact linear operators on normed spaces (cf. [23], II, 1053-1080; cf. also [17], Chap. 8), which he completed in January of 1916, and which was published in 1918; that theory will not concern us here.

3.2 Advances to Unbounded Operators

In his lectures and seminars, Hilbert repeatedly gave examples of kernels $k(s,t)$ of Fredholm integral equations of the second kind for which his spectral theory for compact operators was no longer valid. For instance, in the summer of 1906 he discussed as such a kernel on $[0,1] \times [0,1]$

$$k(s,t) = 1/(s+t). \tag{3.2.1}$$

Integral equations of that kind were later called *singular*, beginning with the Ph.D. thesis *Singular integral equations with special regard to the Fourier integral theorem* of 1908 by H. Weyl (1885-1955), Hilbert's most outstanding student. Weyl wrote at the beginning (cf. [30], I, 1):

> "In one of his lectures Hilbert emphasized that, e.g., the kernel
>
> $$k(s,t) = \cos\ st \tag{3.2.2}$$
>
> on $[0,+\infty) \times [0,+\infty)$ deserves particular interest. From the Fourier integral theorem
>
> $$\phi(s) = \frac{2}{\pi}\int_0^\infty \int_0^\infty \cos(st)\cos(tr)\phi(r)\ dr\ dt \tag{3.2.3}$$
>
> one sees that this kernel has at most two eigenvalues $\pm\sqrt{2/\pi}$."

He confirmed by a simple calculation that these are indeed eigenvalues and each of them has infinite multiplicity.

In his thesis (l.c., 1-87) and a subsequent version of it (l.c., 102-153), Weyl gave a condition sufficient for a kernel on $[0,+\infty) \times [0,+\infty)$ to be such that the corresponding quadratic form is bounded, namely, that

$$\int_0^\infty k(s,t)^2\ dt < \infty \qquad \text{for each } s \geq 0 \tag{3.2.4}$$

and there exists a constant $M > 0$ such that

$$\left|\int_0^\infty \int_0^\infty k(s,t)u(s)v(t)\ ds\ dt\right| \leq M \tag{3.2.5}$$

for all pairs of continuous functions u, v satisfying

$$\int_0^\infty u(s)^2\, ds \le 1, \quad \int_0^\infty v(s)^2\, ds \le 1. \tag{3.2.6}$$

As examples of singular kernels, Weyl considered on $\mathbb{R}^2$

$$k(s,t) = \frac{1}{2}\ \mathrm{sgn}\ (s+t) \exp\Big(-\frac{s^2}{4} - \frac{t^2}{4}\Big) \tag{3.2.7}$$

(l.c., p. 131), which he showed to be closely related to the *Hermite polynomials*, and

$$k(s,t) = \frac{1}{2}\exp(-|s-t|), \tag{3.2.8}$$

which is related to the *Laguerre polynomials* and which he treated in a similar fashion.

In his *Habilitationsschrift* of 1910 (l.c., 248-297) Weyl applied his theory of singular integral equations to the theory of linear ordinary differential equations of second order "which at one end of the real interval have a singularity of a more or less complicated character".

This concerned *singular Sturm-Liouville problems*, two special ones of which had been considered by E. Hilb in his Habilitationsschrift two years earlier. Weyl found out that, based on his theory, he was able to handle the general situation, "that is, without making any restrictive assumption on the nature of the singularity [of the] ... differential equation".

He considered on $[0, +\infty)$

$$L(u) = \frac{d}{ds}\Big(p(s)\frac{du}{ds}\Big) - q(s)u(s) = 0, \tag{3.2.9}$$

$$\begin{aligned} u_1(0) = 1,\ p(0)u_1'(0) = 0, \\ u_2(0) = 0,\ p(0)u_2'(0) = 1, \end{aligned} \tag{3.2.10}$$

and $L(u) + \lambda u = 0$, finding out that it sufficed to take $\lambda = i$; thus

$$L(u) + iu = 0. \tag{3.2.11}$$

From u_1 and u_2 and any fixed $h > 0$ he formed

$$\begin{aligned} v_h &= -u_1 \sin h + u_2 \cos h \\ w_h &= u_1 \cos h + u_2 \sin h, \end{aligned} \tag{3.2.12}$$

so that the solution v_h satisfies the boundary condition

$$u(0) \cos\ h + p(0)u'(0) \sin\ h = 0. \tag{3.2.13}$$

Taking any number $a > 0$ and a number $\beta \geq 0$, he formed

$$u_\mu = v_h + \mu w_h \tag{3.2.14}$$

with complex μ determined by the condition that $u = u_\mu$ satisfy the condition

$$u(a)\cos\beta + p(a)u'(a)\sin\beta = 0. \tag{3.2.15}$$

Weyl showed that for β varying from 0 to 2π the parameter μ describes a circle in the complex plane, which for $a \to +\infty$ may remain a circle (*limit circle case*) or shrink to a point (*limit point case*). From v_h and u_μ he found a Green's function $k_{a,\mu}\ (t,x)$ which in the limiting case always satisfies the conditions in Weyl's thesis. Hence he could apply to the corresponding singular integral equation the Hilbert theory of bounded bilinear forms. In this way he obtained integral formulas as a substitute for the Fourier expansion of the classical Sturm-Liouville theory.

In this very original paper Weyl finally also gave a thorough discussion of the corresponding spectral theory for the operator L. He defined the *spectrum* as the complement of the set in the λ-plane for which $L(u) + \lambda u = g(x)$ has a solution $u \in L^2([0,+\infty))$ satisfying (3.2.14), for every $g \in L^2([0,+\infty))$. He investigated the spectrum under various assumptions on p and q. For instance, he gave an example such that the spectrum is the whole real line $\mathbb{R}$.

In another paper of 1910 (l.c., 222-247), Weyl extended his theory to the case where the equation is considered on the whole real line and is of the limit point type at both ends.

Weyl's work is the earliest study of *unbounded* symmetric operators and of its self-adjoint extensions, and the integral operator defined by the limiting case of $k_{a,\mu}$ with complex values is perhaps the first example of a non-self-adjoint *normal* operator in Hilbert space.

Whereas Weyl's work followed Hilbert's Fourth Paper within a short time, the further development on singular integral equations was slower. The next pioneering advance after Weyl's accomplishments was that by T. Carleman (1892-1945), who since 1924 was H. von Koch's successor at the Technical University of Stockholm.

Beginning in 1920, Carleman investigated singular integral equations

$$\phi(s) - \lambda \int_a^b k(s,t)\phi(t)\, dt = f(s) \tag{3.2.16}$$

with Hermitian kernel (i.e., $k(s,t) = \overline{k(t,s)}$) satisfying (3.2.4) for almost all s (but not necessarily (3.2.5)) – such a kernel is now called a *Carleman kernel.*

Carleman discovered that those equations may be divided into two classes with quite different properties, depending on whether for genuinely complex λ they have a unique solution ("class I") or not ("class II"), and he gave a corresponding criterion. He showed that for $f \in L^2([a, b])$ and every nonreal λ the equation (3.2.16) has at least one solution. On the other hand, for all such λ the corresponding homogeneous equation has the same number ν of linearly independent solutions $\phi \in L^2([a, b])$, where ν may also be 0 (this is class I) or $+\infty$, and any $\nu > 0$ corresponds to class II.

Some of Carleman's results were quite unexpected. They were shown in their proper context in von Neumann's theory, which we shall consider in the next section.

3.3 Impact of Quantum Mechanics

We have seen that Hilbert sequence space l^2 appeared in 1904 in Hilbert's work on integral equations, and Lebesgue space L^2 was defined by F. Riesz in 1906, and the two spaces were found to be isometrically isomorphic by Fischer and Riesz a year later. At that time it was also discovered that continuous linear functionals on l^2 or L^2 can be taken care of by the inner product.

Furthermore, continuous linear operators and their spectral theory were obtained in 1913 by Riesz from Hilbert's results in the Fourth Paper of 1906, as we have discussed in Sec. 3.1.

Hence the beginnings on Hilbert space came from *mathematics* – integral equations –whereas, apart from the two forerunners just considered, the further evolution had to wait for a decade, until finally a demand for further development arose – and this time from *physics*, namely, from *quantum mechanics*. This happened under Hilbert's influence in the following way.

Hilbert's investigation of integral equations was his fourth field of work, after invariant theory, algebraic number theory, and the foundations of geometry. The latter suggested to him:

> "To treat in the same manner, by means of axioms, those physical sciences in which mathematics plays an important part ... "

This quotation is taken from his Sixth Problem of his famous talk of 1900 at the International Congress of Mathematicians.

H. Weyl ([30], IV, 171) mentioned the following.

> "Already before Minkowski's death in 1909, Hilbert had begun a systematic study of theoretical physics, in close collaboration with his friend ... Minkowski's work on relativity theory was the first fruit of these joint studies. Hilbert continued them through the years, and between 1910 and 1930 often lectured and conducted seminars on topics of physics."

Subjects of his interests were:

Kinetic theory of gases (*Math. Ann.* **71** (1912), 562-577),
Radiation theory (*Physik. Zeitschr.* **13** (1912), 1056-1064),
Relativity (*Math. Ann.* **92** (1924), 1-32),
Quantum mechanics (*Math. Ann.* **98** (1927), 1-30).

Quantum mechanics had just been born in a dramatic sequence of events in 1925-26. Indeed, in July 1925 W. Heisenberg (1901-76), then 23 years old and like many other famous physicists a student of A. Sommerfeld (1868-1951) in Munich, had completed his fundamental paper (in German) *On a quantum-theoretical reinterpretation of kinematic and mechanical relations.* Together with M. Born and P. Jordan he developed the new "matrix mechanics" in a paper completed four months later.

P. Dirac (1902-84) in Cambridge, even slightly younger than Heisenberg, had independently developed a different scheme, based on Heisenberg's July paper, namely, his method of q-numbers, in November 1925, in which many-electron atoms and the relativistic Compton effect could be handled, as he showed in Spring of 1926.

In January of the same year, E. Schrödinger (1887-1961) began introducing his wave mechanics. The equivalence of the three approaches was finally proved in December 1926 by Dirac and Jordan.

Hilbert's paper listed above was an outgrowth of his lectures in 1926-27, in which Hilbert proposed to axiomatize quantum mechanics.

Based on papers of 1926 by Schrödinger, Born, and N. Wiener, Hilbert employed operators because:

> "It is known that the connection between Schrödinger's theory and Heisenberg's matrix mechanics is given by operator calculus. Suitably generalized, the latter turned out to be the analytic framework [Gerüst] of general quantum mechanics."

Although "Hilbert's vast plans in physics never matured" (as Weyl expressed it), he had great indirect influence on the development of quantum mechanics, through the appearance of vol. I of R. Courant and D. Hilbert, *Methods of Mathematical Physics* as well as through his contact with the physicists in Göttingen. Heisenberg wrote later:

> "Indirectly Hilbert exerted the strongest influence on the development of quantum mechanics in Göttingen ... It was an especially fortunate coincidence that the mathematical methods of quantum mechanics turned out to be a direct application of Hilbert's theory of integral equations.
>
> ... From Sommerfeld I learned optimism, from the Göttingen people mathematics, and from Bohr physics."

Hilbert's paper of 1927 (reprinted in [27], I, 104-133) was co-authored by L. Nordheim, Hilbert's "physical assistant" from 1922 to 1927, who formulated the final version of the paper, and J. von Neumann (1903-57), who had studied under E. Schmidt in Berlin and had just joined Hilbert's Institute as an assistant. Von Neumann worked out "important mathematical parts" of the paper. The approach as a whole was an attempt of a "not yet complete" *axiomatization* (cf. p. 106), as a counterpart of the *complete* axiomatization of Euclidean geometry (by Pasch, Hilbert, Veronese, etc.). "By the axiomatization somewhat vague concepts, such as *probability*, etc., will lose their mystical character since they will then be defined implicitly by axioms." (Recall that Kolmogoroff's axiomatization of probability occurred only later, in 1933!)

In 1926, M. Born and N. Wiener had recognized that matrices can be regarded as special cases of linear operators, and the paper under discussion was based on operators under the assumption (l.c., p. 111) that these can be represented as integal operators, written

$$(3.3.1) \qquad T^{\binom{x}{y}}(f(y)) = \int_{-\infty}^{\infty} \phi(xy)f(y)\,dy,$$

perhaps with an "improper kernel" (reminiscent of Dirac, who was mentioned on p. 114). *Position and momentum operators*

$$(3.3.2) \qquad qf(x) = xf(x) \quad \text{and} \quad pf(x) = \varepsilon\frac{\partial f(x)}{\partial x}$$

appeared (not under these names) on p. 114, along with the *Heisenberg commutation relation*

$$(3.3.3) \qquad (pq - qp) = \varepsilon 1.$$

As kernels for (3.3.2) the authors gave in terms of *Dirac's delta*

$$(3.3.4) \qquad x\delta(x-y) \quad \text{and} \quad \varepsilon\delta'(x-y).$$

A physical interpretation of the whole operator formalism in terms of probabilities followed on pp. 116-128. *Schrödinger's differential equation* was written (p. 131)

$$(3.3.5) \qquad \left\{H\left(\varepsilon\frac{\partial}{\partial x}, x\right) - W\right\}\phi(xW; qH) = 0,$$

where ϕ is "the probability amplitude for the position of the [one-dimensional physical] system to have a value between x and $x + dx$ when the value W of the energy is given ... [If (3.3.5) has a discrete spectrum] then in ϕ ... there occurs a split into *eigenfunctions* ...

$$(3.3.6) \qquad \phi(xW_n; qH) = \psi_n(x)."$$

Here, W_n is the nth eigenvalue. The paper of 1927 by Hilbert and his two assistants concludes with a summary of the accomplishments and remaining difficulties of the theory and an announcement of von Neumann's forthcoming paper, in which the questions of mathematical justifications were to be discussed further. We shall explain the basic ideas in von Neumann's work in the next section.

3.4 Spectral Theory of Unbounded Operators

Although the Hilbert-Nordheim-von Neumann attempt of axiomatizing quantum mechanics involved unsolved basic difficulties, it advanced the theory and application of linear operators and their integral representation and caused the impression that Hilbert space theory would be adequate for the then most modern part of physics.

This was by and large confirmed in von Neumann's further work published in three papers of 1927 (cf. [27], I, 151-255) and two further papers of 1929 (l.c., II, 3-143), in part written in preparation for von Neumann's well-known book [28] of 1932, *Mathematische Grundlagen der Quantenmechanik.* We shall investigate these papers from the viewpoint of the development of Hilbert space theory.

At the time of Hilbert's lectures (1926-27), von Neumann had begun working on an axiomatic mathematical theory of quantum mechanics of his own. He realized the "unsolvable problems" in the approach which the physicists had taken, e.g., in the diagonalization of infinite matrices in the case of systems that involved a spectrum with a continuous part, and he believed that the *usual* mathematics could be developed to the extent needed for quantum mechanics. He emphasized (in [27], I, 157) that a main difficulty resulted from the fact that even the simplest physical systems required *unbounded operators* for their description, whereas Hilbert's theory concerned bounded operators (i.e., continuous forms).

Unbounded operators had occurred before, in connection with more or less isolated instances, in Hilb's, Weyl's, and Carleman's work, and in a spectral representation of unbounded Jacobi forms by E. Hellinger (*Math. Ann.* **86** (1922), 18-29). (Cf. also [17], where on p. 525 a modern version of the Hellinger-Toeplitz theorem, originally obtained in 1910, is discussed.)

Von Neumann began the development of his spectral theory of unbounded linear operators with a very important step forward, the earliest axiomatic definition of (separable) *Hilbert space* H by the now familiar axioms (p. 165),

A. Axioms of a complex vector space.
B. Axioms of the inner product $Q(f, g)$ and the corresponding metric.
C. dim $H = \infty$.
D. H is separable (has an everywhere dense sequence).
E. H is complete (the Cauchy convergence condition holds).

Thus, twenty years after the Riesz-Fischer theorem, ℓ^2 and L^2 finally became models of the same abstract space.

Von Neumann continued with a discussion of orthonormal sequences (pp. 168-172) and operator calculus. This included *projection operators* (p. 175, called *Einzeloperatoren*, after Hilbert's term) and *position* and *momentum* (p. 185), and culminated with a *spectral representation* of a self-adjoint (possibly unbounded) linear operator T (p. 183; cf. also [17], 559)

$$T = \int_{-\infty}^{\infty} \ell \, dE(\ell). \tag{3.4.1}$$

We can skip the last few pages of this first paper as well as von Neumann's second and third papers of 1927, *Probabilistic development of quantum mechanics* and *Thermodynamics of quantum mechanical systems*, which concern primarily physical aspects of the mathematical theory just developed.

Von Neumann completed his spectral theory of unbounded linear operators in Hilbert space in his two basic papers of 1929 (cf. [27], II, 3-143) mentioned above. In the first of these papers he developed the spectral theory of (in modern terms) *symmetric linear operators* R (his notation) defined on a dense subset of a (separable) Hilbert space H, i.e.,

$$\langle f, Rg \rangle = \langle Rf, g \rangle. \tag{3.4.2}$$

He called this a *Hermitian operator* (abbreviated H. O.). He showed (on p. 44) that any such operator R can be extended to a *maximal symmetric operator*, i.e., one without proper symmetric extensions, and he gave an interesting criterion for none, or some, or all of these extensions to be self-adjoint (he said *hypermaximal*). (Cf. pp. 533-534 of [17] for the distinction.)

The latter concept he owed to E. Schmidt, together with the information that self-adjointness is necessary and sufficient for the existence of a spectral family $F(A)$, which he used in establishing a *spectral representation* (Theorem 36, p. 46)

$$\langle Rf, g \rangle = \int_{-\infty}^{+\infty} d\langle F(\lambda)f, g \rangle. \tag{3.4.3}$$

He based the proof on the Kunstgriff that the *Cayley transform*

$$U = (R - iI)(R + iI)^{-1} \tag{3.4.4}$$

of a self-adjoint operator R is *unitary*, which reduced the possibly unbounded case to the bounded one.

These results and others in the paper showed that the situation had changed drastically, compared to that in the case of bounded operators.

Von Neumann's second paper of 1929 (in [27], II, 86-143) was in its first part the earliest publication on *algebras* of (bounded linear) operators on Hilbert space H. He considered a W^*-algebra B of bounded linear operators on H (he called it a *ring*) and in B rings M that contain A^* for every $A \in M$ and are closed in the weak operator topology (hence the term W^*!). The motivation for this paper came probably more from the Peter-Weyl theory and results by I. Schur than from quantum mechanics. Indeed, the highlight was the *Double commutant theorem* (p. 109, an analog of the Burnside-Frobenius-Schur theorem), stating that $(B')' = B$, where B' is the commutant of B (the family of operators in B that commute with all operators in B).

In the second part of the paper von Neumann investigated properties of (unbounded) *normal operators* (closed densely defined operators R that commute with their Hilbert-adjoint R^*). This culminated with spectral representations of both R and R^* (p. 127) involving the same unique spectral family.

This completed von Neumann's spectral theory of unbounded operators.

In a separate paper ([27], II, 144-172), von Neumann showed that in the case of an unbounded symmetric linear operator R, representations by matrices $A = [a_{mn}]$, where

$a_{mn} = \langle Re_m, e_n \rangle$, (e_m) an orthonormal basis, are no longer adequate, simply because any nonmaximal R and any of its symmetric extensions would correspond to the same matrix.

In his monograph *Linear Transformations in Hilbert Space and Their Applications in Analysis*, M. H. Stone showed (p. 530) how von Neumann's spectral theory could be applied to yield Weyl's results (cf. Sec. 3.2). In particular, L is symmetric, and the defects of L^{**} in the limit circle case and in the limit point case are (2, 2) and (1, 1), respectively.

Similarly, Carleman's results can now be explained in a systematic fashion; cf. Dieudonné [7], 181.

4. Hilbert Spaces and Complex Analysis

Classical complex analysis was created during the 19th century by Cauchy, Riemann, and Weierstrass, who laid the foundations and developed most parts of the basic theory. The further evolution of it was virtually independent of that of functional analysis, except for a few episodes, such as the use of complex integration in spectral theory, for instance, in work on compact operators by F. Riesz. Notice the conspicuous absence of complex analysis in the monumental work by Bourbaki designed to systematize and unify modern mathematics.

However, although complex analysis exerted no leadership in functional analysis, it was brought into contact with Hilbert spaces, mainly through the idea of reproducing kernel functions, as shown in this chapter.

4.1 Hilbert Spaces with Kernel Functions

The theory of kernel functions is a constructive approach to boundary value problems and conformal mapping. This theory furnishes an important bridge between complex analysis and Hilbert spaces.

The idea of using orthogonal systems in spaces of complex functions was very actively pursued at the University of Berlin under the leadership of E. Schmidt and L. Bieberbach. Schmidt became H. A. Schwarz's successor at Berlin in 1917, and Bieberbach joined the Faculty two years later.

Schmidt's important functional-analytic ideas have been discussed in Sec. 2.3. Bieberbach was a leader in modern complex analysis, known, for instance, by the Bieberbach conjecture, which was finally proved by L. de Branges in the early eighties.

Orthogonalization and orthogonal expansions were adopted entirely from real analysis, where they had been used for a long time, and only in connection with convergence questions the principles of complex analysis entered into considerations. It was only relatively recent that the advantages of these expansions for complex analysis were noticed, first by Bieberbach.

Three major works on kernel functions appeared in 1921-22, a paper by G. Szegö (Dozent at Berlin) and two Ph.D. theses, written by S. Bochner and S. Bergman. We shall concentrate on basic ideas in Bergman's and Schiffer's works (cf. [2] and [4]) that are of particular interest from the viewpoint of Hilbert spaces.

Bergman's thesis (*Math. Ann.* **96** (1922), 237-271) was supervised by Schmidt and R. von Mises, who in 1920 was appointed director of the newly founded Institute of Applied Mathematics at the University.

Kernel functions have long been known, in connection with Green's and Neumann's functions. What was new in Bergman's work was the amalgamation of complex analysis and Hilbert space theory in a constructive fashion, an L^2-theory of holomorphic functions.

Two approaches to kernel functions have been pursued. The first defines a kernel function as a series involving the terms of an orthonormal sequence. The second uses the so-called 'reproducing property' explained below.

The first of these two possibilities is Bergman's original approach. In it, one considers the space $HL^2(\Omega)$ consisting of the functions $f(z)$ that are holomorphic and absolutely Lebesgue square integrable on a bounded domain Ω in the complex z-plane, $z = x + iy$, with the inner product defined by

$$(4.1.1) \qquad \langle f, g\rangle = \int_\Omega f(z)\overline{g(z)}\ d\omega \quad (d\omega = dx\ dy),$$

hence with the metric defined by $d(f,g) = \langle f-g, f-g\rangle^{1/2}$. This inner produce space is complete; hence it is a Hilbert space.

For simplicity we assume in this chapter that Ω is simply-connected. (Most of our results extend to bounded domains of arbitrary finite connectivity.)

Since $HL^2(\Omega)$ is separable, it contains a total orthonormal sequence $(\phi_n(z))$. Using it, we define the *Bergman kernel function* k by the series

$$(4.1.2) \qquad k(z,t) = \sum_{n=1}^{\infty} \phi_n(z)\overline{\phi_n(t)}.$$

For *real* orthonormal functions such a series may not converge. Thus, there are Hilbert spaces without kernel functions.

In our case, however, in every closed subregion of Ω the series (4.1.2) converges absolutely, uniformly in z for fixed t and uniformly in t for fixed z.

Bergman showed convergence of the series (4.1.2) essentially by using the inequality

$$(4.1.3) \qquad |f(t)|^2 \leq \|f\|^2/\pi d^2,$$

where d is the distance of t from the boundary of the domain.

Note that a similar inequality between function values and norm does not exist in real! This observation deserves to be considered from the viewpoint of completing inner product spaces. It will turn out to be typical of Hilbert spaces with reproducing kernel.

A pre-Hilbert space X of functions defined on a set E can always be completed (cf. [17], 139). In this embedding of X in a space $\tilde{X}$ of equivalence classes of Cauchy sequences a relation between function values and norm existing in X will get lost, in general. Under suitable conditions this does not happen, as was investigated in 1956 by N. Aronszajn and K. T. Smith (*Ann. Inst. Fourier* **6,** 125-185), with the following result.

We call X *proper* if for every $x \in E$ and $f \in X$ the evaluation functional $F_x(f) = |f(x)|$ is continuous, say,

$$|f(x)| \leq M_x \|f\| \tag{4.1.4}$$

for some positive constant M_x (which may depend on x). We further call $\tilde{X}$ a *proper functional completion* of X if $\tilde{X}$ is a proper function space (and X is dense in $\tilde{X}$, as usual).

It can be shown that such a completion of X exists if and only if for each Cauchy sequence (f_n) in X,

$$\lim f_n(x) = 0 \text{ for all } x \in E \text{ implies } \lim \|f_n\| = 0. \tag{4.1.5}$$

It is not difficult to see that X can have at most one such completion (except for isometry) and that $HL^2(\Omega)$ is proper, so that (4.1.3) is typical.

A property of basic importance is the *reproducing property* of k, that is,

$$\langle f, k(z, .) = f(z) \rangle. \tag{4.1.6}$$

This follows by substituting (4.1.2) and the orthonormal expansion of f in terms of (ϕ_n),

$$f(z) = \sum_{n=1}^{\infty} a_n \phi_n(z), \qquad a_n = \langle f, \phi_n \rangle. \tag{4.1.7}$$

Because of (4.1.6) the function k is often called a *reproducing kernel.*

Bergman pointed out that (4.1.6) implies the independence of k of the particular choice of a total orthonormal system in the space considered, and that it also furnished an immediate proof of the uniqueness of k.

A proof of existence for Bergman's kernel function based on the Riesz representation for continuous linear functionals on a Hilbert space (cf. Sec. 2.4) was given in 1949 by O. Lehto (*Ann. Sci. Fenn. Ser.* A I, **67**), who obtained the boundedness of the functional from Bieberbach's Flächensatz (cf. [5], II, 72).

4.2 Confomal Mapping. Topological Duality

Bergman's kernel function has some remarkable properties that can be used in conformal mapping in two ways,

1. in constructing (approximations to) mapping functions, and
2. in defining a Riemannian metric that is invariant under conformal mapping.

We begin with a very important property, which Bergman ([2], 23) discovered, motivated by Bieberbach's investigations of conformal mapping (cf. [6], 135).

Minimum property. Let $z_0 \in \Omega$. Then among all functions $f \in HL^2(\Omega)$ with $f(z_0) = 1$ the unique function of minimum norm is

$$f(z) = g(z, z_0) = \frac{k(z, z_0)}{k(z_0, z_0)} \tag{4.2.1}$$

and the minimum value of the norm is

$$\min \|f\| = \|g\| = 1/\sqrt{k(z_0, z_0)}. \tag{4.2.2}$$

Interesting in itself, this property is the key to the application of Bergman's kernel function to conformal mapping as a *constructive* method of actually obtaining mapping functions or at least approximations to them. Indeed, the minimum property implies the following

Mapping theorem. The function

$$K(z, t) = \frac{1}{k(t, t)} \int_t^z k(\zeta, t)\, d\zeta \tag{4.2.3}$$

maps Ω (which is assumed to be simply connected and bounded) conformally onto the disk of area $1/k(t, t)$ in such a way that $K(0, 0) = 0$ and $K_z(t, t) = 1$.

A closed representation of the kernel function is rarely possible. Hence, practically one depends on the use of partial sums of (4.1.2). The accuracy can often be increased substantially by including singularity functions into the system of functions before the process of orthonormalization. For recent developments on this method, see [10], 38-42, and the references given there.

We now turn to the second topic in this section, the conformally *invariant Bergman metric* obtained from the kernel function.

This is a Riemann metric whose first fundamental form

$$ds^2 = g_{11}\, dx^2 + 2g_{12}\, dx\, dy + g_{22}\, dy^2 \tag{4.2.4}$$

has the coefficients $g_{11} = g_{22} = k(z, z)$, $g_{12} = 0$. Writing $k = k_\Omega$, Bergman thus had

$$ds^2 = k_\Omega(z, z)\, |dz|^2 = k_\Omega(z, z)(dx^2 + dy^2) \tag{4.2.5}$$

and similarly for a domain $\tilde{\Omega}$ in the ζ -plane, $\zeta = \xi + i\eta$, and its kernel function $k_{\tilde{\Omega}}$,

$$\widetilde{ds}^2 = k_{\tilde{\Omega}}(\zeta, \zeta)|d\zeta|^2 = k_{\tilde{\Omega}}(\zeta, \zeta)(d\xi^2 + d\eta^2). \tag{4.2.6}$$

He assumed that $\tilde{\Omega}$ can be mapped bijectively and conformally onto Ω, say, by $z = h(\zeta)$. Then for the terms of the orthonormal sequence (ϕ_n) in Ω, used in (4.1.2), he obtained

$$\int_\Omega |\phi_n(z)|^2\, d\omega_z = \int_{\tilde{\Omega}} |\phi_n(h(\zeta))|^2 J\, d\omega_\zeta, \tag{4.2.7}$$

the Jacobian J of the mapping being equal to $|h'(\zeta)|^2$ (cf. [19], 862). Hence to $\phi_n(z)$ in (4.1.2) there corresponded

$$\psi_n(\zeta) = \phi_n(h(\zeta))h'(\zeta) \tag{4.2.8}$$

in the series for $k_{\tilde{\Omega}}$. This then implied the invariance $\widetilde{ds}^2 = ds^2$.

Bergman called the Riemann manifold with the metric (4.2.4) the *basic space* of Ω and noted that the mapping $z = h(\zeta)$ corresponds to a transformation of the parameters in the basic space.

For the unit disk he took $\phi_n(z) = \sqrt{n/\pi}z^{n-1}$, $n = 1, 2, \ldots$, so that he was able to sum the series in (4.2.1), and in (4.2.5) he obtained

$$\begin{aligned} g_{11} = g_{22} = k(z, z) &= \pi^{-1}(1 - |z|^2)^{-2} \\ &= \pi^{-1}(1 - x^2 - y^2)^{-2}. \end{aligned} \tag{4.2.9}$$

Thus we see that in this case he obtained the hyperbolic metric

$$ds^2 = \frac{|dz|^2}{\pi(1 - |z|^2)^2}. \tag{4.2.10}$$

Because of the conformal invariance, from this we can read off by a familiar formula for Riemann manifolds of constant curvature – or by calculation from (46.17') on p. 147 of [18] – the following remarkable result.

Gauss curvature. For any simply-connected bounded domain the Gauss curvature of the Bergman metric is constant; it equals -4π .

Bergman's innovative work on the kernel function and the corresponding Hilbert space of analytic functions was followed by a more "abstract" approach sparked by developments on functionals. The latter had begun in 1903 when Hadamard had given his representation (2.4.1) for continuous linear functionals on $C[a, b]$. Important to us in the present connection is the definition of abstract *separable* Hilbert space in 1927 by J. von Neumann and the discovery that separability could be dropped from that definition without great loss, as was realized in 1934 independently by H. Löwig, F. Rellich, and F. Riesz. The latter also gave the general representation

$$F(f) = \langle f, g \rangle \tag{4.2.11}$$

for continuous linear functionals, where g is uniquely determined by F, and $\|g\| = \|F\|$; cf. [17], 188-190, where Riesz's original proof is given.

Beginning around that time, N. Aronszajn used the reproducing property (stated for the Bergman kernel in (4.1.6)) for developing a theory of general reproducing kernels k on a Hilbert space H of functions f defined on a set E (cf. *Trans. Amer. Math. Soc.* **68** (1950), 337-404, and some earlier papers listed in [2], 233).

Aronszajn called k a *reproducing kernel* on H if

1. for every fixed y the function k considered as a function of x is an element of H,

and

2. k has the *reproducing property*

$$\langle f, k(x,.) = f(x)\rangle. \tag{4.2.12}$$

Clearly, such a theory also applies to non-separable spaces, and it can give results common to kernel functions from potential theory, more general elliptic equations, and other fields, in addition to complex analysis, but on the expense of losses of properties typical of the situation in complex analysis.

Existence theorem. H has a reproducing kernel if and only if for every fixed $y \in E$,

$$f(y) = \langle f, k(y,.)\rangle \tag{4.2.13}$$

regarded as a linear functional on H is bounded.

Indeed, existence of k implies boundedness of (4.2.13) by the Schwarz inequality, and, conversely, boundedness implies the existence of a unique $g_y \in H$ such that

$$f(y) = \langle f, g_y\rangle$$

by the Riesz representation theorem; cf. (4.2.11). With $g_y = k(y,.)$ this is (4.2.13).

4.3 Extension to Harmonic Functions

Functional analysis in general, and Hilbert spaces in particular, evolved gradually from the main topic of 19th century analysis, which was the solution of equations whose unknowns are *functions*; this included ordinary and partial differential equations (whose study had been initiated already in the 18th century), integral equations, and other functional equations.

Whereas for *ordinary* differential equations the search for "general solutions" was of practical interest, it was soon recognized, at the very latest at Riemann's times, that for a *partial* differential equation one should primarily study problems consisting of an equation and additional conditions (boundary conditions, initial conditions, etc.), as suggested by physical applications.

Green's famous paper of 1828 containing Green's formulas became known in 1846 through its publication in *Journal für die reine und angewandte Mathematik* by W. Thomson (Lord Kelvin).

In his paper, Green coined the term *potential function* (called *potential* by Gauss in 1840) and introduced *Green's function.* Subsequent work by various mathematicians concerned the existence and properties of solutions of boundary value problems (Dirichlet's, Neumann's, mixed) under various assumptions on the boundary of domains, on properties of the given values, etc.

It is well known that methods and results of complex analysis can be used in two-dimensional potential theory, and we remember in passing that, conversely, it was Riemann who first employed the theory of harmonic functions as a tool in the theory of complex analytic functions.

In connection with kernel functions, Bergman ([2], Chap. V) extended the theory from analytic functions to harmonic functions φ on a domain Ω of (finite) connectivity p; we assume $p = 1$ for simplicity.

For orthonormalization he introduced the inner product

$$\langle \varphi, \psi \rangle = \int_\Omega (\text{grad } \varphi) \cdot (\text{grad } \psi)\, d\omega \tag{4.3.1}$$

and the "*Dirichlet norm*" defined by the corresponding *Dirichlet integral*

$$\|\varphi\| = \langle \varphi, \varphi \rangle^{1/2}. \tag{4.3.2}$$

Using a total orthonormal sequence (φ_n) of harmonic functions on Ω with finite Dirichlet integral, he defined the *kernel function* h by the series

$$h(z,t) = \sum_{n=1}^{\infty} \varphi_n(z)\varphi_n(t). \tag{4.3.3}$$

It is not difficult to see that h enjoys a minimum property similar to that of k, namely, the function

$$q(z,t) = h(z,t)/h(t,t) \tag{4.3.4}$$

minimizes the Dirichlet integral (4.3.2) among all harmonic functions φ under consideration satisfying $\varphi(t) = 1$.

Bergman showed that, although his function h has no immediate physical meaning, it is related to Green's and Neumann's function by a simple and remarkable formula

$$h(z,t) = N(z,t) - G(z,t). \tag{4.3.5}$$

Hence the familiar solution formulas for the Dirichlet and Neumann problems can now be written in terms of h,

$$u(z) = -\int_{\partial\Omega} u(t) \frac{\partial h(z,t)}{\partial n} ds_t + \frac{2\pi}{\ell} \int_{\partial\Omega} u(t)\, ds_t \tag{4.3.6}$$

and

$$u(z) = \int_{\partial\Omega} \frac{\partial u(t)}{\partial n} h(z,t)\, ds_t, \tag{4.3.7}$$

respectively, where n is the *inner* normal and ℓ is the length of $\partial\Omega$.

The basic relation (4.3.5) follows by establishing the reproducing property, where $\partial\Omega$ is assumed to be such that N and G exist. Relation (4.3.6) is then obtained by noting that $\partial N/\partial n = 2\pi/\ell$, and relation (4.3.7) follows because G approaches zero as z approaches any point of the boundary.

These and more general results, also for p-fold connected domains, were obtained by Bergman and Schiffer in papers published between 1947 and 1950, listed in [2].

Orthogonalization over the boundary $\partial\Omega$ of a domain Ω is an alternative idea initiated by G. Szegö (*Math. Zeitschr.* **9,** 218-270) in 1921. For the space of functions U holomorphic in Ω and on $\partial\Omega$ a corresponding *inner product* is

$$\langle f, g\rangle = \int_{\partial\Omega} f(z)\overline{g(z)}\, ds; \tag{4.3.8}$$

similarly for harmonic functions.

For instance, in the case of the unit disk $\Omega = D$ a total orthonormal system in the sense of (4.3.8) is

$$f_n(z) = z^n/\sqrt{2\pi} \qquad (n = 0, 1, \ldots).$$

A function $f \in U$ can then be represented as

$$f(z) = \sum_{n=0}^{\infty} a_n f_n(z) \tag{4.3.9}$$

and has the norm $\|f\|$, where

$$\|f\|^2 = \sum_{n=0}^{\infty} |a_n|^2.$$

For functions f, $g \in U$ with developments (4.3.8) and

$$g(z) = \sum_{n=0}^{\infty} b_n f_n(z) \tag{4.3.10}$$

we can define an *inner product* by

$$\langle f, g\rangle_0 = \sum_{n=0}^{\infty} a_n \bar{b}_n. \tag{4.3.11}$$

This defines a Hilbert space isomorphic with Hilbert sequence space ℓ^2, where f corresponds to (a_n). For functions in U this inner product is equal to that in (4.3.8).

A *reproducing kernel*, named after Szegö, is

$$k_s(z, t) = \sum_{n=0}^{\infty} f_n(z)\overline{f_n(t)} = \frac{1}{2\pi(1 - z\bar{t})}. \tag{4.3.12}$$

If $w = w(z)$ is a holomorphic bijection of D onto a domain D' and $z = z(w)$ is its inverse, then a total orthonormal system in D' consists of the functions

$$g_n(w) = (z'(w)/2\pi)^{1/2} z(w)^n \qquad (n = 0, 1, \ldots)$$

and for the Szegö kernel on D' we obtain

$$k_s(w, v) = \frac{(z'(s)\overline{z'(v)})^{1/2}}{2\pi(1 - z(w)\overline{z(v)})}. \tag{4.3.13}$$

It can be shown that for a simply connected domain the relation to the Bergman kernel k is

$$k_s(w, v)^2 = \frac{1}{4\pi} k(w, v). \tag{4.3.14}$$

4.4 Extension to More General Elliptic Equations

Bergman, Schiffer, and others have shown that Hilbert space methods involving kernel functions can be extended from Laplace's equation to more general elliptic linear equations. This required a norm given by an integral whose Euler-Lagrange equation is the differential equation to be considered.

We explain these ideas in terms of the partial differential equation

$$\Delta u - q(x, y)u(x, y) = 0 \tag{4.4.1}$$

in a bounded domain Ω with piecewise smooth boundary $\partial\Omega$ and a Dirichlet condition on $\partial\Omega$ that has an extension over $\bar{\Omega} = \Omega \cup \partial\Omega$ which has bounded first derivatives on $\bar{\Omega}$. Also, q is assumed to be continuously differentiable and positive on the closure $\bar{\Omega}$ of Ω.

We consider all twice continuously differentiable functions on $\bar{\Omega}$ and define an inner product by

$$\langle u, v \rangle = \int_\Omega [(\text{grad } u).(\text{grad } v) + quv] \, d\omega. \tag{4.4.2}$$

The corresponding norm is

$$\|u\| = \langle u, u \rangle^{1/2} = \left[\int_\Omega (|\text{grad } u|^2 + qu^2) \, d\omega \right]^{1/2} \tag{4.4.3}$$

Obviously, this Dirichlet integral is finite for the functions under consideration. Let W denote the Hilbert space thus obtained, and let U be the subspace of the solutions u of (4.4.1) satisfying that boundary condition. Because of our assumption on the latter, U is closed (cf. [11], Chap 8). By the projection theorem (cf. [17], 146) we can write

$$W = U \oplus V. \tag{4.4.4}$$

Here, the *orthogonal complement* V of U in W consists of the functions in W that are zero on the boundary.

In order to solve the Dirichlet problem for (4.4.1), one can introduce *Green's function* G and then represent the solutions in terms of the given boundary data as

$$u(Q) = \int_{\partial\Omega} u(P) \frac{\partial G(P,Q)}{\partial n}\, ds. \tag{4.4.5}$$

Similarly for Neumann's problem, with *Neumann's function* N,

$$u(Q) = -\int_{\partial\Omega} \frac{\partial u(P)}{\partial n} N(P,Q)\, ds. \tag{4.4.6}$$

Here P and Q are convenient notations for points (ξ, η) and (x, y), respectively, and n is the *inner* normal.

As for the Laplace equation, Bergman and Schiffer ([4] , 275) defined the *reproducing kernel H of the class U with respect to the metric determined by* (4.4.3),

$$H(P,Q) = N(P,Q) - G(P,Q). \tag{4.4.7}$$

This definition is justified by the fact that for $u \in U$ it gives the *reproducing property*

$$\langle H(.,Q), u\rangle = u(Q), \tag{4.4.8}$$

characteristic of a kernel function, whereas for $v \in V$ one obtains

$$\langle H(.,Q), v\rangle = 0 \tag{4.4.9}$$

(proofs below). Together this gives for $w = u + v$

$$\langle H(.,Q), w\rangle = u(Q).$$

This shows that H *is a projector of* W *onto* U *that preserves the boundary values on* $\partial\Omega$.

To prove (4.4.8), we note first that by Green's identity, for any function in W we have from (4.4.2)

$$\langle w, u\rangle = -\int_{\partial\Omega} w \frac{\partial u}{\partial n}\, ds - \int_{\Omega} w(\Delta u - qu)\, d\omega. \tag{4.4.10}$$

For a solution u of (4.4.1) the integral over Ω is zero. Also, $H = N$ on $\partial\Omega$ because G is zero on $\partial\Omega$ in (4.4.7). Thus by (4.4.6),

$$\langle H(.,Q), u\rangle = -\int_{\partial\Omega} H(P,Q) \frac{\partial u(P)}{\partial n}\, ds = u(Q).$$

We prove (4.4.9). From (4.4.7) we have

$$\langle H(.,Q),v\rangle = \langle N(.,Q),v\rangle - \langle G(.,Q),v\rangle.$$

On the right, by (4.4.10), for any $v \in W$, since $\partial N/\partial n$ is zero on $\partial\Omega$,

$$\langle N(.,Q),v\rangle = -\int_{\partial\Omega} \frac{\partial N}{\partial n} v\, ds - \int_{\Omega} N(\Delta v - qv)\, d\omega = -v(Q)$$

and for any $v \in V$, since v is zero on $\partial\Omega$,

$$\langle G(.,Q),v\rangle = -\int_{\partial\Omega} \frac{\partial G}{\partial n} v\, ds - \int_{\Omega} G(\Delta v - qv)\, d\omega = -v(Q).$$

From this, (4.4.9) follows.

The present kernel can be characterized by a minimum property quite similar to that in the previous section. To obtain it, we may proceed as follows. We apply the Schwarz inequality to (4.4.8),

$$u(Q)^2 \leq \|H(.,Q)\|^2 \|u\|^2.$$

Now since $H \in U$, we also get from (4.4.8)

$$\langle H(.,Q), H(.,Q)\rangle = H(Q,Q),$$

so that

(4.4.11) $$u(Q)^2 \leq H(Q,Q)\|u\|^2.$$

For $u(P) = H(P,Q)$ this becomes an equality and gives the following

Minimum property. The function

$$u(P) = H(P,Q)/H(Q,Q)$$

has the smallest norm, namely, $1/\sqrt{H(Q,Q)}$, of all the functions $u \in U$ with $u(Q) = 1$.

We finally turn to the *interpolation problem* in U, following a method by Bergman and Schiffer (cf. [4], 277-279). Inequality (4.4.11) with $u = H(P,R)$ gives

(4.4.12) $$H(Q,R)^2 \leq H(Q,Q)H(R,R).$$

We see that this is an estimate of the kernel with different argument points in terms of kernels at one argument point.

Now (4.4.12) can be generalized. For this we take any points $Q_1, \ldots, Q_n$ and real numbers $x_1, \ldots, x_n$ and consider the function

(4.4.13) $$u_1(P) = \sum_{j=1}^{n} x_j H(P, Q_j).$$

$H \in U$ implies $u_1 \in U$. This function has the norm $\|u_1\|$,

$$\|u_1\|^2 = \sum_{j=1}^{n} x_i x_j H(Q_i, Q_j) \geq 0.$$

Hence $M = [H(Q_i, Q_j)]$ is the coefficient matrix of a positive definite quadratic form. If equality holds, then $\|u_1\| = 0$, hence $u_1 = 0$ for all $P \in \Omega$. By the reproducing property, for any $u \in U$ we then have

$$\langle u_1, u \rangle = \sum_{j=1}^{n} u(Q_j) x_j = 0.$$

Hence if det $M = 0$, then $f_j(u) = u(Q_j)$, $u \in U$, $j = 1, \ldots, n$, are linearly dependent functionals on U.

If det $M \neq 0$, we can construct a function $u_1 \in U$ with given values $u_1(Q_j) = a_j$ at n points, called the *nodes* of the interpolation problem.

Our analysis shows that this interpolation problem can be solved if and only if it can be solved by an expression of the form (4.4.13) involving the kernel.

This exhibits the central role of the latter in interpolation.

Obviously, that problem can be solved by any function

$$v(P) = u_1(P) + u_2(P),$$

where $u_2 \in U$ is zero at the nodes. Because of the reproducing property this function v has the norm $\|v\|$, where

$$\|v\|^2 = \sum_{i=1}^{n} \sum_{j=1}^{n} x_i x_j H(Q_i, Q_j) + \|u_2\|^2 = \|u_1\|^2 + \|u_2\|^2. \tag{4.4.14}$$

This result shows that u_1 is the solution of the interpolation problem of minimum norm, the norm being given by the square root of the double sum in (4.4.14).

For the present equation (4.4.1) the use of orthonormal sequences of functions in connection with the kernel function follows the pattern outlined in the previous sections of this chapter. This will not be considered here.

Similarly, another method for obtaining total orthonormal sequences of particular solutions of a given partial differential equations is based on *Bergman operators*, for which we have listed some references, in particular, [3] and [16], at the end.

References

[1] Bergman, S., Zur Theorie der Funktionen, die eine lineare partielle Differentialgleichung befriedigen. *Mat. Sbornik N. S.* **2** (1937), 1169-1198.

[2] —, *The Kernel Function and Conformal Mapping.* 2nd ed. Providence, RI: American Mathematical Society, 1970.

[3] —, *Integral Operators in the Theory of Linear Partial Differential Equations.* 3rd ed. New York: Springer, 1971.

[4] — and M. Schiffer, *Kernel Functions and Elliptic Differential Equations.* New York: Academic Press, 1953.

[5] Bieberbach, L., *Lehrbuch der Funktionentheorie.* 2 vols. (1934, 1931). New York: Chelsea, 1945.

[6] —, *Conformal Mapping.* (Translation from the German). New York: Chelsea, 1964.

[7] Dieudonné, J., *History of Functional Analysis.* Amsterdam: North-Holland, 1981.

[8] Dirichlet, G. Lejeune, Sur la convergence des séries trigonométriques qui servent à représenter une fonction arbitraire entre des limites données. *J. reine angew. Math.* **4** (1829), 157-169.

[9] Fourier, J.-B. J., *Théorie analytique de la chaleur.* Paris: 1822.

[10] Gaier, D., *Vorlesungen über Approximation im Komplexen.* Basel: Birkhäuser, 1980.

[11] Garabedian, P.R., *Partial Differential Equations.* New York: Wiley, 1964.

[12] Hellinger, E., and O. Toeplitz, *Integralgleichungen und Gleichungen mit unendlichvielen Unbekannten.* (1927). New York: Chelsea, 1953.

[13] Hilbert, D., *Grundzüge einer allgemeinen Theorie der linearen Integralgleichungen.* (1912). New York: Chelsea, 1953.

[14] —, *Gesammelte Abhandlungen.* 3 vols. (1932-35). New York: Chelsea, 1965.

[15] Kline, M., *Mathematical Thought from Ancient to Modern Times.* New York: Oxford University Press, 1972.

[16] Kracht, M., and E. Kreyszig, *Methods of Complex Analysis in Partial Differential Equations with Applications.* New York: Wiley, 1988.

[17] Kreyszig, E., *Introductory Functional Analysis with Applications.* New York: Wiley, 1989.

[18] —, *Differential Geometry.* New York: Dover, 1991.

[19] —, *Advanced Engineering Mathematics.* 7th ed. New York: Wiley, 1993.

[20] Lebesgue, H., *Oeuvres scientifiques.* 5 vols. Geneva: L'Enseignement Mathématique, 1972.

[21] Meschkowski, H., *Hilbertsche Räume mit Kernfunktion.* Berlin: Springer, 1962.

[22] Riemann, B., *Gesammelte mathematische Werke und wissenschaftlicher Nachlass.* (1892). New York: Dover, 1953.

[23] Riesz, F., *Gesammelte Arbeiten.* Budapest: Akadémiai Kiadó, 1960.

[24] Schmidt, E., Über die Auflösung linearer Gleichungen mit unendlich vielen Unbekannten. *Rend. Mat. Palermo* **25** (1908), 53-77.

[25] Sommerfeld, A., *Partielle Differentialgleichungen der Physik.* Leipzig: Akademische Verlagsgesellschaft, 1947.

[26] Volterra, V., *Opere matematiche.* 5 vols. Rome: Accademia Nazionale dei Lincei, 1954-62.

[27] Von Neumann, J., *Collected Works.* 6 vols. New York: Pergamon, 1961-63.
[28] —, *Mathematische Grundlagen der Quantenmechanik.* Berlin: Springer, 1932. Reprinted 1968.
[29] Weierstrass, K., *Mathematische Werke.* 7 vols. (1884-1927). Hildesheim: Olms, no date.
[30] Weyl, H., *Gesammelte Abhandlungen.* 4 vols. Berlin: Springer, 1968.

Manfred Kracht, Universität Düsseldorf, Germany

Erwin Kreyszig, Carleton University, Ottawa, Canada

GRADIMIR V MILOVANOVIĆ

Orthogonal polynomial systems and some applications(*)

1. Introduction and Basic Definitions

The theory of inner product spaces enables an introduction of orthogonal systems which play an important role in many branches of mathematics, physics and other applied and computational sciences. We start with the following definition:

Definition 1.1. Given a real linear space of functions X, an *inner product* (f, g) defined is a mapping of X^2 into $\mathbb{R}$ such that

(a) $(f+g, h) = (f, h) + (g, h)$ (Linearity),

(b) $(\alpha f, g) = \alpha(f, g)$ (Homogeneity),

(c) $(f, g) = (g, f)$ (Symmetry),

(d) $(f, f) > 0,\ (f, f) = 0 \Leftrightarrow f = 0$ (Positivity),

where $f, g, h \in X$ and α is a real parameter. The space X will be called an *inner product space.*

A similar definition can be done for complex spaces. Namely, if X is a complex linear space, then the inner product $(f, g)\colon X^2 \to \mathbb{C}$ is such that the condition (c) is replaced by

(c′) $(f, g) = \overline{(g, f)}$ (Hermitian Symmetry).

The bar in the above line designates the complex conjugate.

Definition 1.2. A system S of elements of an inner product space is called *orthogonal* if $(f, g) = 0$ for every $f \neq g$ $(f, g \in S)$. If $(f, f) = 1$ for each $f \in S$, then the system is called *orthonormal.*

Starting from a linearly independent system of elements of an inner product space and using the well-known *Gram-Schmidt orthogonalizing process* we can construct the corresponding orthogonal (orthonormal) system. In this survey we consider only orthogonal (orthonormal) polynomial systems with respect to different inner product spaces.

Definition 1.3. A system of polynomials $\{\pi_k\}$, where

$$\begin{aligned} &\pi_k(t) = b_k t^k + c_k t^{k-1} + \text{ lower degree terms}, \quad b_k > 0, \\ &(\pi_k, \pi_m) = \delta_{km}, \quad k, m \geq 0, \end{aligned} \tag{1.1}$$

(*) This work was supported in part by the Serbian Scientific Foundation, grant number 04M03.

is called a system of orthonormal polynomials with respect to the inner product $(\,.\,,\,.\,)$.

Orthogonal polynomial systems are very useful in many problems in the approximation theory, mathematical and numerical analysis, and their applications (for example, Gaussian quadrature processes, least square approximation of functions, differential and difference equations, Fourier series, etc.).

This paper is organized as follows. In Section 2 we consider the orthogonality on the real line, with several classes of orthogonal polynomials such as classical, semi-classical and non-classical polynomials, as well as orthogonal polynomials with a Sobolev inner product. Also, an interpretation of s-orthogonality is included. Section 3 is devoted to some important applications of orthogonal polynomials on the real line as Gauss-Christoffel quadrature formulas, moment-preserving spline approximation and summation of slowly convergent series. The case of orthogonality on the unit circle is considered in Section 4, and the cases on the semicircle and a circular arc in Section 5. In Section 6 we deal with some applications of polynomials orthogonal on the semicircle. Finally, in Section 7 we consider orthogonal polynomials on the radial rays in the complex plane.

2. Orthogonality on the Real Line

Let $\lambda\colon \mathbb{R} \to \mathbb{R}$ be a fixed non-decreasing function with infinitely many points of increase for which all moments $\mu_k = \int_{\mathbb{R}} t^k \, d\lambda(t)$, $k = 0, 1, \ldots$, exist and are finite.

Then the improper Stieltjes integral $\int_{\mathbb{R}} P(t)\, d\lambda(t)$ exists for every polynomial P. By the application of the Lebesgue-Stieltjes integral $\int_{\mathbb{R}} f(t)\, d\lambda(t)$ to characteristic functions of sets, the function λ engenders a Lebesgue-Stieltjes measure $d\lambda(t)$, which is known also as *m-distribution* (cf. Freud [26]). Moreover, if $t \mapsto \lambda(t)$ is an absolutely continuous function, then we say that $\lambda'(t) = w(t)$ is a *weight function*. In that case, the measure $d\lambda$ can be express as $d\lambda(t) = w(t)\, dt$, where the weight function $t \mapsto w(t)$ is a non-negative and measurable in Lebesgue's sense for which all moments exists and $\mu_0 = \int_{\mathbb{R}} w(t)\, dt > 0$.

In the general case the function λ can be written in the form $\lambda = \lambda_{\mathrm{ac}} + \lambda_{\mathrm{s}} + \lambda_{\mathrm{j}}$, where λ_{ac} is absolutely continuous, λ_{s} is singular, and λ_{j} is a jump function.

The support of the measure, i.e., the set of points of increase of $t \mapsto \lambda(t)$ we denote by $\operatorname{supp}(d\lambda)$. It is always an infinite and closed set. If $\operatorname{supp}(d\lambda)$ is bounded, then the smallest closed interval containing $\operatorname{supp}(d\lambda)$ we will denote by $\Delta(d\lambda)$. For example, if $\Delta(d\lambda) = [a, b]$ then we say that $d\lambda(t)$ lies in $[a, b]$. In that case we have $\lambda(t) = \lambda(-\infty)$ for $t < a$ and $\lambda(t) = \lambda(+\infty)$ for $t > b$. In addition, if $t \mapsto \lambda(t)$ is absolutely continuous, then the weight function $t \mapsto w(t)$ vanishes outside of $[a, b]$, or more generally, outside of $\operatorname{supp}(d\lambda)$.

2.1. General properties. For any m-distribution $d\lambda(t)$ there exists a unique system of orthonormal polynomials $\pi_k(\cdot) = \pi_k(\,\cdot\,; d\lambda)$, $k = 0, 1, \ldots$, defined by (1.1), where $b_k = b_k(d\lambda)$, $c_k = c_k(d\lambda)$ and the inner product is given by

$$(f, g) = \int_{\mathbb{R}} f(t)g(t)\, d\lambda(t) \qquad (f, g \in X = L^2(\mathbb{R}) \equiv L^2(\mathbb{R}; d\lambda)). \tag{2.1.1}$$

If we have an absolutely continuous function $t \mapsto \lambda(t)$, then instead of $\pi_k(\,\cdot\,; d\lambda)$, $b_k(d\lambda)$, $\text{supp}(d\lambda)$, ..., we usually write $\pi_k(\,\cdot\,; w)$, $b_k(w)$, $\text{supp}(w)$, ..., respectively, where $\lambda'(t) = w(t)$.

If we have $\text{supp}(w) = [a, b]$, where $-\infty < a < b < +\infty$, we say that $\{\pi_k\}$ is a system of orthonormal polynomials in a finite interval $[a, b]$. For (a, b) we say that it is an *interval of orthogonality*.

Now we give a few basic properties of orthogonal polynomials:

Theorem 2.1.1. *The system of orthonormal polynomials $\{\pi_k\}$, associated with the distribution $d\lambda(t)$, satisfy a three-term recurrence relation*

$$t\pi_k(t) = u_{k+1}\pi_{k+1}(t) + v_k\pi_k(t) + u_k\pi_{k-1}(t) \quad (k \geq 0), \tag{2.1.2}$$

where $\pi_{-1}(t) = 0$ and the coefficients $u_k = u_k(d\lambda)$ and $v_k = v_k(d\lambda)$ are given by

$$u_k = \frac{b_{k-1}}{b_k} \qquad \textit{and} \qquad v_k = \int_{\mathbb{R}} t\pi_k(t)^2\, d\lambda(t).$$

Since $\pi_0(t) = b_0 = 1/\sqrt{\mu_0}$ and $b_{k-1} = u_k b_k$ we have that $b_k = b_0/(u_1 u_2 \cdots u_k)$. Notice that $u_k > 0$ for each k.

Contrary, for two given real sequences $\{u_k\}_{k\in\mathbb{N}}$ and $\{v_k\}_{k\in\mathbb{N}_0}$, where $u_k > 0$ for each $k \in \mathbb{N}$, one can construct a sequence of polynomials using the three-term recurrence relation (2.1.2), starting with initial values $\pi_{-1}(t) = 0$ and $\pi_0(t) = 1$. It is well-known by Favard's theorem (cf. Chihara [20]) that there exists a positive measure $d\sigma(t)$ on $\mathbb{R}$ such that

$$\int_{\mathbb{R}} \pi_k(t)\pi_m(t)\, d\sigma(t) = \delta_{km}, \qquad k, m \geq 0.$$

The measure $d\sigma(t)$ is not unique which depends of the fact whether or not the Hamburger moment problem is determined. A sufficient condition for a unique measure is the Carleman's condition given by $\sum_{k=1}^{+\infty} (1/u_k) = +\infty$. Evidently, it holds if $\{u_k\}_{k\in\mathbb{N}}$ is a bounded sequence.

In many considerations and applications we use the monic orthogonal polynomials $\hat{\pi}_k(t) = \pi_k(t)/b_k = t^k +$ lower degree terms. Such polynomials satisfy the following three-term recurrence relation

$$\hat{\pi}_{k+1}(t) = (t - \alpha_k)\hat{\pi}_k(t) - \beta_k\hat{\pi}_{k-1}(t), \qquad k = 0, 1, 2, \ldots, \tag{2.1.3}$$

where $\alpha_k = v_k$ and $\beta_k = u_k^2 > 0$.

Because of orthogonality, we have that

$$\alpha_k = \frac{(t\hat{\pi}_k, \hat{\pi}_k)}{(\hat{\pi}_k, \hat{\pi}_k)} \quad (k \geq 0), \quad \beta_k = \frac{(\hat{\pi}_k, \hat{\pi}_k)}{(\hat{\pi}_{k-1}, \hat{\pi}_{k-1})} \quad (k \geq 1).$$

The coefficient β_0, which multiplies $\hat{\pi}_{-1} = 0$ in three-term recurrence relation may be arbitrary. Sometimes, it is convenient to define it by $\beta_0 = \mu_0 = \int_{\mathbb{R}} d\lambda(t)$. Then the norm of $\hat{\pi}_k$ can be express in the form

$$\|\hat{\pi}_k\| = \sqrt{(\hat{\pi}_k, \hat{\pi}_k)} = \sqrt{\beta_0\beta_1 \cdots \beta_k}\,. \tag{2.1.4}$$

We mention that the existence of a three-term recurrence relation for orthogonal polynomials is a consequence of the property $(tf, g) = (f, tg)$ of the inner product (2.1.1).

Theorem 2.1.2. *All zeros of* $t \mapsto \pi_n(t; d\lambda)$, $n \geq 1$, *are real and distinct and are located in the interior of the interval* $\Delta(d\lambda)$.

Let $\tau_k^{(n)}$, $k = 1, \ldots, n$, denote the zeros of $\pi_n(t; d\lambda)$ in increasing order

$$\tau_1^{(n)} < \tau_2^{(n)} < \cdots < \tau_n^{(n)}.$$

Theorem 2.1.3. *The zeros of* $\pi_n(t; d\lambda)$ *and* $\pi_{n+1}(t; d\lambda)$ *interlace, i.e.,*

$$\tau_k^{(n+1)} < \tau_k^{(n)} < \tau_{k+1}^{(n+1)} \qquad (k = 1, \ldots, n;\ n \in \mathbb{N}).$$

Taking $k = 0, 1, \ldots, n-1$ in (2.1.2), one can obtain the following system of equations

$$t\boldsymbol{p}_n(t) = J_n(d\lambda)\boldsymbol{p}_n(t) + u_n\pi_n(t)\boldsymbol{e}_n,$$

where

$$J_n(d\lambda) = \begin{bmatrix} v_0 & u_1 & & & O \\ u_1 & v_1 & u_2 & & \\ & u_2 & v_2 & \ddots & \\ & & \ddots & \ddots & u_{n-1} \\ O & & & u_{n-1} & v_{n-1} \end{bmatrix}, \quad \boldsymbol{p}_n(t) = \begin{bmatrix} \pi_0(t) \\ \pi_1(t) \\ \pi_2(t) \\ \vdots \\ \pi_{n-1}(t) \end{bmatrix}, \quad \boldsymbol{e}_n = \begin{bmatrix} 0 \\ 0 \\ 0 \\ \vdots \\ 1 \end{bmatrix}.$$

The tridiagonal matrix $J_n = J_n(d\lambda)$ is known as the *Jacobi matrix*. It is clear that $\pi_n(t) = 0$ if and only if $t\boldsymbol{p}_n(t) = J_n\boldsymbol{p}_n(t)$, i.e., the zeros $\tau_k^{(n)}$ of $\pi_n(t)$ are the same as the eigenvalues of the Jacobi matrix J_n. Also, notice that the monic polynomial $\hat{\pi}_n(t)$ can be expressed in the following determinant form

$$\hat{\pi}_n(t) = \det(tI_n - J_n),$$

where I_n is the identity matrix of the order n.

Define the function $(t,x) \mapsto K_n(t,x)$ by

$$K_n(t,x) = \sum_{\nu=0}^{n-1} \pi_\nu(t)\pi_\nu(x) \qquad (n \geq 1), \tag{2.1.5}$$

which plays a fundamental role in the integral representation of partial sums of the orthogonal expansions. Namely, $K_n(t,x)$ represents the kernel of the n-th partial sum. Notice that $K_n(t,x) = K_n(x,t)$. Using the three-term recurrence relation from Theorem 2.1.1 we can prove:

Theorem 2.1.4. *Let $K_n(t,x)$ be defined by* (2.1.5)*. Then we have*

$$K_n(t,x) = u_{n+1}\frac{\pi_n(t)\pi_{n-1}(x) - \pi_{n-1}(t)\pi_n(x)}{t-x}, \tag{2.1.6}$$

where u_{n+1} is defined in Theorem 2.1.1.

Letting $t \to x$ we find

$$K_n(x,x) = u_{n+1}(\pi_n'(x)\pi_{n-1}(x) - \pi_{n-1}'(x)\pi_n(x)).$$

Formula (2.1.6) is known as the *Christoffel-Darboux identity*.

We mention now an important extremal problem for monic polynomials (cf. [110, pp. 416–417]): *Among all polynomials of degree n, with leading coefficient unity, find the polynomial which deviates least from zero in L^2-norm*

$$\|P\| = \sqrt{(P,P)} = \left(\int_{\mathbb{R}} |P(t)|^2 \, d\lambda(t)\right)^{1/2}.$$

The solution is given in the following theorem:

Theorem 2.1.5. *Let $P(t) = \sum_{\nu=0}^{n} a_\nu t^\nu$, with $a_n = 1$, be an arbitrary monic polynomial of degree n. Then*

$$\|P\| \geq \|\hat{\pi}_n\| = b_n^{-1},$$

with equality only if $P(t) = \hat{\pi}_n(t) = \pi_n(t)/b_n$, where $\hat{\pi}_n$ is the monic polynomial orthogonal with respect to the measure $d\lambda(t)$ on $\mathbb{R}$.

A survey on characterization theorems for orthogonal polynomials on the real line was given recently by Al-Salam [6]. In next subsection we consider a special class of orthogonal polynomials so-called *classical orthogonal polynomials*. For some extensions of this polynomial class see Andrews and Askey [8], Askey and Wilson [9], and Atakishiyev and Suslov [10].

2.2. Classical orthogonal polynomials. A very important class of orthogonal polynomials on an interval of orthogonality $(a, b) \in \mathbb{R}$ is constituted by so-called the *classical orthogonal polynomials*. They are distinguished by several particular properties.

Let $\mathcal{P}_n$ be the set of all algebraic polynomials P ($\not\equiv 0$) of degree at most n and the inner product is given by

$$(f, g)_w = \int_a^b w(t)f(t)g(t)\, dt. \tag{2.2.1}$$

Since every interval (a, b) can be transformed by a linear transformation to one of following intervals: $(-1, 1)$, $(0, +\infty)$, $(-\infty, +\infty)$, we will restrict our consideration (without loss of generality) only to these three intervals.

Definition 2.2.1. The orthogonal polynomials $\{Q_k\}$ on (a, b) with respect to the inner product (2.2.1) are called the *classical orthogonal polynomials* if their weight functions $t \mapsto w(t)$ satisfy the differential equation

$$\frac{d}{dt}(A(t)w(t)) = B(t)w(t),$$

where

$$A(t) = \begin{cases} 1 - t^2, & \text{if } (a, b) = (-1, 1), \\ t, & \text{if } (a, b) = (0, +\infty), \\ 1, & \text{if } (a, b) = (-\infty, +\infty), \end{cases}$$

and $B(t)$ is a polynomial of the first degree. For such classical weights we will write $w \in CW$.

We note that if $w \in CW$, then $w \in C^1(a, b)$, and also the following property:

Theorem 2.2.1. *If $w \in CW$ then for each $m = 0, 1, \ldots$ we have*

$$\lim_{t \to a+} t^m A(t)w(t) = 0 \quad \text{and} \quad \lim_{t \to b-} t^m A(t)w(t) = 0.$$

Based on the above definition, the classical orthogonal polynomials $\{Q_k\}$ on (a, b) can be specificated as the *Jacobi polynomials* $P_k^{(\alpha,\beta)}(t)$ $(\alpha, \beta > -1)$ on $(-1, 1)$, the *generalized Laguerre polynomials* $L_k^s(t)$ $(s > -1)$ on $(0, +\infty)$, and finally as the *Hermite polynomials* $H_k(t)$ on $(-\infty, +\infty)$. Their weight functions and the corresponding polynomials $A(t)$ and $B(t)$ are given in Table 2.2.1.

(a, b)	$w(t)$	$A(t)$	$B(t)$	λ_k
$(-1, 1)$	$(1-t)^\alpha(1+t)^\beta$	$1 - t^2$	$\beta - \alpha - (\alpha + \beta + 2)t$	$k(k + \alpha + \beta + 1)$
$(0, +\infty)$	$t^s e^{-t}$	t	$s + 1 - t$	k
$(-\infty, +\infty)$	e^{-t^2}	1	$-2t$	$2k$

TABLE 2.2.1: The classification of the classical orthogonal polynomials

Special cases of the Jacobi polynomials are:

1° The *Legendre polynomials* $P_k(t)$ (for $\alpha = \beta = 0$);
2° The *Chebyshev polynomials of the first kind* $T_k(t)$ (for $\alpha = \beta = -1/2$);
3° The *Chebyshev polynomials of the second kind* $S_k(t)$ (for $\alpha = \beta = 1/2$);
4° The *Chebyshev polynomials of the third kind* $U_k(t)$ (for $\alpha = -\beta = -1/2$);
5° The *Chebyshev polynomials of the fourth kind* $V_k(t)$ (for $\alpha = -\beta = 1/2$);
6° The *Gegenbauer* or *ultraspherical polynomials* $C_k^\lambda(t)$ (for $\alpha = \beta = \lambda - 1/2$).

If $s = 0$, the generalized Laguerre polynomials reduces to the standard Laguerre polynomials $L_k(t)$.

There are many characterizations of the classical orthogonal polynomials. In sequel we give the basic common properties of these polynomials (cf. [110]).

Theorem 2.2.2. *The derivatives of the classical orthogonal polynomials* $\{Q_k\}_{k\in\mathbb{N}_0}$ *form also a sequence of the classical orthogonal polynomials.*

Applying the induction method we can prove a more general result:

Theorem 2.2.3. *The sequence* $\{Q_k^{(m)}\}_{k=m,m+1,\dots}$ *is a classical orthogonal polynomial sequence on* (a,b) *with respect to the weight function* $t \mapsto w_m(t) = A(t)^m w(t)$. *The differential equation for this weight is*

$$(A(t)w_m(t))' = B_m(t)w_m(t),$$

where $B_m(t) = mA'(t) + B(t)$.

Theorem 2.2.4. *The classical orthogonal polynomial* $t \mapsto Q_k(t)$ *is a particular solution of the second order linear differential equation of hyphergeometric type*

$$L[y] = A(t)y'' + B(t)y' + \lambda_k y = 0, \tag{2.2.2}$$

where

$$\lambda_k = -k\left(\tfrac{1}{2}(k-1)A''(0) + B'(0)\right). \tag{2.2.3}$$

The equation (2.2.2) can be written in the Sturm-Liouville form

$$\frac{d}{dt}\left(A(t)w(t)\frac{dy}{dt}\right) + \lambda_k w(t)y = 0. \tag{2.2.4}$$

The coefficients λ_k are also displayed in Table 2.2.1.

Similarly, the m-th derivative of Q_k satisfies the differential equation

$$\frac{d}{dt}\left(A(t)w_m(t)\frac{dy}{dt}\right) + \lambda_{k,m}w_m(t)y = 0,$$

where $\lambda_{k,m} = -(k-m)\left(\frac{1}{2}(k+m-1)A''(0) + B'(0)\right)$. We note that this expression for $\lambda_{k,m}$ reduces to (2.2.3) for $m = 0$, i.e., $\lambda_{k,0} = \lambda_k$.

REMARK 2.2.1. The characterization of the classical orthogonal polynomials by differential equation (2.2.2), i.e. (2.2.4), was proved by Lesky [80], and conjectured by Aczél [2] (see also Bochner [15]). Such a differential equation appears in many mathematical models in atomic physics, electrodynamics and acoustics. As an example we mention the well-known Schrödinger equation.

The classical orthogonal polynomials possess a Rodrigues' type formula (cf. Bateman and Erdélyi [11], Tricomi [139], and Suetin [135]).

Theorem 2.2.5. *The classical orthogonal polynomial $Q_k(t)$ can be expressed in the form*

$$Q_k(t) = \frac{C_k}{w(t)} \cdot \frac{d^k}{dt^k}\left(A(t)^k w(t)\right), \tag{2.2.5}$$

where C_k are constants different from zero.

Using the Cauchy formula for k-th derivative of a regular function, (2.2.5) can be represented in the following integral form

$$Q_k(t) = \frac{C_k}{w(t)} \cdot \frac{k!}{2\pi i} \oint_{\Gamma} \frac{A(z)^k w(z)}{(z-t)^{k+1}}\, dz, \tag{2.2.6}$$

where Γ is a closed contour such that $t \in \operatorname{int} \Gamma$.

The constants C_k in (2.2.5) and (2.2.6) can be chosen in different way (for example, Q_k to be monic, orthonormal, etc.). A historical reason leads to

$$C_k = \begin{cases} \dfrac{(-1)^k}{2^k k!} & \text{for } P_k^{(\alpha,\beta)}(t), \\ 1 & \text{for } L_k^s(t), \\ (-1)^k & \text{for } H_k(t). \end{cases}$$

In addition, the Gegenbauer and the Chebyshev polynomials need

$$C_k^\lambda(t) = \frac{(2\lambda)_k}{\left(\lambda + \frac{1}{2}\right)_k} P_k^{(\alpha,\alpha)}(t) \quad (\alpha = \lambda - 1/2),$$

$$T_k(t) = \frac{k!}{\left(\frac{1}{2}\right)_k} P_k^{(-1/2,-1/2)}(t),$$

$$S_k(t) = \frac{(k+1)!}{\left(\frac{3}{2}\right)_k} P_k^{(1/2,1/2)}(t),$$

where $(s)_k$ is the standard notation for Pochhammer's symbol

$$(s)_k = s(s+1)\cdots(s+k-1) = \frac{\Gamma(s+k)}{\Gamma(s)} \quad (\Gamma \text{ is the gamma function}).$$

For such defined polynomials $Q_k(t) = a_k\big(t^k + r_k t^{k-1} + \cdots\big)$, we give the leading coefficient a_k, the coefficient r_k, and the norm $\|Q_k\|$.

1° *Jacobi case*:

$$a_k = \frac{(k+\alpha+\beta+1)_k}{2^k k!}, \qquad r_k = \frac{k(\alpha-\beta)}{2k+\alpha+\beta},$$

$$\|P_k^{(\alpha,\beta)}\|^2 = \frac{2^{\alpha+\beta+1}\Gamma(k+\alpha+1)\Gamma(k+\beta+1)}{k!(2k+\alpha+\beta+1)\Gamma(k+\alpha+\beta+1)};$$

2° *Gegenbauer case*:

$$a_k = \frac{2^k}{k!}(\lambda)_k, \quad r_k = 0, \quad \|C_k^\lambda\|^2 = \sqrt{\pi}\,\frac{(2\lambda)_k\Gamma(\lambda+1/2)}{(k+\lambda)k!\Gamma(\lambda)};$$

3° *Legendre case*:

$$a_k = \frac{(2k)!}{2^k(k!)^2}, \qquad r_k = 0, \qquad \|P_k\|^2 = \frac{2}{2k+1};$$

4° *Chebyshev case of the first kind*:

$$a_k = 2^{k-1}, \quad r_k = 0, \quad \|T_0\|^2 = \pi, \quad \|T_k\|^2 = \frac{\pi}{2} \quad (k \neq 0);$$

5° *Chebyshev case of the second kind*:

$$a_k = 2^k, \qquad r_k = 0, \qquad \|S_k\|^2 = \frac{\pi}{2};$$

5° *Generalized Laguerre case*:

$$a_k = (-1)^k, \quad r_k = -k(k+s), \quad \|L_k^s\|^2 = k!\Gamma(k+s+1);$$

6° *Hermite case*:

$$a_k = 2^k, \qquad r_k = 0, \qquad \|H_k\|^2 = 2^k k!\sqrt{\pi}.$$

Putting $t = \cos\theta$, $-1 \leq t \leq 1$, Chebyshev polynomials of the first and second kind can be expressed in the forms

$$T_k(x) = T_k(\cos\theta) = \cos k\theta \quad \text{and} \quad S_k(x) = S_k(\cos\theta) = \frac{\sin(k+1)\theta}{\sin\theta},$$

respectively. Thus, trigonometric representations of these polynomials are

$$T_k(t) = \cos(k\arccos t) \quad \text{and} \quad S_k(x) = \frac{\sin\big((k+1)\arccos t\big)}{\sqrt{1-t^2}}.$$

Similarly to the well-known Landau inequality (see Landau [76]) for continuously-differentiable functions and other generalizations (cf. Djordjević and Milovanović [22], Gorny [62], Hille [67], Kolmogoroff [72], Milovanović [94], Schoenberg [129], Steckhin [133]), Agarwal and Milovanović [3] (see also [103]) stated the following characterization of the classical orthogonal polynomials:

Theorem 2.2.6. *Let* $\|f\|^2 = (f, f)_w$, *where* $w \in CW$. *For all* $P \in \mathcal{P}_n$ *the inequality*

$$(2\lambda_n + B'(0))\|\sqrt{A}P'\|^2 \le \lambda_n^2\|P\|^2 + \|AP''\|^2 \tag{2.2.7}$$

holds, with equality if only if $P(t) = cQ_n(t)$, *where* $Q_n(t)$ *is the classical orthogonal polynomial on* (a, b) *with respect to the weight function* $t \mapsto w(t)$, *and* c *is an arbitrary real constant.* λ_n, $A(t)$ *and* $B(t)$ *are given in Table* 2.2.1.

The equality case in (2.2.7) gives a characterization of the classical orthogonal polynomials. For $w(t) = e^{-t^2}$ on $(-\infty, +\infty)$, the inequality (2.2.7) reduces to Varma's result [141]

$$\|P'\|^2 \le \frac{1}{2(2n-1)}\|P''\|^2 + \frac{2n^2}{2n-1}\|P\|^2.$$

Recently, Guessab and Milovanović [65] have considered a weighted L^2 analogues of the well-known Bernstein's inequality, which can be stated in the following form (cf. [110]):

$$\|\sqrt{1-t^2}P'(t)\|_\infty \le n\|P\|_\infty \qquad (P \in \mathcal{P}_n), \tag{2.2.8}$$

where $\|f\|_\infty = \max_{-1\le x\le 1} |f(t)|$. Using the norm $\|f\|^2 = (f, f)_w$, $w \in CW$, they have considered the following problem connected with the Bernstein's inequality (2.2.8): *Determine the best constant* $C_{n,m}(w)$ $(1 \le m \le n)$ *such that the inequality*

$$\|A^{m/2}P^{(m)}\|_w \le C_{n,m}(w)\|P\|_w \tag{2.2.9}$$

holds for all $P \in \mathcal{P}_n$.

Theorem 2.2.7. *For all* $P \in \mathcal{P}_n$ *the inequality* (2.2.9) *holds with the best constant*

$$C_{n,m}(w) = \sqrt{\lambda_{n,0}\lambda_{n,1}\cdots\lambda_{n,m-1}}, \tag{2.2.10}$$

where $\lambda_{n,k} = -(n-k)\left(\frac{1}{2}(n+k-1)A''(0) + B'(0)\right)$. *The equality is attained in* (2.2.10) *if and only if* $P(t)$ *is a constant multiple of the classical polynomial* $Q_n(t)$ *orthogonal with respect to the weight function* $w \in CW$ *on* (a, b).

We list now the coefficients α_k $(k \ge 0)$ and β_k $(k \ge 1)$ in the three-term recurrence relation for the monic classical orthogonal polynomials $\hat{Q}_k(t)$ on (a, b),

$$\begin{aligned} &\hat{Q}_{k+1}(t) = (t - \alpha_k)\hat{Q}_k(t) - \beta_k\hat{Q}_{k-1}(t) \quad (k \ge 0), \\ &\hat{Q}_1(t) = 0, \quad \hat{Q}_0(t) = 1. \end{aligned} \tag{2.2.11}$$

The coefficients u_k and v_k in the corresponding recurrence relation for orthonormal polynomials (see (2.1.3)) are given by $u_k = \sqrt{\beta_k}$ and $v_k = \alpha_k$.

Also, we give the moment $\mu_0 = \int_a^b w(t)\,dt \quad (w \in CW)$.

1° *Jacobi case:* $\hat{P}_k^{(\alpha,\beta)}(t) = 2^k k!/\big((k+\alpha+\beta+1)_k\big) P_k^{(\alpha,\beta)}(t)$,

$$\mu_0 = \frac{2^{\alpha+\beta+1}\Gamma(\alpha+1)\Gamma(\beta+1)}{\Gamma(\alpha+\beta+2)},$$

$$\alpha_k = \frac{\beta^2-\alpha^2}{(2k+\alpha+\beta)(2k+\alpha+\beta+2)},$$

$$\beta_k = \frac{4k(k+\alpha)(k+\beta)(k+\alpha+\beta)}{(2k+\alpha+\beta)^2\big((2k+\alpha+\beta)^2-1\big)}.$$

2° *Gegenbauer case:* $\hat{C}_k^{\lambda}(t) = k!/\big(2^k(\lambda)_k\big) C_k^{\lambda}(t)$,

$$\mu_0 = \frac{\sqrt{\pi}\,\Gamma(\lambda+1/2)}{\Gamma(\lambda+1)}, \quad \alpha_k = 0, \quad \beta_k = \frac{k(k+2\lambda-1)}{4(k+\lambda)(k+\lambda-1)}.$$

3° *Legendre case:* $\hat{P}_k(t) = 2^k(k!)^2/\big((2k)!\big) P_k(t)$,

$$\mu_0 = 2, \quad \alpha_k = 0, \quad \beta_k = \frac{k^2}{4k^2-1}.$$

4° *Chebyshev case of the first kind:* $\hat{T}_0(t) = 1$, $\hat{T}_k(t) = 2^{1-k}T_k(t)$,

$$\mu_0 = \pi, \quad \alpha_k = 0, \quad \beta_1 = \frac{1}{2}, \quad \beta_k = \frac{1}{4} \quad (k \ge 2).$$

5° *Chebyshev case of the second kind:* $\hat{S}_k(t) = 2^{-k}S_k(t)$,

$$\mu_0 = \frac{\pi}{2}, \quad \alpha_k = 0, \quad \beta_k = \frac{1}{4} \quad (k \ge 1).$$

5° *Generalized Laguerre case:* $\hat{L}_k^s(t) = (-1)^k L_k^s(t)$,

$$\mu_0 = \Gamma(s+1), \quad \alpha_k = 2k+s+1, \quad \beta_k = k(k+s).$$

6° *Hermite case:* $\hat{H}_k(t) = 2^{-k}H_k(t)$,

$$\mu_0 = \sqrt{\pi}, \quad \alpha_k = 0, \quad \beta_k = \frac{k}{2}.$$

The norm of monic orthogonal polynomials can be calculated using (2.1.4).

In the case of classical orthogonal polynomials one can express $Q_k'(t)$ in terms of $Q_k(t)$ and $Q_{k-1}(t)$. Such a formula for the monic orthogonal polynomials is

$$A(t)\frac{d}{dt}\hat{Q}_k(t) = (e_k t + f_k)\hat{Q}_k(t) + \omega_k\beta_k\hat{Q}_{k-1}(t), \tag{2.2.12}$$

where $A(t)$ is given in Table 2.2.1 and the coefficient β_k is the same as in the three-term recurrence relation (2.2.11).

In the Jacobi case we have

$$e_k = -k, \qquad f_k = \frac{k(\alpha - \beta)}{2k + \alpha + \beta}, \qquad \omega_k = 2k + \alpha + \beta + 1.$$

In the generalized Laguerre case and the Hermite case, formula (2.2.12) reduces to

$$t\frac{d}{dt}\hat{L}_k^s(t) = k\hat{L}_k^s(t) + k(k+s)\hat{L}_{k-1}^s(t) \quad \text{and} \quad \frac{d}{dt}\hat{H}_k(t) = k\hat{H}_{k-1}(t),$$

respectively.

2.3. Non-classical orthogonal polynomials. As we have seen in the previous subsection the monic Chebyshev polynomials of the second kind

$$\hat{S}_k(t) = \frac{1}{2^k} \cdot \frac{\sin(k+1)\theta}{\sin\theta}, \qquad t = \cos\theta,$$

have a very simple three-term recurrence relation (2.2.11) with $\alpha_k = 0$ and $\beta_k = 1/4$. A system of orthogonal polynomials for which the recursion coefficients satisfy

$$\lim_{k\to+\infty} \alpha_k = 0 \quad \text{and} \quad \lim_{k\to+\infty} \beta_k = \frac{1}{4} \tag{2.3.1}$$

will be said to be a perturbation of the polynomials $\hat{S}_k(t)$. Similarly, one can consider orthogonal polynomials with such coefficients that

$$\lim_{k\to+\infty} \alpha_k = a \quad \text{and} \quad \lim_{k\to+\infty} \beta_k = \frac{b^2}{4} > 0,$$

where $a, b \in \mathbb{R}$. These polynomials are perturbations of $t \mapsto b^k\hat{S}_k((t-a)/b)$ and are said to belong to the class $\mathcal{N}(a,b)$, which was introduced and considered in detail by Nevai [120][1]. Evidently, (2.3.1) holds for polynomials from the class $\mathcal{N}(0,1)$.

There are several classes of orthogonal polynomials which are in certain sense close to the classical orthogonal polynomials. For example, when the weight $t \mapsto W(t)$ is the product of a classical weight $t \mapsto w(t)$ times a polynomial, Ronveaux [126] found the second-order differential equation for the corresponding orthogonal polynomials. Ronveaux and Thiry [128] developed a REDUCE package giving such differential equations. The following cases have been studied by Ronveaux and Marcellán [127]:

1° *Rational Case.* $W(t) = R(t)w(t)$, where R is a rational function with poles and zeros outside the support of w;

---

[1] Originally, Nevai defined this class for orthonormal polynomials.

2° *δ Dirac distribution.*

$$W(t) = w(t) + \sum_{k=1}^{m} \lambda_k \delta(t - t_k),$$

where the positive mass λ_k is located at t_k (t_k outside or inside the support of w).

In both cases, orthogonal polynomials are *semi-classical* (see Maroni [89]).

A nice survey about orthogonal polynomials and spectral theory was given by Everitt and Littlejohn [24].

In many applications of orthogonal polynomials it is very important to know the recursion coefficients α_k and β_k. If $d\lambda(t)$ is one of the classical measures, then α_k and β_k are known explicitly. Furthermore, there are certain non-classical measures when we know also these coefficients. In sequel we mention only a few of them:

1° *Generalized Gegenbauer weight* $w(t) = |t|^\mu (1 - t^2)^\alpha$, $\mu, \alpha > -1$, on $[-1, 1]$. The (monic) generalized Gegenbauer polynomials $W_k^{(\alpha,\beta)}(t)$, $\beta = (\mu - 1)/2$, were introduced by Lascenov [78] (see, also, Chihara [20, pp. 155–156]). These polynomials can be expressed in terms of the Jacobi polynomials,

$$W_{2k}^{(\alpha,\beta)}(t) = \frac{k!}{(k + \alpha + \beta + 1)_k} P_k^{\alpha,\beta)}(2t^2 - 1),$$
$$W_{2k+1}^{(\alpha,\beta)}(t) = \frac{k!}{(k + \alpha + \beta + 2)_k} x P_k^{\alpha,\beta+1)}(2t^2 - 1).$$

Notice that $W_{2k+1}^{(\alpha,\beta)}(t) = tW_{2k}^{(\alpha,\beta+1)}(t)$. Their three-term recurrence relation is

$$W_{k+1}^{(\alpha,\beta)}(t) = tW_k^{(\alpha,\beta)}(t) - \beta_k W_{k-1}^{(\alpha,\beta)}(t), \quad k = 0, 1, \ldots,$$
$$W_{-1}^{(\alpha,\beta)}(t) = 0, \quad W_0^{(\alpha,\beta)}(t) = 1,$$

where

$$\beta_{2k} = \frac{k(k + \alpha)}{(2k + \alpha + \beta)(2k + \alpha + \beta + 1)}, \quad \beta_{2k-1} = \frac{(k + \beta)(k + \alpha + \beta)}{(2k + \alpha + \beta - 1)(2k + \alpha + \beta)},$$

for $k = 1, 2, \ldots$, except when $\alpha + \beta = -1$; then $\beta_1 = (\beta + 1)/(\alpha + \beta + 2)$. Some applications of these polynomials in numerical quadratures and least square approximation with constraint were given in [73] and [107], respectively.

2° *The hyperbolic weight* $w(t) = 1/\cosh t$ on $(-\infty, +\infty)$. The coefficients in three-term recurrence relation are given by

$$\alpha_k = 0, \qquad \beta_0 = \pi, \quad \beta_k = \frac{\pi^2 k^2}{4} \quad (k \geq 1).$$

For details and generalizations see Chihara [20, pp. 191–193].

3° *The logistic weight* $w(t) = e^{-t}/(1+e^{-t})^2$ on $(-\infty, +\infty)$. Here we have

$$\alpha_k = 0, \qquad \beta_0 = 1, \quad \beta_k = \frac{\pi^2 k^4}{4k^2 - 1} \quad (k \geq 1).$$

Recently Gautschi and Li [47] considered a set of orthogonal polynomials induced by a given orthogonal polynomial $\pi_n(\,\cdot\,) = \pi_n(\,\cdot\,; d\lambda)$. Namely, for a fixed $n \geq 1$ and the orthogonal polynomial π_n of degree n relative to some positive measure $d\lambda(t)$ on $\mathbb{R}$, they introduced the polynomial system $\{\tilde{\pi}_{k,n}\}$ which is orthogonal with respect to the modified measure $d\tilde{\lambda}(t) = \pi_n(t)^2 d\lambda(t)$. Such (monic) polynomials satisfy the three-term recurrence relation

$$\tilde{\pi}_{k+1,n}(t) = (t - \tilde{\alpha}_{k,n})\tilde{\pi}_{k,n}(t) - \tilde{\beta}_{k,n}\tilde{\pi}_{k-1,n}(t), \qquad k = 0, 1, 2, \ldots, \tag{2.3.2}$$

with $\tilde{\pi}_{0,n}(t) = 1$ and $\tilde{\pi}_{-1,n}(t) = 0$.

For the Chebyshev measure of the first kind ($d\lambda(t) = (1-t^2)^{-1/2}\,dt$ on $(-1, 1)$, evidently $\tilde{\alpha}_{k,n} = 0$. Gautschi and Li [47] proved:

(i) For $n = 1$,

$$\tilde{\beta}_{0,1} = \frac{\pi}{2}, \qquad \tilde{\beta}_{k,1} = \begin{cases} \dfrac{1}{4}\left(1 - \dfrac{2}{k+1}\right), & k \text{ (even)} \geq 2, \\[2ex] \dfrac{1}{4}\left(1 + \dfrac{2}{k}\right), & k \text{ (odd)} \geq 1; \end{cases}$$

(ii) For any $n \geq 2$,

$$\tilde{\beta}_{k,n} = \begin{cases} \dfrac{\pi}{2^{2n-1}}, & \text{if } k = 0, \\[2ex] \dfrac{1}{4}\left(1 - \dfrac{(-1)^{k/n}}{1 + (k/n)}\right), & \text{if } k \equiv 0 \pmod{n} \ (k \neq 0), \\[2ex] \dfrac{1}{4}\left(1 + \dfrac{(-1)^{(k-1)/n}}{1 + (k-1)/n}\right), & \text{if } k \equiv 1 \pmod{n}, \\[2ex] \dfrac{1}{4}, & \text{otherwise.} \end{cases}$$

The corresponding recursion coefficients in analytic form were also obtained for the Chebyshev measure of the second kind $d\lambda(t) = (1-t^2)^{1/2}\,dt$, as well as for the third and fourth kind $d\lambda(t) = (1-t)^{\mp 1/2}(1+t)^{\pm 1/2}\,dt$.

A system of orthogonal polynomials for which the recursion coefficients are not known explicitly will be said to be *strong non-classical* orthogonal polynomials. In such cases there are a few known approaches to compute the first n coefficients α_k, β_k, $k = 0, 1, \ldots, n-1$. These then allow us to compute all orthogonal polynomials of degree $\leq n$ by a straightforward application of the three-term recurrence relation (2.2.11).

One of approaches for numerical construction of the monic orthogonal polynomials $\{\hat{\pi}_k\}$ is the *method of moments*, or precisely, *Chebyshev* or *modified Chebyshev algorithm*.

The second method makes use of explicit representations

$$\alpha_k = \frac{(t\hat{\pi}_k, \hat{\pi}_k)}{(\hat{\pi}_k, \hat{\pi}_k)} \quad (k \geq 0), \quad \beta_0 = (\hat{\pi}_0, \hat{\pi}_0), \quad \beta_k = \frac{(\hat{\pi}_k, \hat{\pi}_k)}{(\hat{\pi}_{k-1}, \hat{\pi}_{k-1})} \quad (k \geq 1),$$

in terms of the inner product $(\,.\,,\,.\,)$. The method is known as the *Stieltjes procedure*. Using a discretization of the inner product by some appropriate quadrature

$$(f, g) \approx (f, g)_N = \sum_{k=1}^{N} w_k f(x_k) g(x_k), \quad w_k > 0,$$

the corresponding method is called the *discretized Stieltjes procedure*.

For details in numerical construction of orthogonal polynomials see papers of Gautschi [32], [39], [41], [44–45]. A stable computational algorithm for determining the coefficients in the recurrence relation (2.3.2) for an arbitrary measure was given in [47]. In Section 3 we will mention a few non-classical weights for which the recursion coefficients were constructed numerically as well as the corresponding Gaussian formulas.

2.4. On s-orthogonal polynomials. In this subsection we give a short account of so-called *s-orthogonal polynomials* which are connected with Gauss-Turán quadrature formulas (see Turán [140]).

2.4.1. Turán quadratures and s-orthogonal polynomials. In 1950 P. Turán investigated numerical quadratures with multiple nodes,

$$\int_{-1}^{1} f(t)\,dt = \sum_{\nu=1}^{n} \sum_{i=0}^{k-1} A_{i,\nu} f^{(i)}(\tau_\nu) + R_{n,k}(f), \tag{2.4.1}$$

where

$$A_{i,\nu} = \int_{-1}^{1} \ell_{\nu,i}(t)\,dt \qquad (\nu = 1, \ldots, n;\ i = 0, 1, \ldots, k-1)$$

and $\ell_{\nu,i}(t)$ are the fundamental functions of Hermite interpolation. The $A_{i,\nu}$ are Cotes number of higher order. This formula is exact if f is a polynomial at most $kn-1$ and the points $-1 \leq \tau_1 < \tau_2 < \cdots < \tau_n \leq 1$ are arbitrary.

For $k = 1$ this formula can be exact for all polynomials of degree $\leq 2n - 1$ if the nodes τ_ν are the zeros of the Legendre polynomial P_n. That is the well-known Gauss-Legendre quadrature.

Because of the theorem of Gauss it is natural to ask whether knots τ_ν can be chosen so that the quadrature formula (2.4.1) will be exact for polynomials of degree not exceeding $(k+1)n-1$. Turán [140] showed that the answer is negative for $k = 2$, and for $k = 3$ it is positive. He proved that the knots τ_ν should be chosen as the zeros of the monic polynomial $\pi^*(t) = t^n + \cdots$, which minimizes the following integral

$$\int_{-1}^{1} \pi_n(t)^4 \, dt,$$

where $\pi_n(t) = t^n + a_{n-1}t^{n-1} + \cdots + a_1 t + a_0$.

More generally, the answer is negative for even, and positive for odd k, and then τ_ν are the zeros of the polynomial minimizing

$$\int_{-1}^{1} \pi_n(t)^{k+1} \, dt. \tag{2.4.2}$$

For $k = 1$, π_n^* is the monic Legendre polynomial $\hat{P}_n$.

Because of the above, we put $k = 2s + 1$. Instead of (2.4.1) it is interesting to investigate the analogous formula with some positive measure $d\lambda(t)$ on the real line $\mathbb{R}$,

$$\int_{\mathbb{R}} f(t) \, d\lambda(t) = \sum_{i=0}^{2s} \sum_{\nu=1}^{n} A_{i,\nu} f^{(i)}(\tau_\nu) + R(f), \tag{2.4.3}$$

which must be exact for all polynomials of degree at most $2(s+1)n - 1$. The role of the integral (2.4.2) is taken over

$$F_s = \int_{\mathbb{R}} \pi_n(t)^{2s+2} \, d\lambda(t),$$

where $F_s \equiv F_s(a_0, \ldots, a_{n-1})$, $\pi_n(t) = \sum\limits_{k=0}^{n} a_k t^k$, $a_n = 1$. In order to minimize F_s we must have

$$\int_{\mathbb{R}} \pi_n(t)^{2s+1} t^k \, d\lambda(t) = 0, \qquad k = 0, 1, \ldots, n-1. \tag{2.4.4}$$

Usually, instead of $\pi_n(t)$ we write $P_{n,s}(t)$.

The case $d\lambda(t) = w(t)\,dt$ on $[a, b]$ has been considered by the Italian mathematicians Ossicini [123], Ghizzetti and Ossicini [58], Guerra [63–64]. Is known that there exists a unique polynomial $P_{n,s} \in \mathcal{P}_n$ satisfying the conditions (2.4.4), and whose zeros are real, distinct and located in the support interval of the measure $d\lambda(t)$.

These polynomials are known as *s-orthogonal* (or *s-self associated*) *polynomials* with respect to the measure $d\lambda(t)$ on $\mathbb{R}$.

For $s = 0$ we have the standard case of orthogonal polynomials on the real line and the generalized Gauss-Turán formula (2.4.3) reduces to the Gauss-Christoffel quadrature (cf. Section 3).

In the case of the Chebyshev measure $d\lambda(t) = (1-t^2)^{-1/2}\,dt$ on $(-1,1)$, in 1930 Bernstein [14] showed that the monic Chebyshev polynomial $\hat{T}_n(t) = 2^{1-n}T_n(t)$ minimizes the integral F_s for any $s \geq 0$. Therefore, in this case, the nodes in (2.4.3) are the Chebyshev points $\tau_\nu = \cos[(2\nu-1)\pi/2n]$, $\nu = 1,\ldots,n$, for any $s \geq 0$. For other measures the case when $s > 0$ is very difficult.

2.4.2. Construction of s-orthogonal polynomials. G. Vincenti [142] has considered an iterative process to compute the coefficients of s-orthogonal polynomials in a special case when $d\lambda(t) = w(t)\,dt$ with an even weight function on a symmetric interval, say $[-b,b]$. Then $P_{n,s}(-t) = (-1)^n P_{n,s}(t)$. Vincenti applied his process to the Legendre case. When n and s increase, the process becomes ill-conditioned. A stable procedure for such a purpose was given by Milovanović [95] (see also [106]). The main idea was an interpretation of the "orthogonality conditions" (2.4.4) as orthogonality with respect to the positive measure $d\mu(t) = d\mu_{n,s}(t) = (\pi_n^{n,s}(t))^{2s}\,d\lambda(t)$. Thus, these conditions can be interpreted as

$$\int_{\mathbb{R}} \pi_\nu^{n,s}(t)t^i\,d\mu(t) = 0, \quad i = 0,1,\ldots,\nu-1,$$

where $\{\pi_\nu^{n,s}\}$ is a sequence of monic orthogonal polynomials with respect to the new measure $d\mu(t)$. Of course, $P_{n,s}(\,\cdot\,) = \pi_n^{n,s}(\,\cdot\,)$. As we can see, the polynomials $\pi_\nu^{n,s}$, $\nu = 0,1,\ldots$, are implicitly defined, because the measure $d\mu(t)$ depends of $\pi_n^{n,s}(t)$ $(= \pi_n(t))$. The general class of such polynomials was introduced by H. Engels (see [23, pp. 214–226]).

We will write only π_ν instead of $\pi_\nu^{n,s}(\cdot)$. These polynomials satisfy a three-term recurrence relation

$$\begin{aligned} &\pi_{\nu+1}(t) = (t-\alpha_\nu)\pi_\nu(t) - \beta_\nu\pi_{\nu-1}(t), \quad \nu = 0,1,\ldots, \\ &\pi_{-1}(t) = 0, \quad \pi_0(t) = 1, \end{aligned}$$

where, because of orthogonality,

$$\begin{aligned} \alpha_\nu = \alpha_\nu(n,s) &= \frac{(t\pi_\nu,\pi_\nu)}{(\pi_\nu,\pi_\nu)} = \frac{\int_{\mathbb{R}} t\pi_\nu^2(t)\,d\mu(t)}{\int_{\mathbb{R}} \pi_\nu^2(t)\,d\mu(t)}, \\ \beta_\nu = \beta_\nu(n,s) &= \frac{(\pi_\nu,\pi_\nu)}{(\pi_{\nu-1},\pi_{\nu-1})} = \frac{\int_{\mathbb{R}} t\pi_\nu^2(t)\,d\mu(t)}{\int_{\mathbb{R}} \pi_{\nu-1}^2(t)\,d\mu(t)}, \end{aligned}$$

and, for example, $\beta_0 = \int_{\mathbb{R}} d\mu(t)$.

Finding the coefficients α_ν, β_ν ($\nu = 0, 1, \ldots, n-1$) gives us access to the first $n+1$ orthogonal polynomials π_0, π_1, ..., π_n. Of course, we are interested only in the last of them, i.e., $\pi_n \equiv \pi_n^{n,s}$. Thus, for $n = 0, 1, \ldots$, the diagonal (boxed) elements in Table 2.4.1 are our s-orthogonal polynomials $\pi_n^{n,s}$.

n	$d\mu^{n,s}(t)$	Orthogonal Polynomials			
0	$(\pi_0^{0,s}(t))^{2s}\, d\lambda(t)$	$\boxed{\pi_0^{0,s}}$			
1	$(\pi_1^{1,s}(t))^{2s}\, d\lambda(t)$	$\pi_0^{1,s}$	$\boxed{\pi_1^{1,s}}$		
2	$(\pi_2^{2,s}(t))^{2s}\, d\lambda(t)$	$\pi_0^{2,s}$	$\pi_1^{2,s}$	$\boxed{\pi_2^{2,s}}$	
3	$(\pi_3^{3,s}(t))^{2s}\, d\lambda(t)$	$\pi_0^{3,s}$	$\pi_1^{3,s}$	$\pi_2^{3,s}$	$\boxed{\pi_3^{3,s}}$
$\vdots$					

TABLE 2.4.1: Generating of s-orthogonal polynomials

A stable scheme, based on Stieltjes' procedure, that computes $\pi_n^{n,s}$ in an iterative fashion, was developed by Milovanović [95]. For numerical methods for determining the coefficients in the generalized Turán quadrature (2.4.3) see [106] and [116–117].

2.5. Sobolev type orthogonal polynomials. Sobolev type orthogonal polynomials appeared in works of Lewis [81], Althammer [7], Brenner [16], and others. In general, the inner product can be introduced by

$$\langle f, g\rangle = \sum_{k=0}^{m} \int_{\mathbb{R}} f^{(k)}(t) g^{(k)}(t)\, d\lambda_k(t), \tag{2.5.1}$$

where $d\lambda_k(t)$, $k = 0, 1, \ldots, m$ are given positive measures on $\mathbb{R}$.

Let $\{s_n\}$ denote a set of polynomials orthogonal with respect to the inner product (2.5.1). In the case $m = 0$, i.e., when $\langle f, g\rangle$ reduces to the inner product $(f, g) = \int_{\mathbb{R}} f(t)g(t)\, d\lambda_0(t)$, let $\{\pi_k\}$ be the corresponding set of orthogonal polynomials. It is clear that the all zeros of $\pi_n(t)$ lie in the interior of the interval $\Delta(d\lambda_0)$ (see Theorem 2.1.2).

Althammer [7] pointed out that the position of the zeros of $s_n(t)$ can be different from those of $\pi_n(t)$. For example for $m = 1$, if $d\lambda_0(t)$ is the Legendre measure, then the measure $d\lambda_1(t)$ on $(-1, 1)$ can be chosen in such a way that $s_n(t)$ has a zero outside $(-1, 1)$. Brenner [16] also obtained a similar result for $m = 1$ and the Laguerre measure $d\lambda_0(t) = \exp(-t)\, dt$ on $(0, +\infty)$. However, in both cases the measure $d\lambda_1(t)$ can be chosen that $s_n(t)$ has all simple zeros inside of interval of orthogonality. Zero distribution and behaviour of orthogonal polynomials in such cases were considered by Cohen [21].

Recently several authors ([4], [70–71], [85–86], [88], [90–92]) studied polynomials orthogonal with respect to the inner product (2.5.1), where the measures $d\lambda_k(t)$, $k \geq 1$, are concentrated in one point $t = c$ (discrete Sobolev inner product),

$$\langle f, g\rangle = \int_{\mathbb{R}} f(t)g(t)\,d\lambda(t) + \sum_{k=0}^{m} M_k f^{(k)}(c)g^{(k)}(c), \tag{2.5.2}$$

where $M_k \geq 0$, $k = 0, 1, \ldots, m$. Taking $d\lambda(t) = \big(t^\alpha \exp(-t)/\Gamma(\alpha+1)\big)\,dt$, $\alpha > -1$, on $(0, +\infty)$ R. Koekoek [70] proved that the coefficients A_k, $k = 0, 1, \ldots, m+1$, can be chosen in such a way that the polynomials $\big\{L_n^{\alpha, M_0, M_1, \ldots, M_m}(t)\big\}_{n=0}^{+\infty}$, defined by

$$L_n^{\alpha, M_0, M_1, \ldots, M_m}(t) = \sum_{k=0}^{m+1} A_k\, D^k L_n^{(\alpha)}(t),$$

are orthogonal with respect to the inner product (2.5.2). These polynomials satisfy a $(2m+3)$-terms recurrence relation of the form

$$t^{m+1} L_n^{\alpha, M_0, M_1, \ldots, M_m}(t) = \sum_{k=n-m-1}^{n+m+1} E_k^{(n)}\, L_k^{\alpha, M_0, M_1, \ldots, M_m}(t).$$

Also some cases of two symmetric points $t = \pm c$ were investigated by Bavinck and Meijer [12–13].

For more details see papers [4], [68], [85–86], [124], [114], as well as a survey of the results and a complete list of references [84]. Computing orthogonal polynomials of Sobolev type was recently investigated by Gautschi and Zhang [53] (see also [45]).

3. Some Applications of Orthogonal Polynomials on the Real Line

This section is devoted to some important applications of orthogonal polynomials on the real line as Gauss-Christoffel quadrature formulas, moment-preserving spline approximation and summation of slowly convergent series.

3.1. Gaussian type of quadratures. One of the important uses of orthogonal polynomials is in the construction of quadrature formulas of maximum, or nearly maximum, algebraic degree of exactness for integrals involving a positive measure $d\lambda(t)$.

The n-point *Gaussian quadrature formula*

$$\int_{\mathbb{R}} f(t)d\lambda(t) = \sum_{\nu=1}^{n} \lambda_\nu^{(n)} f(\tau_\nu^{(n)}) + R_n(f) \tag{3.1.1}$$

has maximum algebraic degree of exactness $2n-1$, in the sense that $R_n(f)=0$ for all $f \in \mathcal{P}_{2n-1}$. In formula (3.1.1), $\tau_\nu = \tau_\nu^{(n)}$ are the *Gauss nodes*, and $\lambda_\nu = \lambda_\nu^{(n)}$ the *Gauss weights* or *Christoffel numbers*. This formula is also known as *Gauss-Christoffel quadrature formula*. A nice survey on that was given by Gautschi [31].

The nodes τ_ν are the zeros of the n-th orthogonal polynomial $\pi_n(\cdot, d\lambda)$, and the weights λ_ν, which are all positive, can be also expressed in terms of the same orthogonal polynomials. As we have seen in 2.1, the nodes τ_ν are the eigenvalues of the n-th order Jacobi matrix

$$J_n(d\lambda) = \begin{bmatrix} \alpha_0 & \sqrt{\beta_1} & & & \text{O} \\ \sqrt{\beta_1} & \alpha_1 & \sqrt{\beta_2} & & \\ & \sqrt{\beta_2} & \alpha_2 & \ddots & \\ & & \ddots & \ddots & \sqrt{\beta_{n-1}} \\ \text{O} & & & \sqrt{\beta_{n-1}} & \alpha_{n-1} \end{bmatrix},$$

where α_ν and β_ν are the coefficients in three-term recurrence relation for the monic orthogonal polynomials $\pi_n(\cdot, d\lambda)$. The weights λ_ν are given by

$$\lambda_\nu = \beta_0 v_{\nu,1}^2, \qquad \nu = 1, \ldots, n,$$

where $\beta_0 = \int_{\mathbb{R}} d\lambda(t)$ and $v_{\nu,1}$ is the first component of the normalized eigenvector $\boldsymbol{v}_\nu$ corresponding to the eigenvalue τ_ν (cf. Golub and Welsch [59], and Gautschi [29]),

$$J_n(d\lambda)\boldsymbol{v}_\nu = \tau_\nu \boldsymbol{v}_\nu, \qquad \boldsymbol{v}_\nu^T \boldsymbol{v}_\nu = 1, \qquad \nu = 1, \ldots, n.$$

There are well-known and efficient algorithms, such as the QR algorithm with shifts, to compute eigenvalues and eigenvectors of symmetric tridiagonal matrices (cf. the routine GAUSS in the package ORTHPOL given by Gautschi [44]). There are many methods for estimating the remainder term $R_n(f)$ in (3.1.1). Error bounds in the class of analytic functions were investigated by Gautschi and Varga [52].

A simple modification of the previous method can be applied to the construction of Gauss-Radau and Gauss-Lobatto quadrature formulas.

In sequel we mention a few non-classical measures $d\lambda(t) = w(t)\,dt$ for which recursion coefficients $\alpha_k(d\lambda)$, $\beta_k(d\lambda)$, $k = 0, 1, \ldots, n-1$, have been tabulated in the literature and used in the construction of Gaussian quadratures.

1° *One-side Hermite weight* $w(t) = \exp(-t^2)$ on $[0, c]$, $0 < c \le +\infty$. This distribution $w(t)\,d\lambda(t)$ is known as the *Maxwell (velocity) distribution*. The cases

$c = 1$, $n = 10$ and $c = +\infty$, $n = 15$ were considered by Steen, Byrne and Gelbard [134] (see also Gautschi [41]).

2° *Logarithmic weight* $w(t) = t^{\alpha}\log(1/t)$, $\lambda > -1$ on $(0,1)$. Piessens and Branders [125] considered cases when $\alpha = 0, \pm 1/2, \pm 1/3, -1/4, -1/5$ (see also Gautschi [39]).

3° *Airy weight* $w(t) = \exp(-t^3/3)$ on $(0, +\infty)$. The inhomogeneous Airy functions $\mathrm{Hi}(x)$ and $\mathrm{Gi}(x)$, arise in theoretical chemistry (e.g. in harmonic oscillator models for large quantum numbers) and their integral representations (see Lee [79]) are given by

$$\mathrm{Hi}(x) = \frac{1}{\pi}\int_0^{+\infty} w(t)e^{tx}\,dt,$$

$$\mathrm{Gi}(x) = -\frac{1}{\pi}\int_0^{+\infty} w(t)e^{-tx/2}\cos\Big(\frac{\sqrt{3}}{2}\,tx + \frac{2\pi}{3}\Big)\,dt.$$

These functions can be effectively evaluated by Gaussian quadrature relative to the Airy weight $w(t)$. It needs orthogonal polynomials with respect to this weight. Gautschi [35] computed the recursion coefficients for $n = 15$ with 16 decimal digits after the decimal point (D).

3° *Reciprocal gamma function* $w(t) = 1/\Gamma(t)$ on $(0, +\infty)$. Gautschi [34] determined the recursion coefficients for $n = 40$ with 20 significant decimal digits (S). This function could be useful as a probability density function in reliability theory (see Fransén [25]).

4° *Einstein's* and *Fermi's weight functions* on $(0, +\infty)$,

$$w_1(t) = \varepsilon(t) = \frac{t}{e^t - 1} \qquad \text{and} \qquad w_2(t) = \varphi(t) = \frac{1}{e^t + 1}. \tag{3.1.2}$$

These functions arise in solid state physics. Integrals with respect to the measure $d\lambda(t) = \varepsilon(t)^r\,dt$, $r = 1$ and $r = 2$, are widely used in phonon statistics and lattice specific heats and occur also in the study of radiative recombination processes. Similarly, integrals with $\varphi(t)$ are encountered in the dynamics of electrons in metals. For $w_1(t)$, $w_2(t)$, $w_3(t) = \varepsilon(t)^2$ and $w_4(t) = \varphi(t)^2$, Gautschi and Milovanović [48] determined the recursion coefficients α_k and β_k, for $n = 40$ with 25 S, and gave an application of the corresponding Gauss-Christoffel quadratures to summation of slowly convergent series.

5° *The hyperbolic weights* on $(0, +\infty)$,

$$w_1(t) = \frac{1}{\cosh^2 t} \qquad \text{and} \qquad w_2(t) = \frac{\sinh t}{\cosh^2 t}. \tag{3.1.3}$$

The recursion coefficients α_k, β_k, for $n = 40$ with 30 S, were obtained by Milovanović [99]. The discretization was based on the Gauss-Laguerre quadrature rule.

3.2. Moment-preserving spline approximation. In this subsection we give some applications of orthogonal polynomials and Gauss-Christoffel and generalized Gauss-Turán quadratures to the moment-preserving spline approximation of functions. Such a problem of approximation was appeared in physics (see Laframboise and Stauffer [75] and Calder and Laframboise [18]). Namely, it was the problem of finding splines that reproduce as many as possible of the initial moments of a given spherically symmetric function $t \mapsto f(t)$, $t = \|\boldsymbol{x}\|$, $0 \le t < \infty$, in $\mathbb{R}^d$, $d \ge 1$. Gautschi [36] solved this problem of approximating by a piecewise constant function

$$t \mapsto s_n(t) = \sum_{\nu=1}^{n} a_\nu H(\tau_\nu - t) \qquad (a_\nu \in \mathbb{R},\ 0 < \tau_1 < \cdots < \tau_n < +\infty),$$

where H is the Heaviside step function. Also, he considered an approximation by a linear combination of Dirac delta functions. The approximation was to preserve as many moments of f as possible. This work was extended to spline approximation of arbitrary degree by Gautschi and Milovanović [51]. Namely, they considered a spline function of degree $m \ge 0$ on $[0, +\infty)$, vanishing at $t = +\infty$, with $n \ge 1$ positive knots τ_ν $(\nu = 1, \ldots, n)$, which can be written in the form

$$s_{n,m}(t) = \sum_{\nu=1}^{n} a_\nu (\tau_\nu - t)_+^m \qquad (a_\nu \in \mathbb{R},\ 0 \le t < +\infty), \tag{3.2.1}$$

where the plus sign on the right is the cutoff symbol, $u_+ = u$ if $u > 0$ and $u_+ = 0$ if $u \le 0$. Given a function $t \mapsto f(t)$ on $[0, +\infty)$, they determined $s_{n,m}$ such that

$$\int_0^{+\infty} s_{n,m}(t) t^j \, dV = \int_0^{+\infty} f(t) t^j \, dV \qquad (j = 0, 1, \ldots, 2n-1), \tag{3.2.2}$$

where dV is the volume element depending on the geometry of the problem. (For example, $dV = Ct^{d-1}\,dt$ if $d > 1$, where C is some constant, and $dV = dt$ if $d = 1$ were used in [51]. For some details see Gautschi [40].) In any case, the spline $s_{n,m}$ is such to faithfully reproduce the first $2n$ moments of f. Under suitable assumptions on f, it was shown that the problem has a unique solution if and only if certain Gauss-Christoffel quadratures exist corresponding to a moment functional or weight distribution depending on f. Existence, uniqueness and pointwise convergence of such approximation were analyzed. We mention two main results (Gautschi and Milovanović [51]) in the case when $dV = dt$.

Theorem 3.2.1. *Let $f \in C^{m+1}[0, +\infty]$ and*

$$\int_0^{+\infty} t^{2n+m+1} |f^{(m+1)}(t)| \, dt < +\infty .$$

Then a spline function $s_{n,m}$ of the form (3.2.1) *with positive knots τ_ν, that satisfies* (3.2.2), *exists and is unique if and only if the measure*

$$d\lambda(t) = \frac{(-1)^{m+1}}{m!} t^{m+1} f^{(m+1)}(t)\, dt \quad \textit{on} \quad [0, +\infty) \tag{3.2.3}$$

admits an n-point Gauss-Christoffel quadrature formula

$$\int_0^{+\infty} g(t)\, d\lambda(t) = \sum_{\nu=1}^{n} \lambda_\nu^{(n)} g(\tau_\nu^{(n)}) + R_n(g; d\lambda), \tag{3.2.4}$$

with distinct positive nodes $\tau_\nu^{(n)}$, where $R_n(g; d\lambda) = 0$ for all $g \in \mathcal{P}_{2n-1}$. In that event, the knots τ_ν and weights a_ν in (3.2.1) *are given by*

$$\tau_\nu = \tau_\nu^{(n)}, \quad a_\nu = \tau_\nu^{-(m+1)} \lambda_\nu^{(n)} \qquad (\nu = 1, \ldots, n).$$

Theorem 3.2.2. *Given f as in Theorem* 3.2.2, *assume that the measure $d\lambda$ in* (3.2.3) *admits the n-point Gauss-Christoffel quadrature formula* (3.2.4) *with distinct positive nodes $\tau_\nu = \tau_\nu^{(n)}$ and the remainder term $R_n(g; d\lambda)$. Define*

$$\sigma_t(x) = x^{-(m+1)} (x - t)_+^m .$$

Then, for any $t > 0$, we have for the error of the spline approximation (3.2.1), (3.2.2),

$$f(t) - s_{n,m}(t) = R_n(\sigma_t; d\lambda).$$

For example, if f is completely monotonic on $[0, +\infty)$ then $d\lambda(t)$ in (3.2.3) is a positive measure for every m and the Gauss-Christoffel quadrature formula exists uniquely, with n distinct and positive nodes $\tau_\nu^{(n)}$. Theorem 3.2.2 shows that $s_{n,m}$ converges pointwise to f as $n \to +\infty$ if the Gauss-Christoffel quadrature formula (3.2.4) converges for the particular function $x \mapsto g(x) = \sigma_t(x)$ $(x > 0)$.

Similarly, we can consider an approximation of a given function $t \mapsto f(t)$ on $[0, +\infty)$ by defective splines. A spline function of degree $m \geq 2$ and defect k on the interval $0 \leq t < +\infty$, vanishing at $t = +\infty$, with $n \geq 1$ positive knots τ_ν $(\nu = 1, \ldots, n)$, can be written in the form

$$s_{n,m}(t) = \sum_{\nu=1}^{n} \sum_{i=m-k+1}^{m} a_{i,\nu} (\tau_\nu - t)_+^i \tag{3.2.5}$$

where $a_{i,\nu}$ are real numbers.

Under suitable assumptions on f and $k = 2s + 1$, Milovanović and Kovačević [108–109] showed that the approximation problem has a unique solution if and only if certain generalized Turán quadratures exist corresponding to a measure depending on f (cf. Subsection 2.4).

Theorem 3.2.3. *Let* $f \in C^{m+1}[0,+\infty]$ *and*

$$\int_0^{+\infty} t^{2(s+1)n+m+1} |f^{(m+1)}(t)|\, dt < +\infty .$$

Then a spline function $s_{n,m}$ *of the form* (3.2.5) *with* $k = 2s+1$ *and positive knots* τ_ν, *that satisfies* (3.2.2), *with* $j = 0, 1, \ldots, 2(s+1)n-1$, *exists and is unique if and only if the measure*

$$d\lambda(t) = \frac{(-1)^{m+1}}{m!} t^{m+1} f^{(m+1)}(t)\, dt \quad on \quad [0,+\infty)$$

admits a generalized Gauss-Turán quadrature formula

$$\int_0^{+\infty} g(t)\, d\lambda(t) = \sum_{\nu=1}^{n} \sum_{i=0}^{2s} A_{i,\nu}^{(n)} g^{(i)}(\tau_\nu^{(n)}) + R_n^G(g; d\lambda), \tag{3.2.6}$$

with distinct positive nodes $\tau_\nu^{(n)}$, *where* $R_n^G(g; d\lambda) = 0$ *for all* $g \in \mathcal{P}_{2(s+1)n-1}$. *The knots* τ_ν *in* (3.2.5) *are given by* $\tau_\nu = \tau_\nu^{(n)}$, *and coefficients* $a_{i,\nu}$ *by the following triangular system*

$$A_{i,\nu}^{(n)} = \sum_{j=i}^{2s} \frac{(m-j)!}{m!} \binom{j}{i} \left[D^{j-i} t^{m+1}\right]_{t=\tau_\nu} a_{m-j,\nu} \quad (i = 0, 1, \ldots, 2s),$$

where D *is the standard differentiation operator.*

Theorem 3.2.4. *Given* f *as in Theorem* 3.2.3, *assume that the measure* $d\lambda(t)$ *admits the* n*-point generalized Gauss-Turán quadrature formula* (3.2.6) *with distinct positive nodes* $\tau_\nu = \tau_\nu^{(n)}$ *and the remainder term* $R_n^G(g; d\lambda)$. *Then the error of the spline approximation is given by*

$$f(t) - s_{n,m}(t) = R_n^G(\sigma_t; d\lambda) \qquad (t > 0),$$

where $x \mapsto \sigma_t(x) = x^{-(m+1)}(x-t)_+^m$.

Again, if f is completely monotonic on $[0,+\infty)$ then $d\lambda(t)$ is a positive measure for every m and the generalized Gauss-Turán quadrature formula exists uniquely, with n distinct and positive nodes $\tau_\nu^{(n)}$.

Frontini, Gautschi and Milovanović [27] and Frontini and Milovanović [28] considered analogous problems on an arbitrary finite interval, which can be standardized to $[a,b] = [0,1]$. If the approximations exist, they can be represented in terms of generalized Gauss-Lobatto and Gauss-Radau quadrature formulas relative to appropriate measures depending on f, when the defect $k = 1$. Using defective splines with odd defect $k = 2s + 1$, approximation problems reduce to certain generalized Gauss-Turán-Lobatto and Gauss-Turán-Radau quadrature formulas. A more general case with variable defects was considered by Gori and Santi [60] and Kovačević and Milovanović [74]. In that case, approximation problems reduce to Gauss-Turán-Stancu type of quadratures and σ-orthogonal polynomials (cf. Gautschi [31], Gori, Lo Cascio and Milovanović [61], Milovanović and Spalević [117]).

Further extensions of the moment-preserving spline approximation on $[0,1]$ are given by Micchelli [93]. He relates this approximation to the theory of the monosplines.

3.3. Summation of slowly convergent series. We consider convergent series of the type

$$T = \sum_{k=1}^{+\infty} a_k \qquad \text{and} \qquad S = \sum_{k=1}^{+\infty} (-1)^k a_k \tag{3.3.1}$$

and introduce the notation: $T = T^{(m-1)} + T_m^{(\infty)}$, $S = S^{(m-1)} + S_m^{(\infty)}$,

$$T_m^{(n)} = \sum_{k=m}^{n} a_k, \qquad S_m^{(n)} = \sum_{k=m}^{n} (-1)^k a_k,$$

where $T^{(m-1)}$ and $S^{(m-1)}$ are the corresponding partial sums of (3.3.1).

Some methods of summation these series can be found, for example, in the books of Henrici [66], Lindelöf [82], and Mitrinović and Kečkić [118].

Recently, a few new summation/integration procedures for slowly convergent series are developed (see [48], [42–43], [99–101]). Here we give a short account of these methods.

3.3.1. Laplace transform method. Suppose that the general term of T (and S) is expressible in terms of the derivative of a Laplace transform, or in terms of the Laplace transform itself. Namely, let $a_k = F'(k)$, where

$$F(p) = \int_0^{+\infty} e^{-pt} f(t)\, dt, \qquad \operatorname{Re} p \geq 1.$$

Then

$$\sum_{k=1}^{+\infty} F'(k) = -\sum_{k=1}^{+\infty} \int_0^{+\infty} t e^{-kt} f(t)\, dt = -\int_0^{+\infty} \frac{t}{e^t - 1} f(t)\, dt.$$

Similarly, for "alternating" series, one obtains

$$\sum_{k=1}^{+\infty} (-1)^k F'(k) = \int_0^{+\infty} \frac{t}{e^t + 1} f(t)\, dt$$

and

$$\sum_{k=1}^{+\infty}(-1)^k F(k) = -\int_0^{+\infty} \frac{1}{e^t+1} f(t)\, dt.$$

In a joint paper with Gautschi [48] we considered the construction of Gaussian quadrature formulas on $(0, +\infty)$,

$$\int_0^{+\infty} g(t)w(t)\, dt = \sum_{\nu=1}^{n} \lambda_\nu g(\tau_\nu) + R_n(g), \tag{3.3.2}$$

with respect to the weight functions given by (3.1.2). If the series T and S are slowly convergent and the respective function f on the right of the equalities above is smooth, then low-order Gaussian quadrature (3.3.2) applied to the integrals on the right, provides a possible summation procedure. Numerical examples show fast convergence of this procedure (see [48, §4]). A problem which arises with this procedure (*Laplace transform method*) is the determination of the original function f for a given series. For some other applications see [42–43].

3.3.2. Contour integration over a rectangle. Suppose now that $a_k = f(k)$, where $z \mapsto f(z)$ is a holomorphic function in the region

$$G_m = \{z \in \mathbb{C} \mid \operatorname{Re} z \geq \alpha,\ m-1 < \alpha < m\}, \qquad m \in \mathbb{N}.$$

In [99] we derived an alternative summation/integration method for the series (3.3.1) which requires the indefinite integral F of f chosen so as to satisfy the following decay conditions:

(C1) *F is a holomorphic function in the region G_m;*

(C2) $\lim\limits_{|t|\to+\infty} e^{-c|t|} F(x+it/\pi) = 0$, *uniformly for* $x \geq \alpha$;

(C3) $\lim\limits_{x\to+\infty} \int_{-\infty}^{+\infty} e^{-c|t|} \left|F(x+it/\pi)\right| dt = 0$,

where $c = 2$ *or* $c = 1$, *when we consider* $T_m^{(n)}$ *or* $S_m^{(n)}$, *respectively.*

Assume that f and g are analytic functions in a certain domain D of the complex plane with singularities $a_1, a_2, \ldots$ and $b_1, b_2, \ldots$, respectively, in a region $G = \operatorname{int}\Gamma\ (\subset D)$, where Γ is a closed contour. Then by Cauchy's residual theorem, we have

$$\frac{1}{2\pi i}\oint_\Gamma f(z)g(z)\, dz = \sum_\nu \operatorname*{Res}_{z=a_\nu}\Big(f(z)g(z)\Big) + \sum_\nu \operatorname*{Res}_{z=b_\nu}\Big(f(z)g(z)\Big).$$

Taking $\Gamma = \partial G$,

$$G = \Big\{z \in \mathbb{C} \mid \alpha \leq \operatorname{Re} z \leq \beta,\ |\operatorname{Im} z| \leq \frac{\delta}{\pi}\Big\},$$

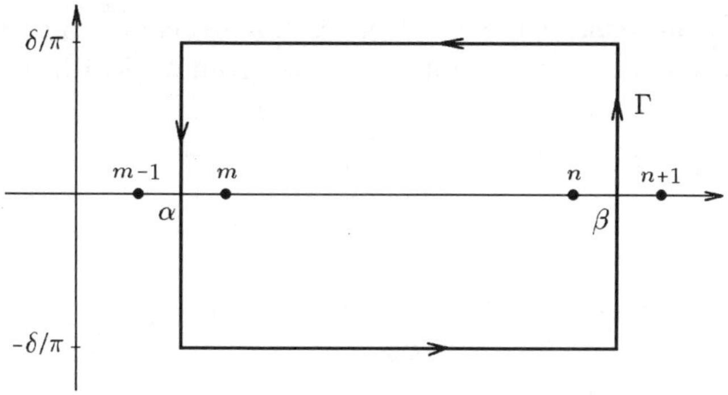

FIGURE 3.3.1: The contour of integration

where $m-1<\alpha<m$, $n<\beta<n+1$ $(m,n\in\mathbb{Z}, m\le n)$ (see Figure 3.3.1), and $g(z)=\pi/\tan\pi z$, then it immediately follows that (cf. [118, p. 212])

$$T_m^{(n)}=\sum_{\nu=m}^{n} f(\nu)=\frac{1}{2\pi i}\oint_\Gamma f(z)\frac{\pi}{\tan\pi z}\,dz-\sum_\nu \operatorname*{Res}_{z=a_\nu}\Big(f(z)\frac{\pi}{\tan\pi z}\Big).$$

Similarly, for $g(z)=\pi/\sin\pi z$ we have

$$S_m^{(n)}=\sum_{\nu=m}^{n}(-1)^\nu f(\nu)=\frac{1}{2\pi i}\oint_\Gamma f(z)\frac{\pi}{\sin\pi z}\,dz-\sum_\nu \operatorname*{Res}_{z=a_\nu}\Big(f(z)\frac{\pi}{\sin\pi z}\Big).$$

For a holomorphic function $z\mapsto f(z)$ in G, we obtain that

$$T_m^{(n)}=\frac{1}{2\pi i}\oint_\Gamma f(z)\frac{\pi}{\tan\pi z}\,dz \quad\text{and}\quad S_m^{(n)}=\frac{1}{2\pi i}\oint_\Gamma f(z)\frac{\pi}{\sin\pi z}\,dz.$$

After integration by parts, these formulas reduce to

$$T_m^{(n)}=\frac{1}{2\pi i}\oint_\Gamma\Big(\frac{\pi}{\sin\pi z}\Big)^2F(z)\,dz \tag{3.3.3}$$

and

$$S_m^{(n)}=\frac{1}{2\pi i}\oint_\Gamma\Big(\frac{\pi}{\sin\pi z}\Big)^2\cos\pi z\,F(z)\,dz, \tag{3.3.4}$$

where $z\mapsto F(z)$ is an integral of $z\mapsto f(z)$.

Taking $\alpha=\alpha_m=m-1/2$, $\beta=\beta_n=n+1/2$, and letting $\delta\to+\infty$ and $n\to+\infty$, under conditions (C1) – (C3), the integrals in (3.3.3) and (3.3.4) over Γ reduce to integrals along the line $z=\alpha_m+iy$ $(-\infty<y<+\infty)$.

After some calculations, we reduce T and S to a problem of quadrature on $(0, +\infty)$ with respect to the hyperbolic weight functions given by (3.1.3). Thus,

$$T = T^{(m-1)} + \int_0^{+\infty} \Phi\left(\alpha_m, t/\pi\right) w_1(t)\, dt$$

and

$$S = S^{(m-1)} + \int_0^{+\infty} \Psi\left(\alpha_m, t/\pi\right) w_2(t)\, dt,$$

where w_1 and w_2 are defined in (3.1.3) and

$$\Phi(x, y) = -\frac{1}{2}\left[F(x + iy) + F(x - iy)\right],$$

$$\Psi(x, y) = \frac{(-1)^m}{2i}\left[F(x + iy) - F(x - iy)\right].$$

Numerical experiments shows that is enough to use only the quadrature with respect to the first weight $w_1(t) = 1/\cosh^2 t$. Namely, in the series S we can include the hyperbolic sine as a factor in the corresponding integrand so that

$$S = S^{(m-1)} + \int_0^{+\infty} \Psi\left(\alpha_m, t/\pi\right) \sinh(t) w_1(t)\, dt.$$

Using this approach we gave an appropriate method for calculating values of the Riemann zeta function $\zeta(z) = \sum\limits_{k=1}^{+\infty} k^{-z}$, which can be transformed to a weighted integral on $(0, +\infty)$ of the function $t \mapsto \exp\bigl(-(z/2)\log(1 + \beta_m^2 t^2)\bigr)\cos(z \arctan(\beta_m t))$, $\beta_m = 2/((2m+1)\pi)$, $m \in \mathbb{N}_0$, involving the hyperbolic weight $w(t) = 1/\cosh^2 t$ (see [100]).

Also some other methods for series with irrational terms were given in [101].

4. Orthogonality on the Unit Circle

Another type of orthogonality is *orthogonality on the unit circle*. The polynomials orthogonal on the unit circle with respect to a given weight function have been introduced and studied by Szegő [136–137] and Smirnov [130–131]. A more general case was considered by Achieser and Kreĭn [1], Geronimus [54–55], Nevai [121–122], Alfaro and Marcellán [5], Marcellán and Sansigre [87], etc. These polynomials are linked with many questions in the theory of time series, digital filters, statistics, image processing, scattering theory, control theory and so on.

In the next three sections we give some basic definitions and properties of polynomials orthogonal on the unit circle.

4.1. Basic definitions and properties of orthogonal polynomials. Let an inner product defined by

$$(f,g) = \frac{1}{2\pi}\int_0^{2\pi} f(e^{i\theta})\overline{g(e^{i\theta})}\,d\mu(\theta), \tag{4.1.1}$$

where $d\mu(\theta)$ is a finite positive measure on the interval $[0, 2\pi]$ whose support is an infinite set. In that case there is a unique system of orthonormal polynomials $\{\varphi_k\}_{k\in\mathbb{N}_0}$ such that

$$(\varphi_k, \varphi_m) = \delta_{km}, \quad \varphi_k(z) = b_k z^k + \cdots, \quad b_k > 0 \ \ (k = 0, 1, \ldots). \tag{4.1.2}$$

If $\theta \mapsto \mu(\theta)$ is an absolutely continuous function on $[0, 2\pi]$, then we say that $\mu'(\theta) = w(\theta)$ is a *weight function.*

The monic polynomials orthogonal on the unit circle with respect to the inner product (4.1.1) we will denote by $\phi_k(z)$. According to (4.1.2) we have that $\phi_k(z) = \varphi_k(z)/b_k$.

Theorem 4.1.1. *The monic orthogonal polynomials $\{\phi_k\}$ on the unit circle $|z| = 1$ satisfy the recurrence relations*

$$\phi_{k+1}(z) = z\phi_k(z) + \phi_{k+1}(0)\phi_k^*(z), \quad \phi_{k+1}^*(z) = \phi_k^*(z) + \overline{\phi_{k+1}(0)}z\phi_k^*(z), \tag{4.1.3}$$

for $k = 0, 1, \ldots,$ where $\phi_k^*(z) = z^k\bar{\phi}_k(1/z)$.

In order to prove the first recurrence relation in (4.1.3) we put

$$\frac{\phi_k(z) - z^k}{z^{k-1}} = \sum_{\nu=0}^{n-1} c_\nu^{(k)}\bar{\phi}_\nu(1/z). \tag{4.1.4}$$

Multiplying this equality by $\phi_m(z)$, where $0 \le m \le k-1$, putting $z = e^{i\theta}$ and integrating, we obtain

$$\frac{1}{2\pi}\int_0^{2\pi} \phi_k(e^{i\theta})\frac{\phi_m(e^{i\theta})}{e^{i(k-1)\theta}}\,d\mu(\theta) - \frac{1}{2\pi}\int_0^{2\pi} e^{i\theta}\phi_m(e^{i\theta})\,d\mu(\theta) = c_m^{(k)}\|\phi_m\|^2,$$

where

$$\|\phi_m\|^2 = \frac{1}{2\pi}\int_0^{2\pi} \phi_\nu(e^{i\theta})\overline{\phi_m(e^{i\theta})}\,d\mu(\theta).$$

Because of orthogonality, the first integral on the left hand side in the previous equality vanish, and we have

$$c_m^{(k)} = \frac{1}{2\pi\|\phi_m\|^2}\int_0^{2\pi} e^{i\theta}\phi_m(e^{i\theta})\,d\mu(\theta) \qquad (m = 0, 1, \ldots, n-1).$$

We note that $c_m^{(k)}$ does not depend on k and therefore we write $c_m^{(k)} = c_m$. Using (4.1.4) we have

$$\phi_k(z) = z^k + z^{k-1} \sum_{\nu=0}^{n-1} c_\nu \bar{\phi}_\nu(1/z),$$

wherefrom we obtain

$$\phi_{k+1}(z) = z\phi_k(z) + z^k c_k \bar{\phi}_k(1/z) = z\phi_k(z) + c_k \phi_k^*(z). \tag{4.1.5}$$

Since $\phi_k^*(0) = 1$, setting $z = 0$ we find $c_k = \phi_{k+1}(0)$. Thus, (4.1.5) reduces to the first relation in (4.1.3).

As we can see recurrence relations (4.1.3) are not three-term relations like (2.3). The values $\phi_k(0)$ are called reflection parameters or Szegő parameters. Defining a sequence of parameters $\{a_k\}$ by $a_k = -\overline{\phi_{k+1}(0)}$, $k = 0, 1, \ldots$, and using (4.1.3), Geronimus [56, Chapter VIII] derived the following three-term recurrence relations:

$$\begin{aligned} \bar{a}_{k-1}\phi_{k+1}(z) &= (\bar{a}_{k-1}z + \bar{a}_k)\phi_k(z) - \bar{a}_k z(1 - |a_{k-1}|^2)\phi_{k-1}(z), \\ a_{k-1}\phi_{k+1}^*(z) &= (a_{k-1}z + a_k)\phi_k^*(z) - a_k z(1 - |a_{k-1}|^2)\phi_{k-1}^*(z), \end{aligned}$$

where $k = 1, 2, \ldots$, and $\phi_0(z) = 1$, $\phi_1(z) = z - \bar{a}_0$.

There is a relation between parameters $\{a_k\}$ and moments $\{c_k\}$, defined by

$$c_k = \frac{1}{2\pi} \int_0^{2\pi} e^{-ik\theta}\, d\mu(\theta), \qquad k = 0, 1, \ldots .$$

Introducing Toeplitz determinants $\Delta_k = |c_{i-j}|_0^k$, $c_{-k} = \bar{c}_k$, $k = 0, 1, \ldots$, then one can express $\phi_k(z)$ in the explicit form

$$\phi_k(z) = \frac{1}{\Delta_{k-1}} \begin{vmatrix} c_0 & c_{-1} & c_{-2} & \cdots & c_{-k} \\ c_1 & c_0 & c_{-1} & & c_{-k+1} \\ \vdots & & & & \\ c_{k-1} & c_{k-2} & c_{k-3} & & c_{-1} \\ 1 & z & z^2 & & z^k \end{vmatrix} = z^k + \text{lower degree terms},$$

where $k = 0, 1, \ldots$ and $\Delta_{-1} = 1$. According to the previous relations, one can get the following relation

$$\frac{\Delta_k}{\Delta_{k-1}} = c_0 \prod_{\nu=0}^{k-1} \left(1 - |a_\nu|^2\right) \qquad (n \in \mathbb{N}). \tag{4.1.6}$$

For a given sequence $\{a_k\}$ such that

$$|a_k| < 1 \qquad (k \in \mathbb{N}_0), \tag{4.1.7}$$

one can construct a sequence of (monic) polynomials $\{\phi_k(z)\}$ and determine a nondecreasing bounded function $\mu(\theta)$ with infinitely many points of increase (see Geronimus [56]). Namely, (4.1.7) is equivalent to the condition $\Delta_k > 0$ for every $k \in \mathbb{N}_0$, which is the necessary and sufficient condition for the existence of $\mu(\theta)$.

For orthonormal polynomials defined in (4.1.2) one can introduce the *kernel polynomial*

$$K_n(z, w) = \sum_{\nu=0}^{n} \overline{\varphi_\nu(z)}\, \varphi_\nu(w),$$

which can be expressed in the form

$$K_n(z, w) = \frac{\overline{\varphi^*_{n+1}(z)}\varphi^*_{n+1}(w) - \overline{\varphi_{n+1}(z)}\varphi_{n+1}(w)}{1 - \bar{z}w}.$$

This analogue of the Christoffel-Darboux formula was proved by Szegő (cf. Freud [26, p. 196]). Some characteristic values of $K_n(z, w)$ are

$$K_n(0, z) = b_n \varphi^*_n(z), \qquad K_n(0, 0) = b_n^2 = \Delta_{n-1}/\Delta_n,$$

where b_n is the coefficient of z^n in $\varphi(z)$.

4.2. Extremal properties of orthogonal polynomials. An analogue of Theorem 2.1.5 is the following result:

Theorem 4.2.1. *Let* $P(z) = \sum_{\nu=0}^{n} a_\nu z^\nu$, *with* $a_n = 1$, *be an arbitrary monic polynomial of degree* n. *Then*

$$\frac{1}{2\pi} \int_0^{2\pi} |P(z)|^2 \, d\mu(\theta) \geq \|\phi_n\|^2 = b_n^{-2} = \frac{\Delta_n}{\Delta_{n-1}}, \qquad z = e^{i\theta},$$

with equality only if $P(z) = \phi_n(z) = \varphi_n(z)/b_n$. *The ratio* Δ_n/Δ_{n-1} *is given by* (4.1.6).

We also mention the following statement (cf. Szegő [138, p. 290]):

Theorem 4.2.2. *Let* $P \in \mathcal{P}_n$ *and*

$$\frac{1}{2\pi} \int_0^{2\pi} |P(z)|^2 \, d\mu(\theta) = 1, \qquad z = e^{i\theta}.$$

For a fixed value of a, *the maximum of* $|P(a)|^2$ *is attained if*

$$P(z) = \frac{\varepsilon K_n(a, z)}{\sqrt{K_n(a, a)}}, \qquad |\varepsilon| = 1.$$

This maximum is given by $K_n(a, a)$.

4.3. Zeros of orthogonal polynomials. For the zeros of orthogonal polynomials on the unit circle one can prove the following result:

Theorem 4.3.1. *For $n \geq 1$, all the zeros of $\phi_n(z)$ lie in $|z| < 1$.*

A short proof of this theorem can be given in the following way (see Landau [77]): Let γ be an arbitrary zero of $\phi_n(z)$, so that $\phi_n(z) = (z - \gamma)Q_{n-1}(z)$, or

$$\phi_n(z) + \gamma Q_{n-1}(z) = zQ_{n-1}(z), \tag{4.3.1}$$

with some $Q_{n-1} \in \mathcal{P}_{n-1}$. Then, since ϕ_n is orthogonal to $\mathcal{P}_{n-1}$, on taking norms in (4.3.1) and using the fact that $(zf, zg) = (f, g)$ we find

$$\|\phi_n\|^2 + |\gamma|^2\|Q_{n-1}\|^2 = \|zQ_{n-1}\|^2 = \|Q_{n-1}\|^2,$$

whence $1 - |\gamma|^2 = \|\phi_n\|^2/\|Q_{n-1}\|^2 > 0$, as required.

For zeros of the kernel polynomials $z \mapsto K_n(a, z)$ we have (see Szegő [138, p. 292]):

Theorem 4.3.2. *For $|a| < 1$ the zeros of $K_n(a, z)$ lie in $|z| > 1$, for $|a| > 1$ in $|z| < 1$, and for $|a| = 1$ on $|z| = 1$.*

At the end of this section, we mention that classical and semiclassical orthogonal polynomials on the unit circle can also be considered (cf. Marcellán [83]). Similarly, orthogonal polynomials on a rectifiable curve or arc lying in the complex plane can be considered (cf. Geronimus [56], Szegő [138]).

5. Orthogonality on the Semicircle and a Circular Arc

Polynomials orthogonal on the semicircle $\Gamma_0 = \{z \in \mathbb{C} \,|\, z = e^{i\theta},\ 0 \leq \theta \leq \pi\}$ have been introduced by Gautschi and Milovanović [49–50]. The inner product is given by

$$(f, g) = \int_\Gamma f(z)g(z)(iz)^{-1}\, dz,$$

where Γ is the semicircle $\Gamma = \{z \in \mathbb{C} \mid z = e^{i\theta}, 0 \leq \theta \leq \pi\}$. Alternatively,

$$(f, g) = \int_0^\pi f(e^{i\theta})g(e^{i\theta})\, d\theta.$$

This inner product is not Hermitian, but the corresponding (monic) orthogonal polynomials $\{\pi_k\}$ exist uniquely and satisfy a three-term recurrence relation of the form

$$\pi_{k+1}(z) = (z - i\alpha_k)\pi_k(z) - \beta_k\pi_{k-1}(z), \qquad k = 0, 1, 2, \ldots,$$
$$\pi_{-1}(z) = 0, \quad \pi_0(z) = 1.$$

Notice that the inner product possesses the property $(zf, g) = (f, zg)$.

The general case of complex polynomials orthogonal with respect to a *complex weight function* was considered by Gautschi, Landau and Milovanović [46]. A generalization of such polynomials on a circular arc was given by de Bruin [17], and further investigations were done by Milovanović and Rajković [112–113].

5.1. Orthogonality on the semicircle. Let $w\colon(-1,1)\mapsto\mathbb{R}_+$ be a weight function which can be extended to a function $w(z)$ holomorphic in the half disc $D_+=\{z\in\mathbb{C}\,|\,|z|<1,\ \operatorname{Im} z>0\}$, and

$$(f,g)=\int_\Gamma f(z)g(z)w(z)(iz)^{-1}\,dz=\int_0^\pi f(e^{i\theta})g(e^{i\theta})w(e^{i\theta})\,d\theta. \tag{5.1.1}$$

Together with (5.1.1) consider the inner product

$$[f,g]=\int_{-1}^1 f(x)\overline{g(x)}w(x)\,dx, \tag{5.1.2}$$

which is positive definite and therefore generates a unique set of real (monic) orthogonal polynomials $\{p_k\}$:

$$[p_k,p_m]=0 \quad \text{for} \quad k\neq m \qquad \text{and} \qquad [p_k,p_m]>0 \quad \text{for} \quad k=m \qquad (k,m\in\mathbb{N}_0).$$

On the other hand, the inner product (5.1.1) is not Hermitian; the second factor g is not conjugated and the integration is not with respect to the measure $|w(e^{i\theta})|\,d\theta$. The existence of corresponding orthogonal polynomials, therefore, is not guaranteed.

We call a system of complex polynomials $\{\pi_k\}$ *orthogonal on the semicircle* if

$$(\pi_k,\pi_m)=0 \quad \text{for} \quad k\neq m \qquad \text{and} \qquad (\pi_k,\pi_m)>0 \quad \text{for} \quad k=m \qquad (k,m\in\mathbb{N}_0).$$

where we assume that π_k is monic of degree k.

The existence of the orthogonal polynomials $\{\pi_k\}$ can be established assuming only that

$$\operatorname{Re}(1,1)=\operatorname{Re}\int_0^\pi w(e^{i\theta})\,d\theta\neq 0. \tag{5.1.3}$$

5.1.1. Existence and representation of π_k. Assume that the weight function w is positive on $(-1,1)$, holomorphic in D_+ and such that the integrals in (5.1.1) and (5.1.2) exist for smooth f and g (possibly) as improper integrals. We also assume that the condition (5.1.3) is satisfied.

Let C_ε, $\varepsilon>0$, denote the boundary of D_+ with small circular parts of radius ε and centres at ±1 spared out and let $\mathcal{P}$ be the set of all algebraic polynomials. Then, by Cauchy's theorem, for any $g\in\mathcal{P}$ we have

$$\begin{aligned} 0&=\int_{C_\varepsilon} g(z)w(z)\,dz \qquad\qquad (5.1.4)\\ &=\left(\int_{\Gamma_\varepsilon}+\int_{C_{\varepsilon,-1}}+\int_{C_{\varepsilon,+1}}\right)g(z)w(z)\,dz+\int_{-1+\varepsilon}^{1-\varepsilon} g(x)w(x)\,dx, \end{aligned}$$

where Γ_ε and $C_{\varepsilon,\pm 1}$ are the circular parts of C_ε (with radii 1 and ε respectively). We assume that w is such that for all $g \in \mathcal{P}$,

$$\lim_{\varepsilon \to 0} \int_{C_{\varepsilon,\pm 1}} g(z)w(z)\,dz = 0. \tag{5.1.5}$$

Then, if $\varepsilon \to 0$ in (5.1.4), we obtain

$$0 = \int_C g(z)w(z)\,dz = \int_\Gamma g(z)w(z)\,dz + \int_{-1}^{1} g(x)w(x)\,dx, \quad g \in \mathcal{P}. \tag{5.1.6}$$

The (monic, real) polynomials $\{p_k\}$, orthogonal with respect to the inner product (5.1.2), as well as the associated polynomials of the second kind,

$$q_k(z) = \int_{-1}^{1} \frac{p_k(z) - p_k(x)}{z - x} w(x)\,dx \qquad (k = 0, 1, 2, \ldots),$$

are known to satisfy a three-term recurrence relation of the form

$$y_{k+1} = (z - a_k)y_k - b_k y_{k-1} \qquad (k = 0, 1, 2, \ldots), \tag{5.1.7}$$

where

$$y_{-1} = 0,\ y_0 = 1 \ \text{ for } \{p_k\} \qquad \text{and} \qquad y_{-1} = -1,\ y_0 = 0 \ \text{ for } \{q_k\}. \tag{5.1.8}$$

Denote by m_k and μ_k the moments associated with the inner products (5.1.1) and (5.1.2), respectively,

$$\mu_k = (z^k, 1), \qquad m_k = [x^k, 1], \qquad k \geq 0,$$

where, in view of (5.1.8), $b_0 = m_0$.

In [46], Gautschi, Landau and Milovanović proved the following result:

Theorem 5.1.1. *Let w be a weight function, positive on $(-1, 1)$, holomorphic in $D_+ = \{z \in \mathbb{C} \mid |z| < 1,\ \operatorname{Im} z > 0\}$, and such that* (5.1.5) *is satisfied and the integrals in* (5.1.6) *exist (possibly) as improper integrals. Assume in addition that*

$$\operatorname{Re}(1, 1) = \operatorname{Re} \int_0^\pi w(e^{i\theta})\,d\theta \neq 0.$$

Then there exists a unique system of (monic, complex) orthogonal polynomials $\{\pi_k\}$ relative to the inner product (5.1.1). *Denoting by $\{p_k\}$ the (monic, real) orthogonal polynomials relative to the inner product* (5.1.2), *we have*

$$\pi_k(z) = p_k(z) - i\theta_{k-1} p_{k-1}(z) \qquad (k = 0, 1, 2, \ldots), \tag{5.1.9}$$

where

$$\theta_{k-1} = \frac{\mu_0 p_k(0) + i q_k(0)}{i\mu_0 p_{k-1}(0) - q_{k-1}(0)} \qquad (k = 0, 1, 2, \ldots). \tag{5.1.10}$$

Alternatively,

$$\theta_k = i a_k + \frac{b_k}{\theta_{k-1}} \quad (k = 0, 1, 2, \ldots); \qquad \theta_{-1} = \mu_0, \tag{5.1.11}$$

where a_k, b_k *are the recursion coefficients in* (5.1.7) *and* $\mu_0 = (1, 1)$. *In particular, all* θ_k *are real* (*in fact, positive*) *if* $a_k = 0$ *for all* $k \geq 0$. *Finally,*

$$(\pi_k, \pi_k) = \theta_{k-1}[p_{k-1}, p_{k-1}] \neq 0 \quad (k = 1, 2, \ldots), \qquad (\pi_0, \pi_0) = \mu_0. \tag{5.1.12}$$

As we can see, relation (5.1.9), with (5.1.10), gives a connection between orthogonal polynomials on the semicircle and the standard polynomials orthogonal on $[-1, 1]$ with respect to the same weight function w. The norms of these polynomials are in relation (5.1.12).

We mention now two interesting examples from [46]:

EXAMPLE 5.1.1. Let $w(z) = 1 + z$. Since $\mu_0 = (1, 1) = \pi + 2i$, $\operatorname{Re}\mu_0 \neq 0$, the orthogonal polynomials $\{\pi_k\}$ exist. Furthermore, $b_0 = m_0 = 2$, $a_k = (2k+1)^{-1}(2k+3)^{-1}$ for $k \geq 0$, and $b_k = k(k+1)(2k+1)^{-2}$ for $k \geq 1$, so that by (5.1.11),

$$\theta_0 = \frac{\pi - 4i}{3(2 - i\pi)}, \quad \theta_1 = \frac{3\pi + 8i}{5(4 + i\pi)}, \quad \ldots .$$

Then, from (5.1.7) and (5.1.8), we find

$$p_0(z) = 1, \quad p_1(z) = z - \frac{1}{3}, \quad p_2(z) = z^2 - \frac{2}{3}z - \frac{1}{5}, \quad \ldots ,$$

and by (5.1.9),

$$\pi_0(z) = 1, \quad \pi_1(z) = z - \frac{2}{2 - i\pi}, \quad \pi_2 = z^2 - \frac{i\pi}{4 + i\pi}z - \frac{4}{3(4 + i\pi)}, \quad \ldots .$$

EXAMPLE 5.1.2. Consider now $w(z) = z^2$. Here $\mu_0 = \int_0^\pi e^{2i\theta}\, d\theta = 0$ so that (5.1.3) is violated and thus the polynomials $\{\pi_k\}$ do not exist, even though $w(x) \geq 0$ on $[-1, 1]$ and the polynomials $\{p_k\}$ do exist. It is easily seen that $q_k(0) = 0$ when k is even, so that θ_{k-1} is zero for k even, and undefined for k odd. For an explanation of this example see Theorem 5.1.3.

5.1.2. Three-term recurrence relation for $\pi_k(z)$. We assume that

$$\operatorname{Re}(1, 1) = \operatorname{Re} \int_0^\pi w(e^{i\theta})\, d\theta \neq 0, \tag{5.1.13}$$

so that orthogonal polynomials $\{\pi_k\}$ exist. Since $(zf, g) = (f, zg)$, it is known that they must satisfy a three-term recurrence relation

$$\begin{aligned} &\pi_{k+1}(z) = (z - i\alpha_k)\pi_k(z) - \beta_k \pi_{k-1}(z), \qquad k = 0, 1, 2, \ldots , \\ &\pi_{-1}(z) = 0, \quad \pi_0(z) = 1. \end{aligned} \tag{5.1.14}$$

Using the representation (5.1.9), we can find a connection between the coefficients in (5.1.14) and the corresponding coefficients in the three-term recurrence relation (5.1.7) for polynomials $\{p_k\}$ (see [46]):

Theorem 5.1.2. *Under the assumption* (5.1.13), *the (monic, complex) polynomials* $\{\pi_k\}$ *orthogonal with respect to the inner product* (5.1.1) *satisfy the recurrence relation* (5.1.14), *where the coefficients* α_k, β_k *are given by*

$$\alpha_k = \theta_k - \theta_{k-1} - ia_k, \qquad \beta_k = \frac{\theta_{k-1}}{\theta_{k-2}} b_{k-1} = \theta_{k-1}(\theta_{k-1} - ia_{k-1}),$$

for $k \geq 1$ *and* $\alpha_0 = \theta_0 - ia_0$, *with the* θ_k *defined in Theorem* 5.1.1.

Alternatively, the coefficients α_k can be expressed in the form

$$\alpha_k = -\theta_{k-1} + \frac{b_k}{\theta_{k-1}}, \qquad k \geq 1, \qquad \alpha_0 = \frac{b_0}{\theta_{-1}} = \frac{m_0}{\mu_0}.$$

5.1.3. Distribution of zeros. It follows from (5.1.14) that the zeros of $\pi_n(z)$ are the eigenvalues of the (complex, tridiagonal) matrix

$$J_n = \begin{bmatrix} i\alpha_0 & 1 & & & O \\ \beta_1 & i\alpha_1 & 1 & & \\ & \beta_2 & i\alpha_2 & \ddots & \\ & & \ddots & \ddots & 1 \\ O & & & \beta_{n-1} & i\alpha_{n-1} \end{bmatrix}, \tag{5.1.15}$$

where α_k and β_k are given in Theorem 5.1.2.

If the weight w is symmetric, i.e.,

$$w(-z) = w(z), \qquad w(0) > 0, \tag{5.1.16}$$

then $\mu_0 = (1,1) = \pi w(0) > 0$, $a_k = 0$, $\theta_k > 0$, for all $k \geq 0$, and

$$\alpha_0 = \theta_0, \quad \alpha_k = \theta_k - \theta_{k-1}, \quad \beta_k = \theta_{k-1}^2, \quad k \geq 1.$$

In that case J_n can be transformed into a real nonsymmetric tridiagonal matrix

$$A_n = -iD_n^{-1} J_n D_n = \begin{bmatrix} \alpha_0 & \theta_0 & & & O \\ -\theta_0 & \alpha_1 & \theta_1 & & \\ & -\theta_1 & \alpha_2 & \ddots & \\ & & \ddots & \ddots & \theta_{n-2} \\ O & & & -\theta_{n-2} & \alpha_{n-1} \end{bmatrix},$$

where $D_n = \operatorname{diag}(1, i\theta_0, i^2\theta_0\theta_1, i^3\theta_0\theta_1\theta_2, \dots) \in \mathbb{C}^{n\times n}$. The eigenvalues η_ν, $\nu = 1, \dots, n$, of A_n can be calculated using the EISPACK subroutine HQR (see [132]). Then all the zeros ζ_ν, $\nu = 1, \dots, n$, of $\pi_n(z)$ are given by $\zeta_\nu = i\eta_\nu$, $\nu = 1, \dots, n$.

In [46] we proved the following result for a symmetric weight (5.1.16):

Theorem 5.1.3. *All zeros of π_n are located symmetrically with respect to the imaginary axis and contained in $D_+ = \{z \in \mathbb{C} \mid |z| < 1,\ \operatorname{Im} z > 0\}$, with the possible exception of a single (simple) zero on the positive imaginary axis.*

If we define the half strip $S_+ = \{z \in \mathbb{C} \mid \operatorname{Im} z > 0,\ -\xi_n \le \operatorname{Re} z \le \xi_n\}$, where ξ_n is the largest zero of the real polynomial p_n, then we can prove that all zeros of π_n are also in S_+ (see [46] and [50]). Thus, all zeros are contained in $D_+ \cap S_+$.

For the Gegenbauer weight $w(z) = (1-z^2)^{\lambda-1/2}$, $\lambda > -1/2$, the exceptional case from Theorem 5.1.3 can only arise if $n = 1$ and $-1/2 < \lambda \le 0$. Likewise, no exceptional cases seem to occur for Jacobi weights $w(z) = (1-z)^\alpha(1+z)^\beta$, $\alpha, \beta > -1$, if $n \ge 2$, as was observed by several numerical computations (see [46]). However, in a general case, Gautschi [38] exhibited symmetric functions w for which $\pi_n(\,\cdot\,; w)$, for arbitrary fixed n, has a zero iy with $y \ge 1$.

5.2. Orthogonality on a circular arc. Recently de Bruin [17] considered the polynomials $\{\pi_k^R\}$ orthogonal on a circular arc with respect to the complex inner product

$$(f, g) = \int_\varphi^{\pi-\varphi} f_1(\theta) g_1(\theta) w_1(\theta)\, d\theta, \tag{5.2.1}$$

where $\varphi \in (0, \pi/2)$, and for $f(z)$ the function $f_1(\theta)$ is defined by

$$f_1(\theta) = f\big(-iR + e^{i\theta}\sqrt{R^2+1}\big), \qquad R = \tan\varphi.$$

Alternatively, the inner product (5.2.1) can be expressed in the form

$$(f, g) = \int_{\Gamma_R} f(z) g(z) w(z) (iz - R)^{-1}\, dz, \tag{5.2.2}$$

where $\Gamma_R = \{z \in \mathbb{C} \,|\, z = -iR + e^{i\theta}\sqrt{R^2+1},\ \varphi \le \theta \le \pi - \varphi,\ \tan\varphi = R\}$.

Under suitable integrability conditions on the weight function w, which is positive on $(-1, 1)$ and is holomorphic in the moon-shaped region

$$M_+ = \left\{z \in \mathbb{C} \mid |z + iR| < \sqrt{R^2+1},\ \operatorname{Im} z > 0\right\},$$

where $R > 0$, the polynomials $\{\pi_k^R\}$ orthogonal on the circular arc Γ_R with respect to the complex inner product (5.2.1) always exist and have similar properties like polynomials orthogonal on the semicircle.

For $R = 0$ the arc Γ_R reduces to the semicircle Γ, and polynomials $\{\pi_k^R\}$ to $\{\pi_k\}$. It is easy to prove that the condition

$$\operatorname{Re} \int_{\Gamma_R} w(z)(iz - R)^{-1}\, dz = \operatorname{Re} \int_\varphi^{\pi-\varphi} w_1(\theta)\, d\theta \ne 0$$

is automatically satisfied for $R > 0$ in contrast to the case $R = 0$ (see condition (5.1.13)).

Quite analogous results to Theorems 5.1.1–5.1.4 were proved by de Bruin [17]. For example, for polynomials $\{\pi_k\}$ (the upper index R is omitted) equalities (5.1.9) and (5.1.12), as well as the three-term recurrence relation (5.1.14) hold, where now the θ_k is given by

$$\theta_k = -R + ia_k + \frac{b_k}{\theta_{k-1}} \quad (k = 0, 1, 2, \ldots); \qquad \theta_{-1} = \mu_0,$$

instead of (5.1.11). Also, for the symmetric weight, $w(z) = w(-z)$, all zeros of π_n are contained in M_+ with the possible exception of just one simple zero situated on the positive imaginary axis.

5.2.1. Dual orthogonal polynomials. Let $\{\pi_n\}$ be the set of polynomials orthogonal on the circular arc Γ_R, with respect to the inner product (5.2.1), i.e., (5.2.2). Milovanović and Rajković [112] introduced the polynomials $\{\pi_n^*\}$ orthogonal on the symmetric down circular arc Γ_R^* with respect to the inner product defined by

$$(f, g)^* = \int_{\Gamma_R^*} f(z)g(z)w(z)(iz + R)^{-1}\, dz, \tag{5.2.3}$$

where $\Gamma_R^* = \{z \in \mathbb{C} \,|\, z = iR + e^{-i\theta}\sqrt{R^2+1},\ \varphi \le \theta \le \pi - \varphi,\ \tan\varphi = R\}$. Such polynomials are called *dual orthogonal polynomials* with respect to polynomials $\{\pi_n\}$.

Let M be a lentil-shaped region with the boundary $\partial M = \Gamma_R \cup \Gamma_R^*$, i.e.,

$$M = \{z \in \mathbb{C} \,|\, |z \pm iR| < \sqrt{R^2+1}\},$$

where $R > 0$.

We assume that w is a weight function, positive on $(-1, 1)$, holomorphic in M, and such that the integrals in (5.2.2), (5.2.3), and (5.1.2) exist for smooth functions f and g (possibly) as improper integrals. Under the same additional conditions on w and f, like previous, we have

$$0 = \int_\Gamma f(z)w(z)\, dz + \int_{-1}^{1} f(x)w(x)\, dx,$$

where $\Gamma = \Gamma_R$ or Γ_R^*. Then both systems of the orthogonal polynomials $\{\pi_n\}$ and $\{\pi_n^*\}$ exist uniquely.

The inner products in (5.2.2) and (5.2.3) define the moment functionals

$$\mathcal{L}z^k = \mu_k, \qquad \mu_k = (z^k, 1) = \int_{\Gamma_R} z^k w(z)(iz - R)^{-1}\, dz$$

and

$$\mathcal{L}^* z^k = \mu_k^*, \qquad \mu_k^* = (z^k, 1)^* = \int_{\Gamma_R^*} z^k w(z)(iz+R)^{-1}\, dz,$$

respectively. Using the moment determinants, we can express the (monic) polynomials π_k and π_k^* as

$$\pi_k(z) = \frac{1}{\Delta_k}\begin{vmatrix} \mu_0 & \mu_1 & \cdots & \mu_k \\ \mu_1 & \mu_2 & & \mu_{k+1} \\ \vdots & & & \\ \mu_{k-1} & \mu_k & & \mu_{2k-1} \\ 1 & z & & z^k \end{vmatrix}, \quad \pi_k^*(z) = \frac{1}{\Delta_k^*}\begin{vmatrix} \mu_0^* & \mu_1^* & \cdots & \mu_k^* \\ \mu_1^* & \mu_2^* & & \mu_{k+1}^* \\ \vdots & & & \\ \mu_{k-1}^* & \mu_k^* & & \mu_{2k-1}^* \\ 1 & z & & z^k \end{vmatrix},$$

respectively, where

$$\Delta_k = \begin{vmatrix} \mu_0 & \mu_1 & \cdots & \mu_{k-1} \\ \mu_1 & \mu_2 & & \mu_k \\ \vdots & & & \\ \mu_{k-1} & \mu_k & & \mu_{2k-2} \end{vmatrix}, \qquad \Delta_k^* = \begin{vmatrix} \mu_0^* & \mu_1^* & \cdots & \mu_{k-1}^* \\ \mu_1^* & \mu_2^* & & \mu_k^* \\ \vdots & & & \\ \mu_{k-1}^* & \mu_k^* & & \mu_{2k-2}^* \end{vmatrix}.$$

In [112] we proved that $\pi_k^*(\overline{z}) = \overline{\pi_k(z)}$, as well as the relation

$$\pi_k^*(z) = p_k(z) - i\theta_{k-1}^* p_{k-1}(z), \qquad k = 0, 1, 2, \ldots,$$

where

$$\theta_{k-1}^* = \frac{(\pi_k^*, \pi_k^*)^*}{[p_{k-1}, p_{k-1}]}, \qquad k = 1, 2, \ldots, \qquad \theta_{-1}^* = \mu_0^*.$$

Here, $\theta_{k-1}^* = -\overline{\theta}_{k-1}$, where θ_{k-1} is the corresponding coefficient in the polynomial π_k.

Also, the following theorem holds:

Theorem 5.2.1. *The dual (monic) orthogonal polynomials $\{\pi_k^*\}$ satisfy the three-term recurrence relation*

$$\begin{aligned} \pi_{k+1}^*(z) &= (z - i\alpha_k^*)\pi_k^*(z) - \beta_k^*\pi_{k-1}^*(z), \qquad k = 0, 1, 2, \ldots, \\ \pi_{-1}^*(z) &= 0, \quad \pi_0^*(z) = 1, \end{aligned}$$

with $\alpha_k^* = -\overline{\alpha}_k$ and $\beta_k^* = \overline{\beta}_k$, where α_k and β_k are the coefficients in the corresponding recurrence relation for the polynomials $\{\pi_k\}$.

Using dual polynomials we can give a short proof of the following result stated in [17]:

Theorem 5.2.2. *Let $w(z) = w(-z)$. Then $\theta_{k-1} > 0$, for $k \geq 0$.*

Namely, since $(\pi_k, \pi_k) = \theta_{k-1}[p_{k-1}, p_{k-1}]$ it is enough to prove that $(\pi_k, \pi_k) > 0$. In this *symmetric case*, θ_{k-1} is real and we have $\theta_{k-1}^* = -\theta_{k-1}$ and

$$(\pi_k, \pi_k) = (\pi_k, \pi_k^*) = \int_{\Gamma_R} G(z)w(z)(iz - R)^{-1}\, dz = -\int_{-1}^{1} G(x)\frac{w(x)}{ix - R}\, dx,$$

where $G(z) = p_k(z)^2 + \theta_{k-1}^2 p_{k-1}(z)^2$. Then

$$(\pi_k, \pi_k) = R\int_{-1}^{1} G(x)\frac{w(x)}{R^2 + x^2}\, dx + i\int_{-1}^{1} xG(x)\frac{w(x)}{R^2 + x^2}\, dx.$$

Since $x \mapsto G(x)$ is an even positive function, the second integral on the right-hand side vanishes and $(\pi_k, \pi_k) > 0$.

5.2.2. Differential equation for the Jacobi weight. We consider the Jacobi weight function

$$w(z) = w^{\alpha,\beta}(z) = (1 - z)^\alpha (1 + z)^\beta, \qquad \alpha, \beta > -1,$$

where fractional powers are understood in terms of their principal branches.

The corresponding (monic) polynomials $\{\pi_k^R\}$ orthogonal on the circular arc Γ_R, with respect to the inner product (5.2.1), i.e., (5.2.2), where $w(z) = w^{\alpha,\beta}(z)$, can be expressed in the form

$$\pi_k^R(z) = \pi_k(z) = p_k(z) - i\theta_{k-1}p_{k-1}(z),$$

where $p_k(z) = \hat{P}_k^{\alpha,\beta}(z)$ are the monic Jacobi polynomials and $\theta_{k-1} = \theta_{k-1}^{\alpha,\beta}$ is given by

$$\theta_{k-1} = \frac{1}{i}\frac{\varrho_k(-iR)}{\varrho_{k-1}(-iR)}, \qquad k \geq 1,$$

where

$$\varrho_k(z) = \int_{-1}^{1} \frac{p_k(x)}{z - x}\, w(x)\, dx, \qquad k \geq 0.$$

The monic polynomials $p_k(z)$ satisfy Jacobi's differential equation

$$A(z)u'' + B(z)u' + \lambda_k u = 0 \qquad \text{(cf. (2.2.2))}$$

and differentiation formula

$$A(z)p_k'(z) = [(k + \alpha + \beta + 1)z + v_k]p_k(z) - (2k + \alpha + \beta + 1)p_{k+1}(z),$$

where (see Table 2.2.1)

$$A(z) = 1 - z^2, \quad B(z) = \beta - \alpha - (\alpha + \beta + 2)z, \quad \lambda_k = k(k + \alpha + \beta + 1)$$

and

$$v_k = (\alpha - \beta)\frac{k+\alpha+\beta+1}{2k+\alpha+\beta+2}.$$

Solving the problem: *Find a function* $z \mapsto \Omega(z)$ *in the form*

$$\Omega(z) = (z-1)^{r_k - it_k}(z+1)^{s_k+it_k} \qquad (r_k, s_k, t_k \in \mathbb{R}), \tag{5.2.4}$$

such that $(z^2-1)[\Omega(z)u(z)]' = \Omega(z)v(z)$, *for* $u(z) = p_{n-1}(z)$ *and* $v(z) = \gamma_k \pi_k^R(z)$, *where* γ_k *is a constant*, Milovanović and Rajković [112] proved the following result:

Theorem 5.2.3. *The polynomial* $\pi_k^R(z)$ *satisfies the differential equation*

$$P(z)y'' + Q(z)y' + R(z)y = 0, \tag{5.2.5}$$

with polynomial coefficients

$$\begin{aligned} P(z) &= -A(z)^2\Omega(z)b(z), \\ Q(z) &= A(z)^2\Omega(z)\big(b'(z) - a(z)b(z)\big), \\ R(z) &= A(z)\big[A(z)\Omega(z)\big(a(z)b'(z) - a'(z)b(z)\big) + \Omega(z)^2 b(z)^2\big], \end{aligned} \tag{5.2.6}$$

where $\Omega(z)$ *defined by* (5.2.4) *with parameters*

$$r_k = \frac{1}{2}\left(k+\alpha+\beta+v_{k-1}\right), \; s_k = \frac{1}{2}\left(k+\alpha+\beta-v_{k-1}\right), \; t_k = \frac{1}{2}(2k+\alpha+\beta-1)\theta_{k-1},$$

and $a(z)$ *and* $b(z)$ *are given by*

$$\begin{aligned} a(z) &= \frac{1}{A(z)}(g(z) + B(z) - A'(z)), \\ b(z) &= -\frac{1}{A(z)\Omega(z)}\Big((\lambda_{k-1} + g'(z))A(z) + g(z)(g(z) + B(z) - A'(z))\Big), \end{aligned} \tag{5.2.7}$$

and $g(z) = (k+\alpha+\beta)z + c_{k-1}$, $c_{k-1} = v_{k-1} - i(2k+\alpha+\beta-1)\theta_{k-1}$.

Notice that all coefficients in (5.2.6) have the factor $A(z)\Omega(z)$, but it is because of this factor that the coefficients turn out to be polynomials.

Using (5.2.6) and (5.2.7) we find $P(z) = (1-z^2)C(z)$, where $z \mapsto C(z)$ is the following polynomial of the first degree

$$C(z) = k(k+\alpha+\beta) + (\beta - \alpha + c_{k-1})c_{k-1} - i(2k+\alpha+\beta)(2k+\alpha+\beta-1)\theta_{k-1}z,$$

where

$$c_{k-1} = (\alpha-\beta)\frac{k+\alpha+\beta}{2k+\alpha+\beta} - i(2k+\alpha+\beta-1)\theta_{k-1}.$$

After some calculation, $C(z)$ can be expressed in the form

$$C(z) = \gamma_0 + i\gamma_1\theta_{k-1} + \gamma_2\theta_{k-1}^2 - \eta_k z, \tag{5.2.8}$$

where

$$\gamma_0 = \frac{4k(k+\alpha)(k+\beta)(k+\alpha+\beta)}{(2k+\alpha+\beta)^2}, \quad \gamma_1 = (\beta^2-\alpha^2)\frac{2k+\alpha+\beta-1}{2k+\alpha+\beta},$$

$$\gamma_2 = -(2k+\alpha+\beta-1)^2, \quad \eta_k = (2k+\alpha+\beta)(2k+\alpha+\beta-1)i\theta_{k-1}.$$

We remark that the differential equation (5.2.5) has regular singular points at 1, -1, ∞, and an additional regular singular point ζ_k which depends on k and is given by

$$\zeta_k = \frac{\gamma_1 - i(\gamma_2\theta_{k-1} + \gamma_0/\theta_{k-1})}{(2k+\alpha+\beta)(2k+\alpha+\beta-1)}.$$

The polynomials P, Q, and R in (5.2.6) can be expressed by the coefficients $A(z)$, $B(z)$, and λ_k in differential equation (2.2.2). Namely,

$$\begin{aligned} P(z) &= A(z)C(z), \\ Q(z) &= B(z)C(z) + A(z)\eta_k, \\ R(z) &= \lambda_k C(z) + A(z)a(z)\eta_k, \end{aligned} \tag{5.2.9}$$

where

$$a(z) = \frac{1}{A(z)}\left(kz + \frac{k(\beta-\alpha)}{2k+\alpha+\beta} - i(2k+\alpha+\beta-1)\theta_{k-1}\right)$$

and $C(z)$ is given by (5.2.8).

We can see that Q and R are the complex polynomials of degree two and one, respectively.

When $\alpha = \beta = \lambda - 1/2$ $(\lambda > -1/2)$ we obtain the Gegenbauer case, which is considered for $R = 0$ by Gautschi, Landau and Milovanović in [46].

It is interesting to consider a case when $R \to +\infty$, i.e., when Γ_R reduces to the interval $[-1, 1]$. Since $\lim_{R\to+\infty} \theta_{k-1} = 0$, we have

$$\lim_{R\to+\infty} \eta_k = 0, \quad \lim_{R\to+\infty} C(z) = \gamma_0, \quad \lim_{R\to+\infty} a(z) = \frac{1}{A(z)}\left(kz + \frac{k(\beta-\alpha)}{2k+\alpha+\beta}\right).$$

Thus, the limit case of (5.2.9) gives

$$\lim_{R\to+\infty} P(z) = \gamma_0 A(z), \quad \lim_{R\to+\infty} Q(z) = \gamma_0 B(z), \quad \lim_{R\to+\infty} R(z) = \gamma_0\lambda_k.$$

In that case, dividing (5.2.5) by γ_0 we obtain Jacobi's differential equation, which is, in fact, the result expected.

5.2.3. Functions of the second kind and some associated polynomials. Let the inner product $(\cdot,\cdot)$ be given by (5.2.1), i.e., (5.2.2). In connection with polynomials $\{\pi_k^R\}$ orthogonal with respect to $(\cdot,\cdot)$ on Γ_R, Milovanović and Rajković [112] introduced the functions, so-called *functions of the second kind*,

$$\varrho_k^R(z) = \int_{\Gamma_R} \frac{\pi_k^R(\zeta)}{z-\zeta} \cdot \frac{w(\zeta)}{i\zeta - R}\, d\zeta, \qquad k = 0, 1, 2, \ldots . \tag{5.2.10}$$

It is easily seen that they also satisfy the same recurrence relation as the polynomials π_k^R. Indeed, from the recurrence relation

$$\pi_{k+1}^R(z) = (z - i\alpha_k)\pi_k^R(z) - \beta_k \pi_{k-1}^R(z), \qquad k = 0, 1, 2, \ldots, \tag{5.2.10}$$

for $z = \zeta$, multiplying by $w(\zeta)/((i\zeta - R)(z-\zeta))$ and integrating, we obtain

$$\varrho_{k+1}^R(z) = (z - i\alpha_k)\varrho_k^R(z) - \int_{\Gamma_R} \pi_k^R(\zeta)\frac{w(\zeta)}{i\zeta - R}\, d\zeta - \beta_k \varrho_{k-1}^R(z).$$

By orthogonality, the integral on the right side in the above equality vanishes if $k \geq 1$, and equals μ_0 if $k = 0$. If we define $\varrho_{-1}^R(z) = 1$ (and $\beta_0 = \mu_0$), we have

$$\varrho_{k+1}^R(z) = (z - i\alpha_k)\varrho_k^R(z) - \beta_k \varrho_{k-1}^R(z), \qquad k = 0, 1, 2, \ldots . \tag{5.2.12}$$

For $|z|$ sufficiently large, we can prove (see [112])

$$\varrho_k^R(z) = \frac{\|\pi_k^R\|^2}{z^{k+1}}\Bigl(1 + O\bigl(\frac{1}{z}\bigr)\Bigr),$$

where $\|\pi_k^R\|^2 = (\pi_k^R, \pi_k^R)$. Based on an idea by Stieltjes (see Monegato [119], Gautschi [31]), Milovanović and Rajković [112] considered an expansion of $1/\varrho_k^R(z)$ into descending powers of z, and obtained

$$\frac{1}{\varrho_k^R(z)} = \frac{z^{k+1}}{\|\pi_k^R\|^2}\left(1 + c_1 z^{-1} + c_2 z^{-2} + \cdots\right) = E_{k+1}^R(z) + d_1 z^{-1} + d_2 z^{-2} + \cdots,$$

where

$$E_{k+1}^R(z) = \frac{1}{\|\pi_k^R\|^2}\left(z^{k+1} + c_1 z^k + \cdots + c_{k+1}\right)$$

and $d_k = c_{k+k+1}/\|\pi_k^R\|^2$, $k = 1, 2, \ldots$. We call E_{k+1}^R the *Stieltjes polynomial associated with polynomials* $\{\pi_k^R\}$ orthogonal with respect to $(\cdot,\cdot)$ on Γ_R. By a residue calculation, this polynomial of exact degree $k+1$, can be expressed in the form

$$E_{k+1}^R(z) = \frac{1}{2\pi i}\oint_C \frac{d\zeta}{(\zeta - z)\varrho_k^R(\zeta)},$$

where C is a sufficiently large contour with z in its interior.

Theorem 5.2.4. *Stieltjes' polynomial E_{k+1}^R is orthogonal to all lower-degree polynomials with respect to the complex measure $d\lambda(z) = \pi_k^R(z)w(z)(iz - R)^{-1}dz$, i.e.,*

$$\int_{\Gamma_R} E_{k+1}^R(z)p(z)\pi_k^R(z)\frac{w(z)}{iz - R}\,dz = 0 \qquad (p \in \mathcal{P}_k),$$

where $\mathcal{P}_k$ is the set of all polynomials of degree at most k.

The quantities $\varrho_k^R(z)/\pi_k^R(z)$, $|z| > 1$, are important in getting error bounds for Gaussian quadrature formulas over Γ_R, applied to analytic functions (cf. Gautschi and Varga [52]). Stieltjes' polynomials appear in quadrature formulas of Gauss-Kronrod's type (cf. [37]).

We can also introduce the polynomials

$$\sigma_k^R(z) = \int_{\Gamma_R} \frac{\pi_k^R(z) - \pi_k^R(\zeta)}{z - \zeta} \cdot \frac{w(\zeta)}{i\zeta - R}\,d\zeta, \qquad k = 0, 1, 2, \dots,$$

which are called the *polynomials associated with the orthogonal polynomials* π_k^R. It is easy to see that

$$\varrho_k^R(z) = \pi_k^R(z)\varrho_0^R(z) - \sigma_k^R(z).$$

The polynomials $\{\sigma_k^R\}$ satisfy the same three-term recurrence relation

$$\sigma_{k+1}^R(z) = (z - i\alpha_k)\sigma_k^R(z) - \beta_k\sigma_{k-1}^R(z), \qquad k = 0, 1, 2, \dots, \tag{5.2.13}$$

where $\sigma_0^R(z) = 0$, $\sigma_1^R(z) = \mu_0$. If we define $\sigma_{-1}^R(z) = -1$ and $\beta_0 = \mu_0$, we can note that (5.2.13) also holds for $k = 0$ (see Gautschi [30]).

Using the recurrence relations for $\{\pi_k^R\}$, $\{\varrho_k^R\}$, and $\{\sigma_k^R\}$ ((5.2.11), (5.2.12), and (5.2.13), respectively), where

1° $\pi_{-1}^R(z) = 0, \quad \pi_0^R(z) = 1$;

2° $\varrho_{-1}^R(z) = 1, \quad \varrho_0^R(z) = F(z)$ (defined by (5.2.10));

3° $\sigma_{-1}^R(z) = -1, \quad \sigma_0^R(z) = 0$,

Milovanović and Rajković [112] proved the following identity of Christoffel-Darboux type:

Theorem 5.2.5. *Let $\{f_k\}$ and $\{g_k\}$ satisfy the three-term recurrence relation of the form* (5.2.13), *and $S_k(z, w) = f_{k+1}(z)g_k(w) - g_{k+1}(w)f_k(z)$. Then the identity*

$$(z - w)\sum_{k=0}^{n} \frac{f_k(z)g_k(w)}{\beta_0\beta_1\cdots\beta_k} = \frac{S_n(z, w)}{\beta_0\beta_1\cdots\beta_n} - S_{-1}(z, w)$$

holds, where β_k ($k = 0, 1, 2, \dots$) are the recursion coefficients in (5.2.13). *Under conditions* 1°–3°, *we have the following special cases*

(a) $f_k := \pi_k^R, \quad g_k := \pi_k^R, \quad S_{-1} = 0$;

(b) $f_k := \pi_k^R, \quad g_k := \varrho_k^R, \quad S_{-1} = 1\,;$

(c) $f_k := \pi_k^R, \quad g_k := \sigma_k^R, \quad S_{-1} = -1\,;$

(d) $f_k := \varrho_k^R, \quad g_k := \varrho_k^R, \quad S_{-1} = F(z) - F(w)\,;$

(e) $f_k := \varrho_k^R, \quad g_k := \sigma_k^R, \quad S_{-1} = -F(z)\,;$

(f) $f_k := \sigma_k^R, \quad g_k := \sigma_k^R, \quad S_{-1} = 0\,.$

5.2.5. Geronimus' version of orthogonality on a contour. In the paper [69], J. W. Jayne considered the Geronimus' concept of orthogonality for recursively generated polynomials. Ya. L. Geronimus proved that a sequence of polynomials $\{p_k\}$, which is orthogonal on a finite interval on real line, is also orthogonal in the sense that there is a weight function $z \mapsto \chi(z)$ having one or more singularities inside a simple curve C and such that

$$\langle p_k, p_m\rangle = \frac{1}{2\pi i}\oint_C p_k(z)p_m(z)\chi(z)\,dz = \begin{cases} 0, & k \neq m, \\ h_m, & k = m. \end{cases} \tag{5.2.14}$$

Following Geronimus [57] and Jayne [69], Milovanović and Rajković [111] determined such a complex weight function $z \mapsto \chi(z)$, for (monic) polynomials $\{\pi_k\}$ orthogonal on the semicircle Γ, and also for the corresponding polynomials $\{\pi_k^R\}$ orthogonal on the circular arc Γ_R $(R > 0)$.

Denoting by C any positively oriented simple closed contour surrounding some circle $|z| = r > 1$, we assume that

$$\chi(z) = \sum_{k=1}^{\infty} \omega_k z^{-k}, \quad \omega_1 = 1, \tag{5.2.15}$$

for $|z| > 1$, and express z^n as a linear combination of the monic polynomials π_m, $m = 0, 1, \ldots, n$, which are orthogonal on the semicircle Γ, with respect to the inner product (5.1.1). Thus,

$$z^n = \sum_{m=0}^{n} \gamma_{n,m}\pi_m(z), \tag{5.2.16}$$

where $(z^n, \pi_m) = \gamma_{n,m}(\pi_m, \pi_m)$, $m = 0, 1, \ldots, n$. Using the inner product (5.2.14) and the representation (5.2.15), we obtain

$$\langle z^n, 1\rangle = \frac{1}{2\pi i}\oint_C z^n\chi(z)\,dz = \frac{1}{2\pi i}\oint_C \sum_{k=1}^{\infty}\omega_k z^{n-k}\,dz = \omega_{n+1}.$$

On the other hand, because of (5.2.16) and the orthogonality condition (5.2.14), we find

$$\langle z^n, 1\rangle = \langle \sum_{m=0}^{n} \gamma_{n,m}\pi_m(z), 1\rangle = \sum_{m=0}^{n} \gamma_{n,m}\langle \pi_m, 1\rangle,$$

i.e., $\langle z^n, 1\rangle = \gamma_{n,0}\langle \pi_0, \pi_0\rangle = \gamma_{n,0}h_0$. Thus, we have $w_{n+1} = \gamma_{n,0}h_0 = \gamma_{n,0}$, because $h_0 = \omega_1 = 1$.

Finally, using the moments $\mu_n = (z^n, 1)$, we obtain $\omega_{n+1} = \mu_n/\mu_0$, $n \geq 0$, and

$$\chi(z) = \frac{1}{\mu_0}\sum_{k=1}^{\infty} \mu_{k-1}z^{-k}, \qquad |z| > 1, \tag{5.2.17}$$

where we need the convergence of this series for $|z| > r > 1$.

Suppose that w be a weight function, nonnegative on $(-1, 1)$, holomorphic in $D_+ = \{z \in \mathbb{C} \,|\, |z| < 1, \operatorname{Im} z > 0\}$, integrable over ∂D_+, and such that (5.1.13) is satisfied. Then the moments μ_k can be expressed in the form

$$\mu_0 = \int_\Gamma w(z)(iz)^{-1}\,dz = \frac{1}{i}\left(i\pi w(0) - \text{v.p.}\int_{-1}^{1} \frac{w(x)}{x}\,dx\right)$$

and

$$\mu_k = \int_\Gamma z^k w(z)(iz)^{-1}\,dz = i\int_{-1}^{1} x^{k-1}w(x)\,dx, \quad k \geq 1.$$

These moments are included in the series (5.2.17).

Supposing that the weight function w has such moments μ_k, which provide the convergence of the series (5.2.17), for all z outside some circle $|z| = r > 1$ lying interior to C, Milovanović and Rajković [111] proved:

Theorem 5.2.6. *The monic polynomials* $\{\pi_k\}$, *which are orthogonal on the semicircle* Γ *with respect to the inner product* (5.1.1), *are also orthogonal in the sense of* (5.2.14), *where*

$$\chi(z) = \frac{1}{z}\left(1 + \frac{i}{\mu_0}\int_{-1}^{1} \frac{w(x)}{z - x}\,dx\right), \qquad |z| > r > 1,$$

and

$$\mu_0 = \pi w(0) + i\,\text{v.p.}\int_{-1}^{1} \frac{w(x)}{x}\,dx.$$

In Gegenbauer case they obtained the following result:

Corollary 5.2.7. *Let* $w(z) = (1 - z^2)^{\lambda - 1/2}$, $\lambda > -1/2$. *The monic polynomials* $\{\pi_k\}$, *which are orthogonal on the unit semicircle with respect to the inner product* (5.1.1), *are also orthogonal in the sense of* (5.2.14), *where*

$$\chi(z) = \frac{1}{z} + \frac{i}{\sqrt{\pi}\,z^2} \cdot \frac{\Gamma(\lambda + \frac{1}{2})}{\Gamma(\lambda + 1)}\, F\left(1, \frac{1}{2}, \lambda + 1; \frac{1}{z^2}\right),$$

where F is the Gauss hypergeometric series and Γ is the gamma function.

EXAMPLE 5.2.1. In Legendre case ($\lambda = 1/2$) we have

$$\chi(z) = \frac{1}{z} + \frac{i}{\pi z} \log \frac{z+1}{z-1},$$

where the interval from -1 to 1 on the real axis is considered as a branch cut.

The corresponding complex weight for polynomials $\{\pi_k^R\}$ ($R > 0$) orthogonal on the circular arc Γ_R was also derived in [111] in the form

$$\chi(z) = \frac{1}{\mu_0} \int_{-1}^{1} \frac{(R+ix)w(x)}{(R^2+x^2)(z-x)}\, dx, \qquad |z| > r > 1,$$

where

$$\mu_0 = \int_{-1}^{1} \frac{R+ix}{R^2+x^2} w(x)\, dx.$$

5.3. Orthogonality with the Hermite weight. In this subsection we consider an orthogonality on a growing semicircle $\Gamma_{(\sqrt{\lambda})}$ with radius $\lambda > 1$, and especially a limit case when λ tends to infinity.

Let $w_\lambda(z)$ be the Gegenbauer weight function,

$$w_\lambda(z) = (1-z^2)^{\lambda-1/2} \qquad (\lambda > 0),$$

and D_+ the half disc $D_+ = \{z \in \mathbb{C} \mid |z| \le 1,\ \mathrm{Im}(z) > 0\}$, bounded by the semicircle Γ and the interval $[-1,1]$.

Following (5.1.1) we define the inner product on the semicircle Γ with respect to the weight function $w_\lambda(z)$,

$$(f,g)_\lambda = \int_\Gamma f(z)g(z)w_\lambda(z)(iz)^{-1}\, dz.$$

Then, there exists an unique sequence of monic polynomials $\{\pi_n^\lambda(z)\}$ such that is

$$(\pi_m^\lambda, \pi_n^\lambda) = \delta_{mn} \|\pi_n^\lambda\|_\lambda^2 \qquad (m,n = 0,1,\ldots).$$

A connection with the monic Gegenbauer polynomials $\widehat{C}_n^\lambda(z)$ can be given in the form (see Theorem 5.1.1)

$$\pi_n^\lambda(z) = \widehat{C}_n^\lambda(z) - i\theta_{n-1}^\lambda \widehat{C}_{n-1}^\lambda(z),$$

where

$$\theta_{-1}^\lambda = \pi, \quad \theta_0^\lambda = \frac{\Gamma(\lambda+1/2)}{\sqrt{\pi}\,\Gamma(\lambda+1)}, \quad \theta_k^\lambda = \frac{1}{\lambda+k} \frac{\Gamma((k+2)/2)\Gamma(\lambda+(k+1)/2)}{\Gamma((k+1)/2)\Gamma(\lambda+k/2)}.$$

The norm of such polynomials is given by

$$\|\pi_n^\lambda\|_\lambda^2 = \pi(\theta_0^\lambda \theta_1^\lambda \cdots \theta_{n-1}^\lambda)^2 > 0 \qquad (n \geq 1).$$

Introduce now a new variable u by $u = z\sqrt{\lambda}$. Then the semicircle Γ becomes a new one (see Figure 5.3.1)

$$\Gamma_{(\sqrt{\lambda})} = \{u \in \mathbb{C} \mid |u| \leq \sqrt{\lambda},\ \mathrm{Im}(u) > 0\}.$$

The orthogonality condition becomes

$$\int_{\Gamma_{(\sqrt{\lambda})}} \pi_m^\lambda(u/\sqrt{\lambda})\pi_n^\lambda(u/\sqrt{\lambda})\left(1 - u^2/\lambda\right)^{\lambda-1/2} \frac{du}{iu} = \delta_{mn}\|\pi_n^\lambda\|_\lambda^2.$$

We can define a new sequence of polynomials $\{X_n^\lambda(u)\}$ by

$$X_n^\lambda(u) = \sqrt{\lambda^n}\,\pi_n^\lambda(u/\sqrt{\lambda}),$$

which are orthogonal with respect to the weight

$$\widetilde{w}_\lambda(u) = \left(1 - \frac{u^2}{\lambda}\right)^{\lambda-1/2}$$

on $\Gamma_{(\sqrt{\lambda})}$. Thus,

$$\langle X_m^\lambda, X_n^\lambda\rangle_\lambda = \int_{\Gamma_{(\sqrt{\lambda})}} X_m^\lambda(u)X_n^\lambda(u)\widetilde{w}_\lambda(u)\,\frac{du}{iu} = \delta_{mn}\|X_n^\lambda\|^2. \tag{5.3.1}$$

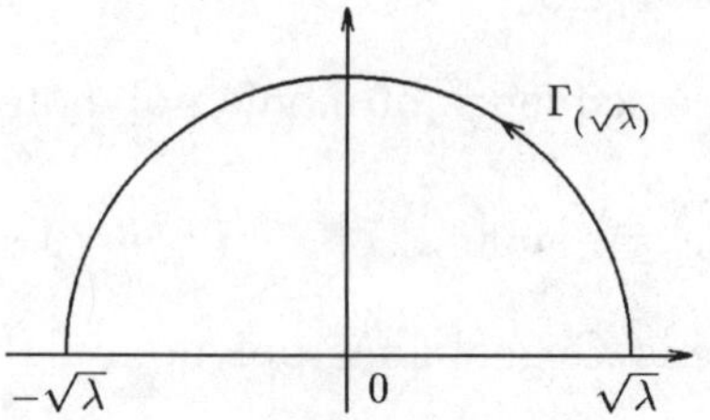

FIGURE 5.3.1: The growing semicircle $\Gamma_{(\sqrt{\lambda})}$

Putting $P_n^\lambda(t) = \sqrt{\lambda^n}\,\widehat{C}_n^\lambda(t/\sqrt{\lambda})$ $(n \in \mathbb{N}_0)$, we see that this sequence satisfies the following three-term recurrence relation

$$P_{k+1}^\lambda(t) = tP_k^\lambda(t) - b_k^\lambda P_{k-1}^\lambda(t), \quad P_{-1}^\lambda(t) = 0, \quad P_0^\lambda(t) = 1,$$

where

$$b_0^\lambda = \int_{-\sqrt{\lambda}}^{+\sqrt{\lambda}} \left(1 - \frac{t^2}{\lambda}\right)^{\lambda-1/2} dt = \sqrt{\lambda\pi}\,\frac{\Gamma(\lambda+1/2)}{\Gamma(\lambda+1)}$$

and

$$b_k^\lambda = \frac{k\lambda(k+2\lambda-1)}{4(k+\lambda)(k+\lambda-1)}, \qquad k \geq 1.$$

These polynomials are orthogonal on $(-\sqrt{\lambda}, \sqrt{\lambda})$ with respect to $\widetilde{w}_\lambda(t)$. Then, we yield

$$X_n^\lambda(u) = P_n^\lambda(u) - i\vartheta_{n-1}^\lambda P_{n-1}^\lambda(u), \tag{5.3.2}$$

where

$$\vartheta_{-1}^\lambda = \theta_{-1}^\lambda, \qquad \vartheta_{n-1}^\lambda = \theta_{n-1}^\lambda\sqrt{\lambda} \qquad (n \geq 1).$$

We can prove these polynomials satisfy a three-term recurrence relation and a second order differential equation. Their zeros lie in the region bounded by $\Gamma_{(\sqrt{\lambda})}$ and $[-\sqrt{\lambda}, \sqrt{\lambda}]$.

Using Cauchy's theorem the inner product $\langle \cdot, \cdot \rangle_\lambda$ can be expressed in the form

$$\langle f, g \rangle_\lambda = \pi f(0)g(0) + i \text{ v.p.} \int_{-\sqrt{\lambda}}^{+\sqrt{\lambda}} f(x)g(x)\frac{w_\lambda(x)}{x}\,dx. \tag{5.3.3}$$

On the other hand, it is known (see Szegő [138, p. 107]) that is

$$\lim_{\lambda\to+\infty} P_k^\lambda(t) = \widehat{H}_k(t). \tag{5.3.4}$$

where $\widehat{H}_k(t)$, $k = 0, 1, \ldots$, are the monic Hermite polynomials which satisfy

$$\widehat{H}_{k+1}(t) = t\widehat{H}_k(t) - b_k\widehat{H}_{k-1}(t), \quad \widehat{H}_{-1}(t) = 0, \quad \widehat{H}_0(t) = 1,$$

with $b_0 = \sqrt{\pi}$ and $b_k = k/2$ $(k \geq 1)$. Defining $\vartheta_k = \lim_{\lambda\to+\infty} \theta_k^{(\lambda)}$, we find

$$\vartheta_{-1} = \pi, \qquad \vartheta_n = \frac{\Gamma((n+2)/2)}{\Gamma((n+1)/2)} \qquad (n = 0, 1, \ldots).$$

Knowing that

$$\lim_{x\to+\infty} \frac{\Gamma(x+a)}{x^a\Gamma(x)} = 1 \qquad (a > 0),$$

we conclude that $\lim_{n\to+\infty} \vartheta_n = \lim_{n\to+\infty} \sqrt{(n+1)/2} = +\infty$.

Now, we can define a sequence of complex polynomials $\{X_n(z)\}$ by

$$X_n(z) = \lim_{\lambda\to+\infty} X_n^\lambda(z).$$

Using (5.3.4) and (5.3.2) we conclude that these polynomials satisfy the following recurrence relation

$$X_n(z) = \widehat{H}_n(z) - i\vartheta_{n-1}\widehat{H}_{n-1}(z), \quad X_{-1}(z) = 0, \quad X_0(z) = 1.$$

From (5.3.1) and (5.3.3) we obtain the inner product

$$\langle f, g\rangle = \pi f(0)g(0) + i \text{ v.p.} \int_{-\infty}^{+\infty} f(x)g(x)\frac{e^{-x^2}}{x}\, dx. \tag{5.3.5}$$

Theorem 5.3.1. *The sequence of polynomials* $\{X_n(z)\}$ *is orthogonal with respect the inner product* (5.3.5), *i.e.,*

$$\langle X_m, X_n\rangle = \delta_{mn}\|X_n\|^2 \qquad (m, n = 0, 1, \ldots).$$

Theorem 5.3.2. *The sequence* $\{X_n(z)\}$ *satisfies the three-term recurrence relation*

$$X_{n+1}(z) = (z - i\alpha_n)X_n(z) - \beta_n X_{n-1}(z), \qquad X_{-1}(z) = 0, \quad X_0(z) = 1,$$

where

$$\alpha_0 = \vartheta_0, \quad \beta_0 = \vartheta_{-1}, \quad \alpha_n = \vartheta_n - \vartheta_{n-1}, \quad \beta_n = \vartheta_{n-1}^2 \quad (n \geq 1).$$

The norm of polynomials is given by

$$\|X_n\|^2 = \langle X_n, X_n\rangle = \beta_0\beta_1\cdots\beta_n = \Gamma\Big(\frac{n+1}{2}\Big)^2 \qquad (n \geq 0).$$

EXAMPLE 5.3.1. A few values of ϑ_n and $X_n(z)$ are given by

$$\begin{aligned}
\vartheta_0 &= \frac{1}{\sqrt{\pi}}, & X_0(z) &= 1,\\
\vartheta_1 &= \frac{\sqrt{\pi}}{2}, & X_1(z) &= z - i\frac{1}{\sqrt{\pi}},\\
\vartheta_2 &= \frac{2}{\sqrt{\pi}}, & X_2(z) &= z^2 - i\frac{\sqrt{\pi}}{2}z - \frac{1}{2},\\
\vartheta_3 &= \frac{3\sqrt{\pi}}{4}, & X_3(z) &= z^3 - i\frac{2}{\sqrt{\pi}}z^2 - \frac{3}{2}z + i\frac{1}{\sqrt{\pi}},\\
\vartheta_4 &= \frac{8}{3\sqrt{\pi}}, & X_4(z) &= z^4 - i\frac{3\sqrt{\pi}}{4}z^3 - 3z^2 + i\frac{9\sqrt{\pi}}{8}z + \frac{3}{4}.
\end{aligned}$$

Notice that $\vartheta_{-1} = \pi$.

REMARK 5.3.1. Introducing

$$\langle f, g\rangle_\lambda = \pi f(0)g(0) + i \text{ v.p.} \int_{-\sqrt{\lambda}}^{+\sqrt{\lambda}} f(x)g(x)|x|^s\Big(1 - \frac{x^2}{\lambda}\Big)^{\lambda-1/2}\, dx,$$

when $\lambda \to +\infty$, we obtain the inner product whose corresponding orthogonal polynomials are known as *the generalized Hermite polynomials*.

Since zeros of the monic Hermite polynomials $\widehat{H}_n(z)$ $(n = 1, 2, \ldots)$ are real, simple and satisfy the separation theorem, for the polynomials $X_n(z)$ we have:

Theorem 5.3.3. *All zeros of* $X_n(z)$ *are contained in the rectangle*

$$R_+ = \{z \in \mathbb{C} \mid -\xi_n \le \mathrm{Re}(z) \le \xi_n,\ 0 < \mathrm{Im}(z) < \vartheta_{n-1}/2\},$$

where ξ_n *is the largest zero of* $\widehat{H}_n(z)$.

Similarly as in 5.2.2 we can prove the following result:

Theorem 5.3.4. *The polynomial* $X_n(z)$ *satisfies the differential equation*

$$P(z)X_n''(z) - 2[zP(z) + i\vartheta_{n-1}]X_n'(z) + 2[nP(z) - 2\vartheta_{n-1}^2]X_n(z) = 0, \tag{5.3.6}$$

where $P(z) = 2i\vartheta_{n-1}z + 2\vartheta_{n-1}^2 - n$.

Dividing (5.3.6) by $P(z)$ we obtain the equation

$$X_n''(z) - 2\left[z + \frac{i\vartheta_{n-1}}{P(z)}\right]X_n'(z) + 2\left[n - \frac{2\vartheta_{n-1}^2}{P(z)}\right]X_n(z) = 0,$$

which is more similar to the HERMITE equation.

For $\vartheta_{n-1} = 0$, the polynomial $X_n(z)$ becomes the polynomial $\widehat{H}_n(z)$ and the differential equation is the corresponding one.

6. Some Applications of Orthogonality on the Semicircle

Several interesting properties of polynomials orthogonal on the semicircle and applications in numerical analysis, especially for Gegenbauer weight, were given in [50] and [97]. Also, differentiation formulas for higher derivatives of analytic functions, using quadratures on the semicircle, were considered in [19].

6.1. Gaussian quadratures. In this section we consider Gauss-Christoffel quadrature formula over the semicircle $\Gamma = \{z \in \mathbb{C} \mid z = e^{i\theta}, 0 \le \theta \le \pi\}$,

$$\int_0^\pi f(e^{i\theta})w(e^{i\theta})\,d\theta = \sum_{\nu=1}^n \sigma_\nu f(\zeta_\nu) + R_n(f), \tag{6.1.1}$$

with Gegenbauer weight $w(z) = (1 - z^2)^{\lambda - 1/2}$, $\lambda > -1/2$, which is exact for all algebraic polynomials of degree at most $2n - 1$.

In this case, (5.1.13) reduces to $\mathrm{Re}(1,1) = \pi \ne 0$, so that the corresponding orthogonal polynomials exist and they can be expressed in terms of monic Gegenbauer polynomials $\hat{C}_k^\lambda(z)$,

$$\pi_k(z) = \hat{C}_k^\lambda(z) - i\theta_{k-1}\hat{C}_{k-1}^\lambda(z),$$

where the sequence $\{\theta_k\}$ is given by

$$\theta_k = \frac{1}{\lambda + k} \cdot \frac{\Gamma((k+2)/2)\Gamma(\lambda + (k+1)/2)}{\Gamma((k+1)/2)\Gamma(\lambda + k/2)}, \qquad k \ge 0.$$

It was shown [46, Sect. 6.3] that all zeros of $\pi_n(z)$, $n \geq 2$, are simple and contained in the upper unit half disc $D_+ = \{z \in \mathbb{C} \,|\, |z| < 1, \operatorname{Im} z > 0\}$. The nodes $\zeta_\nu = \zeta_\nu^{(n)}$ in (6.1.1) are precisely the zeros of the polynomial π_n, i.e., the eigenvalues of the Jacobi matrix J_n given by (5.1.15). The weights $\sigma_\nu = \sigma_\nu^{(n)}$ can be obtained by an adaptation of the procedure of Golub and Welsch [59] (see [31] and [33]).

Following [52], we gave error bounds for the Gaussian quadratures (6.1.1), applied to analytic functions, using a contour integral representation of the remainder term (see [98]).

The Gaussian quadrature (6.1.1) can be applied to calculation of the Cauchy principal value integral

$$I_\lambda(\xi; f) = \text{v.p.} \int_{-1}^{1} \frac{w(t) f(t)}{t - \xi}\, dt,$$

where $-1 < \xi < 1$ and $w(t) = (1 - t^2)^{\lambda - 1/2}$, $\lambda > -1/2$. Firstly, using the linear fractional transformation $t = (x + \xi)/(x\xi + 1)$ we find

$$I_\lambda(\xi; f) = w(\xi)\, \text{v.p.} \int_{-1}^{1} w(x) \frac{g(\xi; x)}{x}\, dx,$$

where $g(\xi; x) = f\Big(\dfrac{x + \xi}{x\xi + 1}\Big)/(x\xi + 1)^{2\lambda}$.

Let f be a meromorphic function with poles p_ν, $\nu = 1, \ldots, m$, in D_+ and $\psi(z) = w(z) g(\xi; z)/z$. In [97] we proved that

$$I_\lambda(\xi; f) \approx w(\xi)\, \operatorname{Im}\Big\{ \sum_{\nu=1}^{n} \sigma_\nu g(\xi; \zeta_\nu) - 2\pi \sum_{\nu=1}^{m} \operatorname*{Res}_{z = p_\nu} \psi(z) \Big\}.$$

6.2. Numerical differentiation. Let f be an analytic function on some domain containing the point a and a circular neighborhood of a with radius r. Using the central difference operator δ_h defined by

$$\delta_h f(a) = \frac{1}{h}\Big(f\Big(a + \frac{h}{2}\Big) - f\Big(a - \frac{h}{2}\Big)\Big),$$

we can find $\delta_h^m f(a) = \delta_h\left(\delta_h^{m-1} f(a)\right)$, i.e.,

$$\delta_h^m f(a) = \frac{1}{h^m} \sum_{k=0}^{m} (-1)^k \binom{m}{k} f\Big(a + \frac{m - 2k}{2} h\Big). \tag{6.2.1}$$

Putting $he^{i\theta}$ instead of h, where h is such that $|a + \frac{mh}{2} e^{i\theta}| < r$, and integrating (6.2.1) over the semicircle, we obtain

$$\int_0^\pi \delta_{he^{i\theta}}^m f(a) w(e^{i\theta})\, d\theta = \pi f^{(m)}(a). \tag{6.2.2}$$

Applying the Gauss-Christoffel quadrature formula on the semicircle (6.1.1) to the integral on the left side in (6.2.2), we obtain the following differentiation formula to higher derivatives

$$f^{(m)}(a) \approx D_{n,h}^{m} f(a) = \frac{1}{\pi} \sum_{\nu=1}^{n} \sigma_\nu \delta_{h\zeta_\nu}^{m} f(a)\,,$$

i.e.,

$$D_{n,h}^{m} f(a) = \frac{1}{\pi h^m} \sum_{\nu=1}^{n} \frac{\sigma_\nu}{\zeta_\nu^m} \sum_{k=0}^{m} (-1)^k \binom{m}{k} f\Big(a + \frac{m-2k}{2} h\zeta_\nu\Big). \tag{6.2.3}$$

Regarding to the truncation error we can give the following result (see [19]):

Theorem 6.2.1. *The error of the differentiation formula* (6.2.3) *for analytical functions is given by*

$$R_{n,h}^{m} f(a) = f^{(m)}(a) - D_{n,h}^{m} f(a) = \frac{1}{\pi} \sum_{p=n}^{\infty} \frac{f^{(m+2p)}(a)}{(m+2p)!} S_{m+2p}^{(m)} R_n(z^{2p}) h^{2p},$$

where

$$S_j^{(m)} = \sum_{k=0}^{m} (-1)^k \binom{m}{k} \Big(\frac{m-2k}{2}\Big)^j$$

and $R_n(z^{2p})$ *is defined by* (6.1.1). *The dominant error term is*

$$\frac{S_{m+2n}^{(m)}}{\pi(m+2n)!} \left(\frac{\Gamma\left((n+1)/2\right)\Gamma\left(\lambda+n/2\right)}{\Gamma(\lambda+n)}\right)^2 f^{(m+2n)}(a) h^{2n}.$$

Some considerations for $n = 2$ and $m = 1$ regarding to λ are given in [97].

For real-valued analytic functions the formula (6.2.3) can be simplified. Namely, when n is even and $\operatorname{Re}\zeta_\nu > 0$, for $\nu = 1, 2, \ldots, n/2$, one finds

$$D_{n,h}^{m} f(a) = \frac{2}{\pi h^m} \sum_{\nu=1}^{n/2} \operatorname{Re}\left\{ \frac{\sigma_\nu}{\zeta_\nu^m} \sum_{k=0}^{m} (-1)^k \binom{m}{k} f\Big(a + \frac{m-2k}{2} h\zeta_\nu\Big) \right\}. \tag{6.2.4}$$

In the simplest case when $n = 2$, we have

$$\zeta_{1,2} = \frac{1}{4}\left(\pm\sqrt{3} + i\right), \qquad \sigma_{1,2} = \frac{\pi}{2}\left(1 \pm i\frac{\sqrt{3}}{3}\right).$$

Then, the corresponding differentiation formula (6.2.4) reduces to

$$D_{2,h}^{m} f(a) = \frac{2}{\pi h^m} \operatorname{Re}\left\{ \frac{\sigma_1}{\zeta_1^m} \sum_{k=0}^{m} (-1)^k \binom{m}{k} f\Big(a + \frac{m-2k}{2} h\zeta_1\Big) \right\}.$$

Its error is $O(h^4)$.

The formula (6.2.3) for real-valued analytic functions can be improved with a little change. Namely, if we put $he^{i\alpha}$ instead of h in (6.2.2), where α is an arbitrary real parameter, and applying again Gauss-Christoffel formula (6.1.1), we obtain the following differentiation formula

$$f^{(m)}(a) \approx D^m_{n,h,\alpha} f(a) = \frac{1}{\pi} \sum_{\nu=1}^{n} \sigma_\nu \delta^m_{he^{i\alpha}\zeta_\nu} f(a).$$

Similar to the above investigation we find an expression for the error, depending on the real parameter α

$$R^m_{n,h,\alpha} f(a) = f^{(m)}(a) - D^m_{n,h,\alpha} f(a) = \frac{1}{\pi} \sum_{p=n}^{\infty} \frac{f^{(m+2p)}(a)}{(m+2p)!} S^{(m)}_{m+2p} R_n(z^{2p}) e^{i2p\alpha} h^{2p}.$$

Since the derivative $f^{(m)}(a)$ is real for real a and real-valued functions the parameter α can be chosen such that the dominant error term in the last expressions be purely imaginary. Then, for such functions, the dominant error term in $R^m_{n,h,\alpha} f(a)$, i.e.,

$$\frac{1}{\pi(m+2n)!} f^{(m+2n)}(a) S^{(m)}_{m+2n} R_n(z^{2n}) e^{i2n\alpha} h^{2n},$$

becomes purely imaginary. This can be achieved for $\alpha = \pi/4n$. In that case, the dominant error term for real-valued functions becomes the real part of the term in $R^m_{n,h,\alpha} f(a)$ for $p = n+1$. So we have the following result:

Theorem 6.2.2. *The dominant error term of the differentiation formula*

$$f^{(m)}(a) \approx \mathrm{Re}\left\{ D^m_{n,h,\pi/4n} f(a) \right\}, \qquad a \in \mathbb{R},$$

for real-valued analytic functions is given by

$$-\frac{\sin(\pi/2n)}{\pi(m+2n+2)!} S^{(m)}_{m+2n+2} R_n(z^{2n+2}) f^{(m+2n+2)}(a) h^{2n+2}, \tag{6.2.5}$$

where $R_n(g)$ is defined in (6.1.1).

Remark 6.2.1. With $\alpha = 3\pi/4n$ we also obtain a rule of degree precision $2n+2$. Then, in the dominant error term (6.2.5), the factor $-\sin(\pi/2n)$ should be replaced by $\sin(3\pi/2n)$.

Several numerical experiments were done in [19], [50], and [96–97].

7. Polynomials Orthogonal on the Radial Rays in the Complex Plane

Recently, we have introduced [104–105] (see also [102] and [115]) a class of polynomials orthogonal on some radial rays in the complex plane and investigated their existence, uniqueness and many other properties. An analogue of the Jacobi polynomials and the corresponding problem with the generalized Laguerre polynomials, have been also treated. In this section we give a short account of these polynomials.

7.1. Definition, existence and uniqueness of polynomials. We start with a nonstandard type of orthogonality on some radial rays in the complex plane. Suppose that we have M points in the complex plane, $z_s = a_s e^{i\varphi_s} \in \mathbb{C}$, $s = 0, 1, \ldots, M-1$, with different arguments φ_s. Some of a_s (or all) can be ∞. The case $M = 5$ is shown in Figure 7.1.1. We can define an inner product on these radial rays ℓ_s in the complex plane which connect the origin $z = 0$ and the points z_s, $s = 0, 1, \ldots, M-1$. Namely,

$$(f, g) = \sum_{s=0}^{M-1} e^{-i\varphi_s} \int_{\ell_s} f(z)\overline{g(z)}\,|w(z)|\,dz,$$

where $z \mapsto w(z)$ is a suitable function (complex weight).

Since, this product can be expressed in the form

$$(f, g) = \sum_{s=0}^{M-1} \int_0^{a_s} f\big(xe^{i\varphi_s}\big)\overline{g\big(xe^{i\varphi_s}\big)}\,\big|w\big(xe^{i\varphi_s}\big)\big|\,dx,$$

we see that

$$(f, f) = \sum_{s=0}^{M-1} \int_0^{a_s} \big|f\big(xe^{i\varphi_s}\big)\big|^2 \big|w\big(xe^{i\varphi_s}\big)\big|\,dx > 0,$$

except when $f(z) = 0$.

We will consider here only the case when M is an even number and $\varphi_s = \pi s/m$, $s = 0, 1, \ldots, 2m-1$. Thus, let $m \in \mathbb{N}$ and $\varepsilon_0, \varepsilon_1, \ldots, \varepsilon_{2m-1}$ be $(2m)$th roots of unity, i.e., $\varepsilon_s = \exp(i\pi s/m)$, $s = 0, 1, \ldots, 2m-1$. We will study orthogonal polynomials relative to the inner product

$$(f, g) = \sum_{s=0}^{2m-1} \varepsilon_s^{-1} \int_{\ell_s} f(z)\overline{g(z)}|w(z)|\,dz. \tag{7.1.1}$$

Suppose that $a_s = 1$ for each s and let $z \mapsto w(z)$ be a holomorphic function such that

$$|w(x\varepsilon_s)| = w(x), \quad s = 0, 1, \ldots, 2m-1,$$

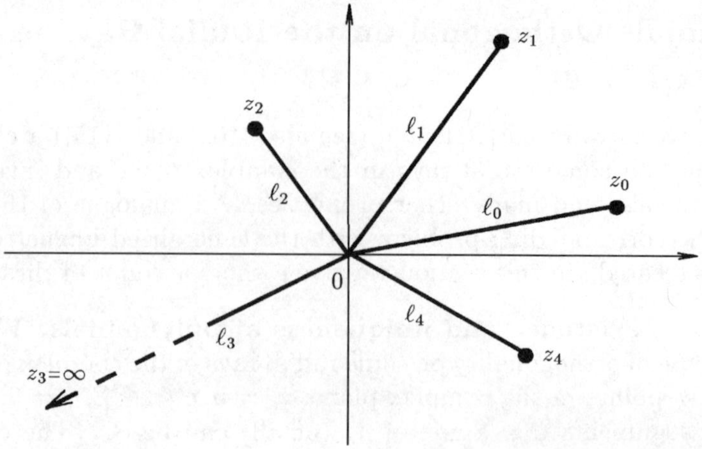

FIGURE 7.1.1: Radial rays in the complex plane ($M = 5$)

and $x \mapsto w(x)$ be a weight function on $(0, 1)$ (nonnegative on the interval $(0, 1)$ and $\int_0^1 w(x)\,dx > 0$). Then, (7.1.1) can be written in the form

$$(f, g) = \int_0^1 \left(\sum_{s=0}^{2m-1} f(x\varepsilon_s)\overline{g(x\varepsilon_s)} \right) w(x)\,dx. \tag{7.1.2}$$

In the case $m = 1$, (7.1.2) becomes

$$(f, g) = \int_{-1}^{1} f(x)\overline{g(x)}w(x)\,dx,$$

so we have the standard case of polynomials orthogonal on $(-1, 1)$ with respect to the weight function $x \mapsto w(x)$.

It is easy to prove that the inner product (7.1.2) has the following property:

Lemma 7.1.1. $(z^m f, g) = (f, z^m g)$.

The moments are given by

$$\mu_{p,q} = (z^p, z^q) = \left(\sum_{s=0}^{2m-1} \varepsilon_s^{p-q} \right) \int_0^1 x^{p+q} w(x)\,dx, \quad p, q \geq 0. \tag{7.1.3}$$

If $p = 2mn + \nu$, $n = [p/(2m)]$, and $0 \leq \nu \leq 2m - 1$, it is easy to verify that

$$\sum_{s=0}^{2m-1} \varepsilon_s^p = \sum_{s=0}^{2m-1} \varepsilon_s^\nu = \begin{cases} 2m & \text{if } \nu = 0, \\ 0 & \text{if } 1 \leq \nu \leq 2m - 1. \end{cases}$$

Thus, $\mu_{p,q}$ in (7.1.3) is different from zero only if $p \equiv q \pmod{2m}$; otherwise $\mu_{p,q} = 0$. Using the moment determinants

$$\Delta_0 = 1, \quad \Delta_N = \begin{vmatrix} \mu_{00} & \mu_{10} & \cdots & \mu_{N-1,0} \\ \mu_{01} & \mu_{11} & \cdots & \mu_{N-1,1} \\ \vdots & & & \\ \mu_{0,N-1} & \mu_{1,N-1} & \cdots & \mu_{N-1,N-1} \end{vmatrix}, \quad N \geq 1,$$

we can prove the following existence result for the (monic) orthogonal polynomials $\{\pi_N(z)\}_{N=0}^{+\infty}$ with respect to the inner product (7.1.2) (see Milovanović [105]):

Theorem 7.1.2. *If $\Delta_N > 0$ for all $N \geq 1$ the monic polynomials $\{\pi_N(z)\}_{N=0}^{+\infty}$, orthogonal with respect to the inner product (7.1.2), exist uniquely.*

7.2. Recurrence relation. It is well known that an orthogonal sequence of polynomials satisfies a three-term recurrence relation if the inner product has the property $(zf, g) = (f, zg)$. In our case the corresponding property is given by $(z^m f, g) = (f, z^m g)$ (see Lemma 7.1.1) and the following result holds:

Theorem 7.2.1. *Let the inner product $(\cdot,\cdot)$ be given by (7.1.2) and let the corresponding system of monic orthogonal polynomials $\{\pi_N(z)\}_{N=0}^{+\infty}$ exists. These polynomials satisfy the recurrence relation*

$$\begin{aligned} \pi_{N+m}(z) &= z^m \pi_N(z) - b_N \pi_{N-m}(z), \quad N \geq m, \\ \pi_N(z) &= z^N, \quad N = 0, 1, \ldots, 2m-1, \end{aligned} \tag{7.2.1}$$

where

$$b_N = \frac{(\pi_N, z^m \pi_{N-m})}{(\pi_{N-m}, \pi_{N-m})} = \frac{\|\pi_N\|^2}{\|\pi_{N-m}\|^2}. \tag{7.2.2}$$

In a simple case when $m = 2$ and $w(x) = 1$, i.e., when the inner product $(\cdot,\cdot)$ is given by

$$(f, g) = \int_0^1 \left[f(x)\overline{g(x)} + f(ix)\overline{g(ix)} + f(-x)\overline{g(-x)} + f(-ix)\overline{g(-ix)} \right] dx, \tag{7.2.3}$$

we can calculate directly the coefficient b_N in the recurrence relation (7.2.1). The moments are given by

$$\mu_{p,q} = (z^p, z^q) = \begin{cases} \dfrac{4}{p+q+1}, & p \equiv q \pmod 4, \\ 0, & \text{otherwise.} \end{cases}$$

Thus, if $p = 4i + \nu$ and $q = 4j + \nu$, $\nu \in \{0, 1, 2, 3\}$, we have

$$\mu_{4i+\nu,4j+\nu} = \frac{4}{4(i+j) + 2\nu + 1}, \qquad i, j \geq 0.$$

In [105] we expressed the moment determinants

$$\Delta_N = \begin{vmatrix} \mu_{00} & \mu_{10} & \cdots & \mu_{N-1,0} \\ \mu_{01} & \mu_{11} & \cdots & \mu_{N-1,1} \\ \vdots & & & \\ \mu_{0,N-1} & \mu_{1,N-1} & \cdots & \mu_{N-1,N-1} \end{vmatrix}, \quad N \geq 1,$$

in terms of the determinants $E_0^{(\nu)} = 1$ and

$$E_n^{(\nu)} = \begin{vmatrix} \mu_{\nu,\nu} & \mu_{4+\nu,\nu} & \cdots & \mu_{4(n-1)+\nu,\nu} \\ \mu_{\nu,4+\nu} & \mu_{4+\nu,4+\nu} & \cdots & \mu_{4(n-1)+\nu,4+\nu} \\ \vdots & & & \\ \mu_{\nu,4(n-1)+\nu} & \mu_{4+\nu,4(n-1)+\nu} & \cdots & \mu_{4(n-1)+\nu,4(n-1)+\nu} \end{vmatrix},$$

where $\nu = 0, 1, 2, 3$.

Lemma 7.2.2. *We have*

$$\begin{aligned} \Delta_{4n} &= E_n^{(0)} E_n^{(1)} E_n^{(2)} E_n^{(3)}, \\ \Delta_{4n+1} &= E_{n+1}^{(0)} E_n^{(1)} E_n^{(2)} E_n^{(3)}, \\ \Delta_{4n+2} &= E_{n+1}^{(0)} E_{n+1}^{(1)} E_n^{(2)} E_n^{(3)}, \\ \Delta_{4n+3} &= E_{n+1}^{(0)} E_{n+1}^{(1)} E_{n+1}^{(2)} E_n^{(3)}. \end{aligned}$$

We note, first of all, that $\Delta_N > 0$ for all $N \geq 1$, and therefore, the orthogonal polynomials $\{\pi_N(z)\}_{N=0}^{+\infty}$ with respect to the inner product (7.2.3) exist uniquely, and

$$(\pi_N, \pi_N) = \|\pi_N\|^2 = \frac{\Delta_{N+1}}{\Delta_N} > 0.$$

Using (7.2.1) and (7.2.2) we obtain:

Theorem 7.2.3. *The (monic) polynomials* $\{\pi_N(z)\}_{N=0}^{+\infty}$*, orthogonal with respect to the inner product* (7.2.3)*, satisfy the recurrence relation*

$$\begin{aligned} \pi_{N+2}(z) &= z^2 \pi_N(z) - b_N \pi_{N-2}(z), \quad N \geq 2, \\ \pi_N(z) &= z^N, \quad N = 0, 1, 2, 3, \end{aligned} \tag{7.2.4}$$

where

$$b_{4n+\nu} = \begin{cases} \dfrac{16n^2}{(8n+2\nu-3)(8n+2\nu+1)} & \text{if } \nu = 0, 1, \\[2ex] \dfrac{(4n+2\nu-3)^2}{(8n+2\nu-3)(8n+2\nu+1)} & \text{if } \nu = 2, 3. \end{cases} \tag{7.2.5}$$

From (7.2.5) we conclude that $b_N \to 1/4$ as $N \to +\infty$, just like in Szegő's theory for orthogonal polynomials on the interval $(-1, 1)$.

Taking $\pi_{-2}(z) = \pi_{-1}(z) = 0$, $\pi_0(z) = 1$, and $\pi_1(z) = z$, it is easy to see that the recurrence relation (7.2.4) holds for every $N \geq 0$.

The norms of the polynomials $\{\pi_N(z)\}$ can be expressed in the form (see [105])

$$\|\pi_N\|^2 = \frac{4}{2N+1}, \quad 0 \leq N \leq 3,$$

$$\|\pi_N\|^2 = \|\pi_{4n+\nu}\|^2 = \frac{4}{8n+2\nu+1}\left(\prod_{k=n}^{2n-1} \frac{4(k-n+1)}{4k+2\nu+1}\right)^2, \quad N \geq 4.$$

7.3. A representation of $\pi_N(z)$ and zeros. Now, we again consider the general case of the inner product (7.1.2) for which the corresponding system of the monic orthogonal polynomials $\{\pi_N(z)\}_{N=0}^{+\infty}$ exists and satisfies the recurrence relation (7.2.1). Based on this recurrence relation, we can conclude and easily prove that $\pi_N(z)$ are incomplete polynomials with the following representation (see Milovanović [105]):

Theorem 7.3.1. *The polynomials from Theorem* 7.1.2 *can be expressed in the form*

$$\pi_{2mn+\nu}(z) = z^\nu q_n^{(\nu)}(z^{2m}), \quad \nu = 0, 1, \ldots, 2m-1;\ n = 0, 1, \ldots, \tag{7.3.1}$$

where $q_n^{(\nu)}(t)$, $\nu = 0, 1, \ldots, 2m-1$, are monic polynomials of exact degree n, which satisfy the three-term recurrence relation

$$\begin{aligned} &q_{n+1}^{(\nu)}(t) = (t - a_n^{(\nu)})q_n^{(\nu)}(t) - b_n^{(\nu)} q_{n-1}^{(\nu)}(t), \quad n = 0, 1, \ldots, \\ &q_0^{(\nu)}(t) = 1, \quad q_{-1}^{(\nu)}(t) = 0. \end{aligned} \tag{7.3.2}$$

The recursion coefficients $a_n^{(\nu)}$ and $b_n^{(\nu)}$ are given in terms of the b-coefficients as

$$a_n^{(\nu)} = b_N + b_{N+m}, \quad b_n^{(\nu)} = b_{N-m} b_N, \quad N = 2mn + \nu.$$

The three-term recurrence relation (7.3.2) shows that the monic polynomial systems $\{q_n^{(\nu)}(t)\}_{n=0}^{+\infty}$, $\nu = 0, 2, \ldots, 2m-1$, are orthogonal. The following theorem gives this orthogonality:

Theorem 7.3.2. *Let $x \mapsto w(x)$ be a weight function in the inner product* (7.1.2) *which guarantees the existence of the polynomials $\pi_N(z)$, i.e., $q_n^{(\nu)}(t)$, $\nu = 0, 1, \ldots$, $2m-1$, determined by* (7.3.1). *For any $\nu \in \{0, 1, \ldots, 2m-1\}$, the sequence of polynomials $\{q_n^{(\nu)}(t)\}_{n=0}^{+\infty}$ is orthogonal on $(0, 1)$ with respect to the weight function $t \mapsto w_\nu(t) = t^{(2\nu+1-2m)/2m} w(t^{1/2m})$.*

As we can see the question of the existence of the polynomials $\pi_N(z)$ is reduced to the existence of polynomials $q_n^{(\nu)}(t)$, orthogonal on $(0, 1)$ with respect to the weight function $w_\nu(t)$, for every $\nu = 0, 1, \ldots, 2m-1$.

The next result gives the zero distribution of the polynomials $\pi_N(z)$ (see [105]):

Theorem 7.3.3. *Let $N = 2mn + \nu$, $n = [N/2m]$, $\nu \in \{0, 1, \dots, 2m-1\}$. All zeros of the polynomial $\pi_N(z)$ are simple and located symmetrically on the radial rays l_s, $s = 0, 1, \dots, 2m-1$, with the possible exception of a multiple zero of order ν at the origin $z = 0$.*

7.4. Some analogues of Jacobi and Laguerre polynomials. Consider now the polynomials $\{\pi_N(z)\}_{N=0}^{+\infty}$ orthogonal with respect to the inner product (7.1.2), where the weight function is given by

$$w(x) = (1 - x^{2m})^{\alpha} x^{2m\gamma}, \qquad \alpha > -1,\ \gamma > -1/(2m). \tag{7.4.1}$$

Theorem 7.4.1. *The monic polynomials $\{\pi_N(z)\}_{N=0}^{+\infty}$ orthogonal with respect to the inner product (7.1.2), where the weight function is given by (7.4.1), can be expressed in the form*

$$\pi_N(z) = 2^{-n} z^{\nu} \hat{P}_n^{(\alpha,\beta_\nu)}(2z^{2m} - 1), \quad N = 2mn + \nu,\ n = [N/2m],$$

where $\nu \in \{0, 1, \dots, 2m-1\}$, $\beta_\nu = \gamma + (2\nu + 1 - 2m)/(2m)$, and $\hat{P}_n^{(\alpha,\beta)}(x)$ denotes the monic Jacobi polynomial orthogonal with respect to the weight $x \mapsto (1-x)^{\alpha}(1+x)^{\beta}$ on $(-1, 1)$. The polynomials $\pi_N(z)$ satisfy the recurrence relation (7.2.1), where

$$b_{2mn+\nu} = \begin{cases} \dfrac{n(n+\alpha)}{(2n+\alpha+\beta_\nu)(2n+\alpha+\beta_\nu+1)} & \text{if } 0 \le \nu \le m-1, \\[2ex] \dfrac{(n+\beta_\nu)(n+\alpha+\beta_\nu)}{(2n+\alpha+\beta_\nu)(2n+\alpha+\beta_\nu+1)} & \text{if } m \le \nu \le 2m-1. \end{cases} \tag{7.4.2}$$

For $\alpha = \gamma = 0$ and $m = 2$, (7.4.2) reduces to (7.2.5).

Using the same method as before, we can investigate the corresponding problem on $(0, +\infty)$ with the inner product

$$(f, g) = \int_0^{+\infty} \left(\sum_{s=0}^{2m-1} f(x\varepsilon_s)\overline{g(x\varepsilon_s)} \right) w(x)\, dx, \tag{7.4.3}$$

where

$$w(x) = x^{2m\gamma} \exp(-x^{2m}), \quad \gamma > -1/(2m). \tag{7.4.4}$$

Theorem 7.4.2. *The monic polynomials $\{\pi_N(z)\}_{N=0}^{+\infty}$ orthogonal with respect to the inner product (7.4.3), where the weight function is given by (7.4.4), can be expressed in the form*

$$\pi_N(z) = z^{\nu} \hat{L}_n^{(\alpha_\nu)}(z^{2m}), \quad N = 2mn + \nu,\ n = [N/2m],$$

where $\nu \in \{0, 1, \dots, 2m-1\}$, $\alpha_\nu = \gamma + (2\nu + 1 - 2m)/(2m)$, and $\hat{L}_n^{(s)}(t)$ denotes the monic generalized Laguerre polynomial orthogonal with respect to the weight $t \mapsto$

$t^s e^{-t}$ *on* $(0, +\infty)$. *The polynomials* $\pi_N(z)$ *satisfy the recurrence relation* (7.2.1), *where*

$$b_{2mn+\nu} = \begin{cases} n + 1 + \alpha_\nu & \text{if } 0 \le \nu \le m-1, \\ n & \text{if } m \le \nu \le 2m-1. \end{cases}$$

Like the generalized Laguerre polynomial $L_n^{(s)}(t)$, the polynomial $\pi_N(z)$ satisfies a second order linear homogeneous differential equation (see Milovanović [102]).

Theorem 7.4.3. *The polynomial* $\pi_N(z)$ *orthogonal with respect to the inner product* (7.4.3) *satisfies the differential equation*

$$z^2 y'' + B(z)y' + C(z)y = 0, \tag{7.4.5}$$

where

$$B(z) = 2z\big[1 + m(\gamma - 1 - z^{2m})\big], \quad C(z) = 2mNz^{2m} - \nu(\nu + 2m(\gamma - 1) + 1), \tag{7.4.6}$$

and $N = 2mn + \nu$, $n = [N/(2m)]$, $\nu \in \{0, 1, \ldots, 2m-1\}$.

Notice that for $m = 1$, the equation (7.4.5) reduces to the well-known Hermite equation $y'' - 2zy' + 2Ny = 0$.

REMARK 7.4.1. A simple case could be if we choose the parameter γ in the weight function (7.4.4) such that the coefficient $B(z)$ in (7.4.6) reduces to a monomial of degree $2m + 1$. Namely, if $\gamma = (m-1)/m$, the equation (7.4.5) reduces to

$$y'' - 2mz^{2m-1}y' + \left[2mNz^{2m-2} - \frac{\nu(\nu - 1)}{z^2}\right] y = 0.$$

References

[1] N. I. Achieser and M. G. Kreĭn, *On some problems in the moment theory*, GONTI, Har'kov, 1938.

[2] J. Aczél, *Eine Bemerkung über die Charakterisierung der "klassischen" Orthogonalpolynome*, Acta Math. Acad. Sci. Hung. **4** (1953), 315–321.

[3] R. P. Agarwal and G. V. Milovanović, *A characterization of the classical orthogonal polynomials*, Progress in Approximation Theory (P. Nevai and A. Pinkus, eds.), Academic Press, New York, 1991, pp. 1–4.

[4] M. Alfaro, F. Marcellán, M. L. Rezola and A. Ronveaux, *On orthogonal polynomials of Sobolev type: Algebraic properties and zeros*, SIAM J. Math. Anal. **23** (1992), 737–757.

[5] M. Alfaro and F. Marcellán, *Recent trends in orthogonal polynomials on the unit circle*, IMACS Annals on Computing and Applied Mathematics, Vol. 9: Orthogonal Polynomials and Their Applications (C. Brezinski, L. Gori and A. Ronveaux, eds.), IMACS, Baltzer, Basel, 1991, pp. 3–14.

[6] W. A. Al-Salam, *Characterization theorems for orthogonal polynomials*, Orthogonal Polynomials – Theory and Practice (P. Nevai, ed.), NATO ASI Series, Series C; Mathematical and Physical Sciences, Vol. 294, Kluwer, Dordrecht, 1990, pp. 1–24.

[7] P. Althammer, *Eine Erweiterung des Orthogonalitätsbegriffes bei Polynomen und deren Anwendung auf die beste Approximation*, J. Reine Angew. Math. **211** (1962), 192–204.

[8] G. E. Andrews and R. Askey, *Classical orthogonal polynomials*, Polynômes Orthogonaux et Applications (C. Brezinski, A. Draux, A. P. Magnus, P. Maroni, A. Ronveaux, eds.), Lect. Notes Math. No. 1171, vol. 1985, Springer Verlag, Berlin – Heidelberg – New York, pp. 36–62.

[9] R. Askey and J. Wilson, *Some basic hypergeometric orthogonal polynomials that generalize Jacobi polynomials*, Memoirs Amer. Mat. Soc. **319** (1985).

[10] N. M. Atakishiyev and S. K. Suslov, *The Hahn and Meixner polynomials of an imaginary argument and some of their applications*, J. Phys. A: Math. Gen. **18** (1985), 1583–1596.

[11] H. Bateman and A. Erdélyi, *Higher Transcendental Functions*, Vol. 2, McGraw–Hill, New York, 1953.

[12] H. Bavinck and H. G. Meijer, *Orthogonal polynomials with respect to a symmetric inner product involving derivatives*, Appl. Anal. **33** (1989), 103–117.

[13] H. Bavinck and H. G. Meijer, *On orthogonal polynomials with respect to an inner product involving derivatives: zeros and recurrence relations*, Indag. Math. (N.S.) **1** (1990), 7–14.

[14] S. Bernstein, *Sur les polinomes ortogonaux relatifs à un segment fini*, J. Math. Pures Appl. **9(9)** (1930), 127–177.

[15] S. Bochner, *Über Sturm-Liouvillesche Polynomsysteme*, Math. Z. **29** (1929), 730–736.

[16] J. Brenner, *Über eine Erweiterung des Orthogonalitäts-begriffes bei Polynomen*, Constructive Theory of Functions (G. Alexits and S. B. Stechkin, eds.), Akadémiai Kiadó, Budapest, 1972, pp. 77–83.

[17] M. G. de Bruin, *Polynomials orthogonal on a circular arc*, J. Comput. Appl. Math. **31** (1990), 253–266.

[18] A. C. Calder and J. G. Laframboise, *Multiple-water-bag simulation of inhomogeneous plasma motion near an electrode*, J. Comput. Phys. **65** (1986), 18–45.

[19] F. Calio', M. Frontini and G. V. Milovanović, *Numerical differentiation of analytic functions using quadratures on the semicircle*, Comput. Math. Appl. **22** (1991), 99–106.

[20] T. S. Chihara, *An Introduction to Orthogonal Polynomials*, Gordon and Breach, New York, 1978.

[21] E. A. Cohen, *Zero distribution and behavior of orthogonal polynomials in the Sobolev space* $W^{1,2}[-1,1]$, SIAM J. Math. Anal. **6** (1975), 105–116.

[22] R. Ž. Djordjević and G. V. Milovanović, *A generalization of E. Landau's theorem*, Univ. Beograd. Publ. Elektrotehn. Fak. Ser. Mat. Fiz. No 498–No 541 (1975), 91–96.

[23] H. Engels, *Numerical quadrature and cubature*, Academic Press, London, 1980.

[24] W. N. Everitt and L. L. Littlejohn, *Orthogonal polynomials and spectral theory: a survey*, IMACS Annals on Computing and Applied Mathematics, Vol. 9, Orthogonal Polynomials and Their Applications (C. Brezinski, L. Gori, and A. Ronveaux, eds.), J.C. Baltzer AG, Scientific Publ. Co., Basel, 1991, pp. 21–55.

[25] A. Fransén, *Accurate determination of the inverse gamma integral*, BIT **19** (1979), 137–138.

[26] G. Freud, *Orthogonal Polynomials*, Akadémiai Kiadó/Pergamon, Budapest, 1971.

[27] M. Frontini, W. Gautschi and G. V. Milovanović, *Moment-preserving spline approximation on finite intervals*, Numer. Math. **50** (1987), 503–518.

[28] M. Frontini and G. V. Milovanović, *Moment-preserving spline approximation on finite intervals and Turán quadratures*, Facta Univ. Ser. Math. Inform. **4** (1989), 45–56.

[29] W. Gautschi, *On generating Gaussian quadrature rules*, Numerische Integration (G. Hämmerlin, ed.), ISNM Vol. 45, Birkhäuser, Basel, 1979, pp. 147–154.

[30] W. Gautschi, *Minimal solutions of three-term recurrence relations and orthogonal polynomials*, Math. Comp. **36** (1981), 457–554.

[31] W. Gautschi, *A survey of Gauss-Christoffel quadrature formulae*, E. B. Christoffel (P.L. Butzer and F. Fehér, eds.), Birkhäuser, Basel, 1981, pp. 72–147.

[32] W. Gautschi, *On generating orthogonal polynomials*, SIAM J. Sci. Statist. Comput. **3** (1982), 289–317.

[33] W. Gautschi, *An algorithmic implementation of the generalized Christoffel theorem*, Numerische Integration. Internat. Ser. Numer. Math. vol. **57** (G. Hämmerlin, ed.), Birkhäuser, Basel, 1982, pp. 89–106.

[34] W. Gautschi, *Polynomials orthogonal with respect to the reciprocal gamma function*, BIT **22** (1982), 387–389.

[35] W. Gautschi, *How and how not to check Gaussian quadrature formulae*, BIT **23** (1983), 209–216.

[36] W. Gautschi, *Discrete approximations to spherically symmetric distributions*, Numer. Math. **44** (1984), 53–60.

[37] W. Gautschi, *Gauss-Kronrod quadrature – a survey*, Numerical Methods and Approximation Theory III (G. V. Milovanović, ed.), University of Niš, Niš, 1988, pp. 39–66.

[38] W. Gautschi, *On the zeros of polynomials orthogonal on the semicircle*, SIAM J. Math. Anal. **20** (1989), 738–743.

[39] W. Gautschi, *Computational aspects of orthogonal polynomials*, Orthogonal Polynomials (Columbus, OH, 1989) (P. Nevai, ed.), NATO Adv. Sci. Inst. Ser. C: Math. Phys. Sci., 294, Kluwer Acad. Publ., Dordrecht, 1990, pp. 181–216.

[40] W. Gautschi, *Spline approximation and quadrature formulae*, Atti Sem. Mat. Fis. Univ. Modena **29** (1991), 47–60.

[41] W. Gautschi, *Computational problems and applications of orthogonal polynomials*, IMACS Annals on Computing and Applied Mathematics, Vol. 9: Orthogonal Polynomials and Their Applications (C. Brezinski, L. Gori and A. Ronveaux, eds.), IMACS, Baltzer, Basel, 1991, pp. 61–71.

[42] W. Gautschi, *A class of slowly convergent series and their summation by Gaussian quadrature*, Math. Comp. **57** (1991), 309–324.

[43] W. Gautschi, *On certain slowly convergent series occurring in plate contact problems*, Math. Comp. **57** (1991), 325–338.

[44] W. Gautschi, *Algorithm 726: ORTHPOL – a package of routines for generating orthogonal polynomials and Gauss-type quadrature rules*, ACM Trans. Math. Software **20** (1994), 21–62.

[45] W. Gautschi, *Orthogonal polynomials: applications and computation*, Acta Numerica (Cambridge) (1996), 45–119.

[46] W. Gautschi, H. Landau, and G. V. Milovanović, *Polynomials orthogonal on the semicircle, II*, Constr. Approx. **3** (1987), 389–404.

[47] W. Gautschi and S. Li, *A set of orthogonal polynomials induced by a given orthogonal polynomial*, Aequationes Math. **46** (1993), 174–198.

[48] W. Gautschi and G. V. Milovanović, *Gaussian quadrature involving Einstein and Fermi functions with an application to summation of series*, Math. Comp. **44** (1985), 177-190.

[49] W. Gautschi and G. V. Milovanović, *Polynomials orthogonal on the semicircle*, Rend. Sem. Mat. Univ. Politec. Torino (Special Functions: Theory and Computation) (1985), 179–185.

[50] W. Gautschi and G. V. Milovanović, *Polynomials orthogonal on the semicircle*, J. Approx. Theory **46** (1986), 230–250.

[51] W. Gautschi and G. V. Milovanović, *Spline approximations to spherically symmetric distributions*, Numer. Math. **49** (1986), 111–121.

[52] W. Gautschi and R. S. Varga, *Error bounds for Gaussian quadrature of analytic functions*, SIAM J. Numer. Anal. **20** (1983), 1170–1186.

[53] W. Gautschi and M. Zhang, *Computing orthogonal polynomials in Sobolev spaces*, Numer. Math. **71** (1995), 159–183.

[54] Ya. L. Geronimus, *Polynomials orthogonal on a circle and their applications*, Zap. Nauč.-issled. Inst. Mat. Mech. HMO **19** (1948), 35–120. (Russian)

[55] Ya. L. Geronimus, *On some properties of generalized orthogonal polynomials*, Mat. Sb. **9** (51): 1 (1941), 121–135. (Russian)

[56] Ya. L. Geronimus, *Polynomials orthogonal on a circle and interval*, Gos. Izd. Fiz.-Mat. Lit., Moscow, 1958. (Russian)

[57] Ya. L. Geronimus, *Orthogonal polynomials*, Amer. Math. Soc. Transl. **108** (1977), 37–130.

[58] A. Ghizzetti and A. Ossicini, *Su un nuovo tipo di sviluppo di una funzione in serie di polinomi*, Rend. Accad. Naz. Lincei **8(43)** (1967), 21–29.

[59] G. H. Golub and J. H. Welsch, *Calculation of Gauss quadrature rules*, Math. Comp. **23** (1969), 221–230.

[60] L. Gori-N. Amati and E. Santi, *On a method of approximation by means of spline functions*, Approximation, Optimization and Computing – Theory and Application (A. G. Law and C. L. Wang, eds.), IMACS, Dalian, China, 1990, pp. 41–46.

[61] L. Gori, M. L. Lo Cascio and G. V. Milovanović, *The σ-orthogonal polynomials: a method of construction*, IMACS Annals on Computing and Applied Mathematics, Vol. 9, Orthogonal Polynomials and Their Applications (C. Brezinski, L. Gori, and A. Ronveaux, eds.), J.C. Baltzer AG, Scientific Publ. Co., Basel, 1991, pp. 281–285.

[62] A. Gorny, *Contribution à l'étude des fonctions dérivables d'une variable réelle*, Acta Math. **71** (1939), 317–358.

[63] S. Guerra, *Polinomi generati da successioni peso e teoremi di rappresentazione di una funzione in serie di tali polinomi*, Rend. Ist. Mat. Univ. Trieste **8** (1976), 172–194.

[64] S. Guerra, *Su un determinate collegato ad un sistema di polinomi ortogonali*, Rend. Ist. Mat. Univ. Trieste **10** (1978), 66–79.

[65] A. Guessab and G. V. Milovanović, *Weighted L^2-analogues of Bernstein's inequality and classical orthogonal polynomials*, J. Math. Anal. Appl. **182** (1994), 244–249.

[66] P. Henrici, *Applied and Computational Complex Analysis*, Vol. 1, Wiley, New York, 1984.

[67] E. Hille, *Remark on the Landau-Kallman-Rota inequality*, Aequationes Math. **4** (1970), 239–240.

[68] A. Iserles, P. E. Koch, S. P. Nørsett, and J. M. Sanz-Serna, *On polynomials orthogonal with respect to certain Sobolev inner products*, J. Approx. Theory **65** (1991), 151–175.

[69] J. W Jayne, *Recursively generated polynomials and Geronimus' version of orthogonality on a contour*, SIAM J. Math. Anal. **19** (1988), 676–686.

[70] R. Koekoek, *Generalizations of Laguerre polynomials*, J. Math. Anal. Appl. **153** (1990), 576–590.

[71] R. Koekoek and H. G. Meijer, *A generalization of Laguerre polynomials*, SIAM J. Math. Anal. **24** (1993), 768–782.

[72] A. Kolmogoroff, *On inequalities between upper bounds of the successive derivatives of an arbitrary function on an infinite interval*, Uchen. Zap. Moskov. Gos. Univ. Mat. **30** (1939), 3–16. (Russian)

[73] M. A. Kovačević and G. V. Milovanović, *Lobatto quadrature formulas for generalized Gegenbauer weight*, 5th Conference on Applied Mathematics (Z. Bohte, ed.), University of Ljubljana, Ljubljana, pp. 81–88.

[74] M. A. Kovačević and G. V. Milovanović, *Spline approximation and generalized quadratures*, Portugal. Math. **53** (1996), 355–366.

[75] J. G. Laframboise and A. D. Stauffer, *Optimum discrete approximation of the Maxwell distribution*, AIAA J. **7** (1969), 520–523.

[76] E. Landau, *Einige Ungleichungen für zweimal differenzierbare Funktionen*, Proc. London. Math. Soc. (2) **13** (1913), 43–49.

[77] H. J. Landau, *Maximum entropy and the moment problem*, Bull. Amer. Math. Soc. **16** (1987), 47–77.

[78] R. V. Lascenov, *On a class of orthogonal polynomials*, Učen. Zap. Leningrad. Gos. Ped. Inst. **89** (1953), 191–206. (Russian)

[79] S.-Y. Lee, *The inhomogeneous Airy functions* $\mathrm{Hi}(z)$ *and* $\mathrm{Gi}(z)$, J. Chem. Phys. **72** (1980), 332–336.

[80] P. Lesky, *Die Charakterisierung der klassischen orthogonalen Polynome durch Sturm-Liouvillesche Differentialgleichungen*, Arch. Rat. Mech. Anal. **10** (1962), 341–351.

[81] D. C. Lewis, *Polynomial least square approximations*, Amer. J. Math. **69** (1947), 273–278.

[82] E. Lindelöf, *Le calcul des résidus*, Gauthier-Villars, Paris, 1905.

[83] F. Marcellán, *Orthogonal polynomials and Toeplitz matrices: some applications*, Rational Approximation and Orthogonal Polynomials (M. Alfaro, ed.), Seminario Matematico "Garcia de Galdeano", Universidad de Zaragoza, 1989, pp. 31–57.

[84] F. Marcellán, M. Alfaro, and M. L. Rezola, *Orthogonal polynomials on Sobolev spaces: Old and new directions*, J. Comput. Appl. Math. **48** (1993), 113–131.

[85] F. Marcellán, T. E. Pérez and M. A. Piñar, *On zeros of Sobolev-type orthogonal polynomials*, Rend. Mat. Appl. (7) **12** (1992), 455–473.

[86] F. Marcellán and A. Ronveaux, *On a class of polynomials orthogonal with respect to a discrete Sobolev inner product*, Indag. Math. (N.S.) **1** (1990), 451–464.

[87] F. Marcellán and G. Sansigre, *Orthogonal polynomials on the unit circle: symmetrization and quadratic decomposition*, J. Approx. Theory **65** (1991), 109–119.

[88] F. Marcellán and W. Van Assche, *Relative asymptotics for orthogonal polynomials with a Sobolev inner product*, J. Approx. Theory **72** (1992), 192–209.

[89] P. Maroni, *Une caractérisation des polynômes orthogonaux semi-classiques*, C.R. Acad. Sci. Paris **301,** Série 1 (1985), 269–272.

[90] H. G. Meijer, *On real and complex zeros of orthogonal polynomials in a discrete Sobolev space*, J. Comput. Appl. Math. **49** (1993), 179–191.

[91] H. G. Meijer, *Laguerre polynomials generalized to a certain discrete Sobolev inner product space*, J. Approx. Theory **73** (1993), 1–16.

[92] H. G. Meijer, *Zero distribution of orthogonal polynomials in a certain discrete Sobolev space*, J. Comput. Appl. Math. **172** (1993), 520–532.

[93] C. A. Micchelli, *Monosplines and moment preserving spline approximation*, Numerical Integration III (H. Brass and G. Hämmerlin, eds.), Birkhäuser, Basel, 1988, pp. 130–139.

[94] G. V. Milovanović, *On some functional inequalities*, Univ. Beograd. Publ. Elektrotehn. Fak. Ser. Mat. Fiz. No 599 (1977), 1–59. (Serbian)

[95] G. V. Milovanović, *Construction of s-orthogonal polynomials and Turán quadrature formulae*, Numerical Methods and Approximation Theory III (Niš, 1987) (G. V. Milovanović, ed.), Univ. Niš, Niš, 1988, pp. 311–328.

[96] G. V. Milovanović, *Some applications of the polynomials orthogonal on the semicircle*, Numerical Methods, (Miskolc, 1986), Colloq. Math. Soc. János Bolyai, Vol. 50, North-Holland, Amsterdam – New York, 1987, pp. 625–634.

[97] G. V. Milovanović, *Complex orthogonality on the semicircle with respect to Gegenbauer weight: Theory and applications*, Topics in Mathematical Analysis (Th.M. Rassias, ed.), World Scientific Publ., Singapore, 1989, pp. 695–722.

[98] G. V. Milovanović, *On polynomials orthogonal on the semicircle and applications*, J. Comput. Appl. Math. **49** (1993), 193–199.

[99] G. V. Milovanović, *Summation of series and Gaussian quadratures*, Approximation and Computation (R. V. M. Zahar, ed.), ISNM Vol. 119, Birkhäuser Verlag, Basel – Boston – Berlin, 1994, pp. 459–475.

[100] G. V. Milovanović, *Summation of series and Gaussian quadratures, II*, Numer. Algorithms **10** (1995), 127–136.

[101] G. V. Milovanović, *Summation of slowly convergent series via quadratures*, Advances in Numerical Methods and Applications – $O(h^3)$ (I. T. Dimov, Bl. Sendov, and P. S. Vassilevski, eds.), World Scientific, Singapore, 1994, pp. 154–161.

[102] G. V. Milovanović, *Generalized Hermite polynomials on the radial rays in the complex plane*, Theory of Functions and Applications (H. B. Nersessian, ed.), Collection of Works Dedicated to the Memory of M. M. Djrbashian, Louys Publishing House, Yerevan, 1995, pp. 125–129.

[103] G. V. Milovanović, *Orthogonal polynomials and applications*, Scientific Review **14** (1995), 5–28.

[104] G. V. Milovanović, *Some nonstandard types of orthogonality (A survey)*, FILOMAT (Formerly: Zb. Rad.) **9** (1995), 517–542.

[105] G. V. Milovanović, *A class of orthogonal polynomials on the radial rays in the complex plane*, J. Math. Anal. Appl. **206** (1997), 121–139.

[106] G. V. Milovanović, *S-orthogonality and generalized Turán quadratures: Construction and applications*, Internat. Conf. Approx. Optim. (Romania) – ICAOR (Cluj-Napoca, 1996) (to appear).

[107] G. V. Milovanović and M. A. Kovačević, *Least squares approximation with constraint: generalized Gegenbauer case*, Facta Univ. Ser. Math. Inform. **1** (1986), 73–81.

[108] G. V. Milovanović and M. A. Kovačević, *Moment-preserving spline approximation and Turán quadratures*, Numerical Mathematics (Singapore, 1988) INSM Vol. 86 (R. P. Agarwal, Y. M. Chow and S. J. Wilson, eds.), Birkhäuser, Basel, 1988, pp. 357–365.

[109] G. V. Milovanović and M. A. Kovačević, *Moment-preserving spline approximation and quadratures*, Facta Univ. Ser. Math. Inform. **7** (1992), 85–98.

[110] G. V. Milovanović, D. S. Mitrinović and Th. M. Rassias, *Topics in Polynomials: Extremal Problems, Inequalities, Zeros*, World Scientific, Singapore – New Jersey – London – Hong Kong, 1994.

[111] G. V. Milovanović and P. M. Rajković, *Geronimus concept of orthogonality for polynomials orthogonal on a circular arc*, Rend. di Matematica, Serie VII (Roma) **10** (1990), 383–390.

[112] G. V. Milovanović and P. M. Rajković, *On polynomials orthogonal on a circular arc*, J. Comput. Appl. Math. **51** (1994), 1–13.

[113] G. V. Milovanović and P. M. Rajković, *Complex orthogonal polynomials with the Hermite weight*, Univ. Beograd Publ. Elektrotehn. Fak. Ser. Mat. **6** (1995), 63–73.

[114] G. V. Milovanović, P. M. Rajković and Z. M. Marjanović, *On properties of some nonclassical orthogonal polynomials*, FILOMAT (Formerly: Zb. Rad.) **9** (1995), 101–111.

[115] G. V. Milovanović, P. M. Rajković and Z. M. Marjanović, *A class of orthogonal polynomials on the radial rays in the complex plane, II*, Facta Univ. Ser. Math. Inform. **11** (1996), 29–47.

[116] G. V. Milovanović and M. M. Spalević, *A numerical procedure for coefficients in generalized Gauss-Turán quadratures*, FILOMAT (Formerly: Zb. Rad.) **9** (1995), 1–8.

[117] G. V. Milovanović and M. M. Spalević, *Construction of Chakalov-Popoviciu's type quadrature formulae*, Supl. Rend. Circ Mat. Palermo (to appear).

[118] D. S. Mitrinović and J. D. Kečkić, *The Cauchy Method of Residues – Theory and Applications*, Reidel, Dordrecht, 1984.

[119] G. Monegato, *Stieltjes polynomials and related quadrature rules*, SIAM Review **24** (1982), 137–158.

[120] P. Nevai, *Orthogonal Polynomials*, Memoirs Amer. Math. Soc., Vol. 213, Amer. Mat. Soc., Providence, R.I., 1979.

[121] P. Nevai, *Géza Freud, orthogonal polynomials and Christoffel functions, A case study*, J. Approx. Theory **48** (1986), 3–167.

[122] P. Nevai, *Orthogonal polynomials, measures and recurrence relations on the unit circle*, Trans. Amer. Math. Soc. **300** (1987), 175–189.

[123] A. Ossicini, *Construzione di formule di quadratura di tipo Gaussiano*, Ann. Mat. Pura Appl. **(4) 72** (1966), 213–238.

[124] T. E. Pérez and M. A. Piñar, *Global properties of zeros for Sobolev-type orthogonal polynomials*, J. Comput. Appl. Math. **49** (1993), 225–232.

[125] R. Piessens and M. Branders, *Tables of Gaussian quadrature formulas*, Appl. Math. Progr. Div. University of Leuven, Leuven (1975).

[126] A. Ronveaux, *Sur l'équation différentielle du second ordre satisfaite par une classe de polynômes orthogonaux semi-classiques*, C.R. Acad. Sci. Paris **305,** Série 1 (1987), 163–166.

[127] A. Ronveaux and F. Marcellán, *Differential equation for classical-type orthogonal polynomials*, Canad. Math. Bull. **32** (4) (1989), 404–411.

[128] A. Ronveaux and G. Thiry, *Differential equations of some orthogonal families in REDUCE*, J. Symb. Comp. **8** (1989), 537–541.

[129] I. J. Schoenberg, *The elementary cases of Landau's problem of inequalities between derivatives*, Amer. Math. Monthly **80** (1973), 121–158.
[130] I. Smirnov, *Sur la thèorie des polynomes orthogonaux à une variable complèxe*, Ž. Leningrad. Fiz.-Mat. Ob. **2** (1928), 155–179.
[131] I. Smirnov, *Sur les valeurs limites des fonctions regulières à l'intérieur d'un cercle*, Ž. Leningrad. Fiz.-Mat. Ob. **2** (1928), 22–37.
[132] B.T. Smith et al., *Matrix Eigensystem Routines – EISPACK Guide*, Lect. Notes Comp. Sci. Vol. 6, Springer-Verlag, New-York, 1974.
[133] S. B. Steckhin, *Inequalities between the upper bounds of the derivatives of an arbitrary function on the half-line*, Mat. Zametki **1** (1967), 665–674. (Russian)
[134] N. M. Steen, G. D. Byrne and E. M. Gelbard, *Gaussian quadratures for the integrals* $\int_0^\infty \exp(-x^2)f(x)\,dx$ *and* $\int_0^b \exp(-x^2)f(x)\,dx$, Math. Comp. **23** (1969), 661–671.
[135] P. K. Suetin, *Classical Orthogonal Polynomials*, Nauka, Moscow, 1976. (Russian)
[136] G. Szegő, *Beiträge zur Theorie der Toeplitzschen Formen*, Math. Z. **6** (1920), 167–202.
[137] G. Szegő, *Beiträge zur Theorie der Toeplitzschen Formen. II*, Math. Z. **9** (1921), 167–190.
[138] G. Szegő, *Orthogonal Polynomials*, Amer. Math. Soc. Colloq. Publ., vol. 23, 4th ed., Amer. Math. Soc., Providence, R. I., 1975.
[139] F. Tricomi, *Serie orthogonali di Funzioni*, Torino, 1948.
[140] P. Turán, *On the theory of the mechanical quadrature*, Acta Sci. Math. (Szeged) **12** (1950), 30–37.
[141] A. K. Varma, *A new characterization of Hermite polynomials*, Acta Math. Hung. **49** (1987), 169–172.
[142] G. Vincenti, *On the computation of the coefficients of s-orthogonal polynomials*, SIAM J. Numer. Anal. **23** (1986), 1290–1294.

Gradimir V. Milovanović
University of Niš
Faculty of Electronic Engineering
Department of Mathematics, P.O. Box 73
18000 Niš, Serbia – Yugoslavia

E-mail: grade@efnis.elfak.ni.ac.yu

MUHAMMAD ASLAM NOOR

Multivalued variational inequalities

1. Introduction

It is well known that the Riesz-Frèchet representation theorem plays an immportant and fundamental part in functioal analysis and theory of differential equations. This theorem has been geeneralized and extended in many directions by using innovative and novel ideas. The history of the Riesz-Frèchet representation theorem goes back as early as the beginning of the twentieth century. In 1906, Hadamard introduced the notion of a linear operator. At the same time, Lebesgue defined the square summable integrals known as the Lebesgue integrals. Keeping in view, the definitions of linear operators as well as the square summable integrals, Riesz [50, p. 1410] discovered the following result for square summable functions, "*For any continuous linear operation, there exists a function k such that the value of the operation for any function f is denoted by the product integral of the function f and k.*"

By "*continuous linear operation*" Riesz meant what is now termed a continuous linear functional. The same result was also discovered by Frèchet [16] independently for square summable functions on the interval $[0, 2\pi]$. In 1909, Hilbert [21] used a class of spaces now known as Hilbert spaces; but it was Von Neumann [31], who actually formulated the axioms for separable Hilbert spaces in 1929. After the definition of an abstract Hilbet space, Riesz [51] gave a nice proof of the Riesz-Frèchet representation theorem. The modern statement of this theorem asserts two important facts about Hilbert spaces, the existence of many continuous linear functionals and a representation for all of them. Both these facts have been extended to more general spaces, the first being easier to generalize and being one of the consequences of the Hahn-Banach theorem. The second fact is the one which has found the greatest concrete use, for it asserts the existence of a function (that is, element of the Hilbert space) with prescribed properties. By choosing the Hilbert space and inner product appropriately, this theorem furnishes one of the major existence theory tools for the theory of partial differential equations. To be more precise, let H be a real Hilbert space whose inner product and norm are denoted by $< \cdot, \cdot >$ and $\| \cdot \|$ respectively. For a given continuous linear functional $f \in H$, consider the functional $I[v]$, where

$$I[v] = \frac{1}{2} < v, v > -2 < f, v >, \quad \text{for all } v \in H. \tag{1.1}$$

The functional $I[v]$, defined by (1.1) is known as the energy (potential) functional. It can be shown that the minimum u of $I[v]$, defined by (1.1), on the whole Hilbert space H can be characterized by the equations

$$< u, v > = < f, v >, \quad \text{for all } v \in H. \tag{1.2}$$

The problem (1.2) is known as the Riesz-Frèchet representation theorem. We remark that the relation (1.1) is called the variational formulation of the Riesz-Frèchet representation theorem. This fact has not been noted in the literature explicitly. This aspect shows that the direct method of the calculus of variations can be used to give another proof of the Riesz-Frèchet representation theorem. Consequently, we have the variational formulation for the Riesz-Frèchet representation theorem. This shows that the origin of the Riesz-Frèchet theorem can be traced back to Fermat, Newton, Leibniz, Bournalli, Euler and Lagrange.

Since the inner product in a Hilbert space is a bilinear continuous form; then naturally the question arises, whether a version of this theorem is still valid for any arbitrary bilinear continuous (not necessarily symmetric) form. The answer to this question is in affirmative. Lax and Milgram [27] announced such a representation theorem in 1952, now known as the Lax-Milgram lemma.

"*For a given continuous coercive bilinear form $a(u, v)$ and a continuous linear function f on H, there exists a unique $u \in H$ such that*

$$a(u, v) = < f, v >, \quad \textit{for all } v \in H. \text{"} \tag{1.3}$$

We note that if $a(u, v)$ is a symmetric positive bilinear form then it can be shown easily that the function $u \in H$ satisfying (1.3) is the minimum of the functional $F[v]$ on H, where

$$F[v] \equiv a(v, v) - 2f(v) = a(v, v) - 2 < f, v >, \quad \text{for all } v \in H, \tag{1.4}$$

and the converse is also true.

In the Lax-Milgram lemma, the bilinear form is defined on the entire Hilbert space. If one tries to minimize the fuctional $F[v]$ defined by (1.4) on a convex set K in a Hilbert space, then one can prove that the minimum of $F[v]$ can be characterized by

$$a(u, v - u) \geq < f, v - u >, \quad \text{for all } v \in K. \tag{1.5}$$

The inequality of the type (1.5) is called the variational inequality, which is originally due to Stampacchia [57]. See also Fichera [15].

Note that for $K = H$, the problem (1.5) is equivalent to the Lax-Milgram lemma and consequently to the Riesz-Frèchet theorem. In recent years, the variational inequalities have been generalized and extended in many direction using innovative and novel techniques. It is worth mentioning that in many important applications, the convex set K also depends upon the solution implicitly or explicitly. In this case, the variational inequality (1.5) is known as the quasi variational inequality. This generalization is due to Bensoussan and Lions [3]. In fact, we have the problem of finding $u \in K(u)$ such that

$$a(u, v-u) \geq < f, v-u >, \quad \text{for all } v \in K(u). \tag{1.6}$$

For a nonlinear differentiable function f, Noor [33] has proved that the minimum of the functional $F[v]$, defined by (1.4) can be characterized by the inequality

$$a(u, v-u) \geq < f'(u), v-u >, \quad \text{for all } v \in K, \tag{1.7}$$

where $f'(u)$ is the Frèchet differential of f at u. The inequalities of the type (1.7) are known as the mildly nonlinear variational inequalities. It has been shown by Noor [33,34,37] that many obstacle, unilateral and free boundary value problems can be studied in the framework of mildly nonlinear variational inequalities. Similarly if the convex set K also depends upon the solution u, then we consider the problem of finding $u \in K(u)$ such that

$$a(u, v-u) \geq < f'(u), v-u > \quad \text{for all } v \in K(u), \tag{1.8}$$

which is known as the mildly nonlinear quasi variational inequality, see Noor [52,19,30,44,5] for formulation and iterative methods.

In many important applications, the problems of the type (1.2), (1.3), (1.5), and (1.7) arise, which cannot be derived from the minimization of the functional on the whole space or on a convex set. This motivates us to consider more general variational inequalities of which the problems (1.1), (1.3), (1.5) and (1.7) are special cases.

For given nonlinear continuous operators $T, A : K \rightarrow H$, consider the problem of finding $u \in K$ such that

$$< T(u) + A(u), v-u > \geq 0, \quad \text{for all } v \in K. \tag{1.9}$$

The inequality of the type (1.9) is known as the strongly nonlinear variational inequality, which is mainly due to Noor [37,33]. For its applications, generalizations and numerical methods, see [34,35,37,41,42,43,44,45] and the reference cited therein.

In 1982, Fang and Peterson [14] considered the multivalued version of the variational inequality (1.5). To be more specific, for a given multivalued operator $T : H \rightarrow 2^H$, we consider the problem of finding $u \in K$ such that $w \in T(u)$ and

$$< w, v-u > \geq < f, v-u >, \quad \text{for all } v \in K, \tag{1.10}$$

which is known as the multivalued variatonal inequality.

We remark that for $K = H$, the problem (1.10) is equivalent to finding $u \in H$ such that $w \in T(u)$ and

$$< w, v > = < f, v >, \quad \text{for all } v \in H \tag{1.11}$$

The problem (1.11) is the multivalued version of the Riesz-Frèchet representation theorem and constitute an interesting subject for future research.

We would like to point out that the theory of variational inequalities can be regarded as the natural development of the 19th and 20th problems of Hilbert, which he formulated in its famous Paris lecture in 1900. This theory allows us to prove the existence of solutions for nonlinear equations and inequalities, which are not necessarily the Euler-Lagrange equations corresponding to variational problems.

In this paper, we consider and study the multivalued extension of the strongly nonlinear (quasi) variational inequalities (1.9). Using essentially the projection technique, we prove the unique existence of a solution of the multivalued strongly nonlinear variational inequalities. This technique is used to establish the equivalence between the multivalued variational inequalities and fixed point problems. This alternative formulation enables us to suggest a number of iterative algorithms for solving variational inequalities. We also study the convergence criteria for these algorithms. It is well known that the scope of the iterative projection algorithms is quite limited. In some cases, one cannot even find the projection of the solution. These facts motivated Noor [35,39,41,42], Glowinski, Lions and Tremolieres [18] and Cohen [7] to develop an alternate method, which does not use the projection. This technique is known as the auxiliary principle technique. This technque has been used quite effectively to study the existence of a solution of the variational inequalities and also to suggest a number of iterative methods, which include Newton, descent, projection and decomposition methods as special cases, see, for example, [1-58] and the references therein.

In Section 2, we formulate the problem and review some basic notations and concepts. Existence results are considered in Section 3. In Section 4, we prove that the multivalued variational inequalities are equivalent to the fixed point problem. This fixed point formulation is used to suggest an iterative algorithm. Using the projection technique, we establish the equivalence between the variational inequalities and the multivalued Wiener-Hopf equations in Section 5. This equivalence enables us to suggest and analyze a number of iterative algorithms. In Section 6, we use the auxiliary principle technique to study the existence of a solution of the multivalued variational inequalities. It has been shown that the auxiliary principle technique can be used to suggest innovative iterative algorithms for variational inequalities. The multivalued quasi variational inequalities are considered and studied in Section 7. It is shown that the projection and the Wiener-Hopf equations techniques can be used to suggest a number of iterative algorithms for solving the multivalued quasi variational inequalities. The results proved in this paper represent a refinement and improvement of the previously known results in these areas.

2. Formulation and Basic Results.

Let H be a real Hilbert space, whose norm and inner product are denoted by $\|.\|$ and $< .,. >$ respectively. Let K be a nonempty closed convex set in H. For given multivalued operators $T : H \to 2^H$ and $V : H \to C(H)$, where $C(H)$ is the family of all nonempty compact subset of H, we consider the problem of finding $u \in K$, $w \in T(u)$, $y \in V(u)$ such that

$$< w + y,\ v - u > \geq 0, \quad \text{for all } v \in K. \tag{2.1}$$

The problem (2.1) is called the multivalued strongly nonlinear variational inequality problem, which is the most general and includes many previously known problems as special cases. For related work, see Ding [12].

I . If V, T are single-valued operators, then the problem (2.1) reduce to the problem of finding $u \in K$ such that

$$< Tu + V(u), v - u > \geq 0 \quad \text{for all } v \in K. \tag{2.2}$$

The problem (2.2) is called the strongly (mildly) nonlinear variational inequality problem, which was introduced by Noor [33] to study the obstacle and unilateral mildly nonlinear elliptic boundary problem in 1975. For the iterative methods, sensitivity analysis, finite element error estimates and generalization of the problem (2.2), see [6,12,23,34-43] and the references therein.

II. If $V \equiv 0$, then problem collapse to the variational inequality problem of finding $u \in K$, $w \in T(u)$ such that

$$< w, v - u > \geq 0 \quad \text{for all } v \in K, \tag{2.3}$$

which is known as the generalized variational inequality, see [14,55].

III. If $K^* = \{u \in H; < u, v > \geq 0 \text{ for all } u \in K\}$ is a polar cone of the convex cone K in H, then the problem (2.1) is equivalent to finding $u \in K$, $w \in T(u), y \in V(u)$ such that

$$w + y \in K^* \quad \text{and} \quad < w + y, u >= 0. \tag{2.4}$$

The problem (2.4) is called the multivalued strongly nonlinear complementarity problem.

IV. If $V \equiv 0$ and $T : H \to H$ is a single-valued operator, then the problem (2.1) is equivalent to the problem of finding $u \in K$ such that

$$< Tu, v - u > \geq 0, \quad \text{for all } v \in K, \tag{2.5}$$

which is known as the standard variational inequality problem. The problem (2.5) is mainly and originally due to Stampacchia [57]. For the recent applications, generalizations and numerical techniques, see [2,3,4,5,9,13,14,18,19,24,30,32 - 46, 52,55,58].

For appropriate and suitable choice of the operators T, V and the convex set K, one can obtain various classes of variational inequalities and complementarity problems as special cases from the problem (2.1). In fact, the multivalued strongly nonlinear variational inequality (2.1) is the unifying one.

We also need the following concepts and results.

Definition 2.1. For all $u_1, u_2 \in H$, the multivalued operator $T : H \to 2^H$ is said to be **strongly monotone** if there exists a constant $\alpha > 0$ such that

$$< w_1 - w_2, u_1 - u_2 > \geq \alpha \|u_1 - u_2\|^2, \quad \text{for all } w_1 \in T(u_1) \text{ and } w_2 \in T(u_2).$$

Note that for the single-valued operator T, the definition 2.1 reduces to the usual definition of strongly monotonicity of a nonlinear operator.

Definition 2.2. The multivalued operator $V : H \to C(H)$ is said to be M-Lipschitz continuous if there exists a constant $\eta > 0$ such that

$$M(V(u), V(v)) \leq \eta \|u - v\|, \quad \text{for all } v, u \in H,$$

where $C(H)$ is the family of all nonempty compact subsets of H and $M(\cdot, \cdot)$ is the Hausdorff metric on $C(H)$.

Lemma 2.1. Given $z \in H$, $u \in K$ satisfies the inequality

$$< u - z, v - u > \geq 0, \quad \text{for all } v \in K,$$

if and only if

$$u = P_K z,$$

where P_K is the projection of H into K. Furthermore, P_K is nonexpansive, that is,

$$\|P_K u - P_K v\| \leq \|u - v\|, \quad \text{for all } u, v \in H.$$

Using Lemma 2.1 and technique of Noor [36], we can prove:

Lemma 2.2. Let K be a closed convex set in H. Then (u, w, y) is a solution of (2.1) if and only if (u, w, y) satisfies the relation

$$u = P_K[u - \rho(w + y)], \tag{2.6}$$

where $\rho > 0$ is a constant and P_K is the projection of H into K.

Lemma 2.2 implies that the multivalued variational inequality (2.1) is equivalent to the fixed point problem. This alternate formulation is very important from both theoretical and numerical point of views. This formulation enables us to study the existence of the solution of the multivalued variational inequality (2.1) and to suggest the iterative algorithm for solving variational inequalities and complementarity problems.

3. Existence Results

In this section, we study those conditions under which there exists a unique solution of the variational inequalities (2.1).

Theorem 3.1. Let the multivalued operator $T : H \to 2^H$ be a strongly monotone with constant $\alpha > 0$ and M-Lipschitz continuous with constant $\beta > 0$. Let $V : H \to C(H)$ be a M-Lipschitz continuous operator with constant $\eta > 0$. If $\eta < \alpha$, then there exists a unique solution $u \in K, w \in T(u),\ y \in V(u)$ such that (2.1) holds.

Proof. Uniqueness. Let $u_1, u_2 \in K,\ u_1 \neq u_2$ be two solutions of (2.1), then

$$< w_1 + y_1, v - u_1 >\geq 0 \quad \text{for all } v \in K \tag{3.1}$$

and

$$< w_2 + y_2, v - u_2 >\geq 0 \quad \text{for all } v \in K. \tag{3.2}$$

Taking $v = u_2$ in (3.1) and $v = u_1$ in (3.2), and adding the resultant inequalities, we have

$$< w_1 - w_2, u_1 - u_2 >\leq< y_1 - y_2, u_2 - u_1 > .$$

Since the multivalued operator T is a strongly monotone so there exists a constant $\alpha > 0$, such that

$$\alpha\|u_1 - u_2\|^2 \leq< w_1 - w_2, u_1 - u_2 >\leq< y_2 - y_1, u_1 - u_2 >\leq \|y_1 - y_2\|\|u_1 - u_2\|. \tag{3.3}$$

Further, since V is M-Lipschitz continuous, we have

$$\|y_1 - y_2\| \leq (M(u_1), M(u_2)) \leq \eta\|u_1 - u_2\|. \tag{3.4}$$

From (3.3) and (3.4) we have

$$\alpha\|u_1 - u_2\| \leq \eta\|u_1 - u_2\|^2,$$

which implies that

$$(\alpha - \eta)\|u_1 - u_2\|^2 \leq 0.$$

Since $\eta < \alpha$, so we conclude that $u_1 = u_2$, the uniqueness of the solution of the multivalued variational inequality (2.1).

Existence. From Lemma 2.2, we know that the variational inequality (2.1) is equivalent to the fixed point problem.

$$F(u) = P_K[u - \rho(w + y)]. \tag{3.5}$$

In order to prove the existence of a solution of (2.1), it is enough to show that the problem (3.5) has a fixed point. Thus for all $u_1, u_2 \in K$, we have

$$\begin{aligned} \|F(u_1) - F(u_2)\| &= \|P_K[u_1 - \rho(w_1 + y_1)] - P_K[u_2 - \rho(w_2 + y_2)]\| \\ &\leq \|u_1 - u_2 - \rho(w_1 - w_2) - \rho(y_1 - y_2)\| \\ &\leq \|u_1 - u_2 - \rho(w_1 - w_2)\| + \rho\|y_1 - y_2\|. \end{aligned} \tag{3.6}$$

Since T is a strongly monotone M-Lipschitz continuous multivalued operator, so

$$\begin{aligned} \|u_1 - u_2 - \rho(w_1 - w_2)\|^2 &= \|u_1 - u_2\|^2 - 2\rho < w_1 - w_2, u_1 - u_2 > +\rho^2\|w_1 - w_2\|^2 \\ &\leq (1 - 2\rho\alpha)\|u_1 - u_2\|^2 + \rho^2\{M(T(u_1), T(u_2)\}^2 \\ &\leq (1 - 2\rho\alpha + \rho^2\beta^2)\|u_1 - u_2\|^2. \end{aligned} \quad (3.7)$$

Combining (3.4), (3.6) and (3.7), we have

$$\begin{aligned} \|F(u_1) - F(u_2)\| &\leq \{\sqrt{(1 - 2\rho\alpha + \beta^2\rho^2)} + \rho\eta\}\|u_1 - u_2\| \\ &= \theta\|u_1 - u_2\|, \end{aligned}$$

where $\theta = \sqrt{1 - 2\rho\alpha + \beta^2\rho^2} + \rho\eta < 1$ for $0 < \rho < \frac{\alpha-\eta}{\beta^2-\eta^2}$, $\rho\eta < 1$ and $\eta < \alpha$.

Since $\theta < 1$, so the map $F(u)$ defined by (3.5) is a contraction mapping and hence has a fixed point $u \in K$, a closed convex set in H, satisfying the multivalued variational inequality (2.1). This completes the proof. □

4. Iterative Algorithms

In this section, we invoke Lemma 2.2 to suggest an iterative algorithm for solving variational inequality (2.1) and its various special cases. We note that (2.6) can be written as

$$u = (1 - \lambda)u + \lambda P_K[u - \rho(w + y)], \quad (4.1)$$

where $0 < \lambda < 1$ is a parameter and $\rho > 0$ is a constant. This formulation is used to suggest the following unified algorithm.

Algorithm 4.1. Assume that $T : H \to 2^H$, $V : H \to C(H)$ and K is a closed convex set in H. For a given $u_0 \in K$, let us consider $w_0 \in T(u_0)$, $y_0 \in V(u_0)$ and

$$u_1 = (1 - \lambda)u_0 + \lambda P_K[u_0 - \rho(w_0 + y_0)].$$

Since $y_0 \in V(u_0), w_0 \in T(u_0)$, there exist $y_1 \in V(u_1), w_1 \in T(u_1)$ such that

$$\begin{aligned} \|y_0 - y_1\| &\leq M(V(u_0), V(u_1)) \\ \|w_0 - w_1\| &\leq M(T(u_0), T(u_1)), \end{aligned}$$

where $M(\cdot,\cdot)$ is the Hausdorff metric on $C(H)$. Let

$$u_2 = (1 - \lambda)u_1 + \lambda P_K[u_1 - \rho(w_1 + y_1)].$$

Continuing this way, we can compute the sequences $\{u_n\}, \{w_n\}$ and $\{y_n\}$ from the iterative schemes

$$w_n \in T(u_n) : \ \|w_n - w_{n+1}\| \leq M(T(u_n), T(u_{n+1})) \quad (4.2)$$

$$y_n \in V(u_n) : \ \|y_n - y_{n+1}\| \leq M(V(u_n), V(u_{n+1})) \quad (4.3)$$

$$u_{n+1} = (1 - \lambda)u_n + \lambda P_K[u_n - \rho(w_n + y_n)], \quad n = 0, 1, 2, \cdots. \quad (4.4)$$

If $V \equiv I$, is the identity operator, then Algorithm 4.1 reduces to:
Algorithm 4.2. For a given $u_0 \in K$ $w_0 \in T(u_0)$, compute the sequences $\{u_n\}$, and $\{w_n\}$ from the iterative schemes

$$w_n \in T(u_n): \ \|w_n - w_{n+1}\| \leq M(T(u_n), T(u_{n+1}))$$
$$u_{n+1} = (1-\lambda)u_n + \lambda P_K[u_n - \rho(w_n + y_n)], \quad n = 0, 1, 2, \cdots.$$

If $T, V : H \to H$ are the single-valued operators, then Algorithm 4.1 collapses to:
Algorithm 4.3. For a given $u_0 \in K$, compute $\{u_n\}$ by the iterative schemes

$$u_{n+1} = (1-\lambda)u_n + \rho P_K[u_n - \rho(T(u_n) + V(u_n))], \ n = 0, 1, 2, \cdots.$$

For the convergence analysis of Algorithm 4.3, see Noor [34].

For appropriate and suitable choice of the operators T, V and the convex set K, one can obtain various iterative algorithms for solving variational inequalities and complementarity problems as special cases from Algorithm 4.1.

We now study those conditions under which the approximate solution computed by Algorithm 4.1 converges to the exact solution of the multivalued strongly nonlinear variational inequality (2.1).

Theorem 4.1. Let the operator $T : H \to 2^H$ be strongly monotone with constant $\alpha > 0$ and M-Lipschitz continuous with constant $\beta > 0$. Let $V : H \to C(H)$ be M-Lipschitz continuous with constant $\eta > 0$. If $\alpha < \eta$, then there exists $u \in K$, $w \in T(u)$, $y \in V(u)$, which are the solutions of the multivalued strongly nonlinear variational inequality (2.1), and the sequences $\{u_n\}, \{w_n\}$ and $\{y_n\}$ generated by Algorithm 4.1 converge strongly to u, w, y in H respectively.

Proof. From Algorithm 3.1, we have

$$\begin{aligned} \|u_{n+1} - u_n\| &\leq (1-\lambda)\|u_n - u_{n-1}\| + \lambda\|P_K[u_n - \rho(w_n - \rho(w_n + y_n)] \\ &\quad - P_K[u_{n-1} - \rho(w_{n-1} + y_{n-1})]\| \\ &\leq (1-\lambda)\|u_n - u_{n-1}\| + \lambda\|u_n - u_{n-1} - \rho(w_n - w_{n-1})\| + \lambda\rho\|y_n - y_{n-1}\| \\ &\leq \{(1-\lambda) + \lambda(\sqrt{1 - 2\alpha\rho + \beta^2\rho^2} + \rho\eta)\}\|u_n - u_{n-1}\|, \\ &= \{1 - \lambda(1-\theta)\}\|u_n - u_{n-1}\|, \end{aligned} \tag{4.5}$$

where

$$\theta = \sqrt{1 - 2\alpha\rho + \beta^2\rho^2} + \rho\eta. \tag{4.6}$$

Thus

$$\|u_{n+1} - u_n\| \leq h\|u_n - u_{n-1}\|, \tag{4.7}$$

where $h = 1 - \lambda(1-\theta)$.

Since $\theta = \sqrt{1 - 2\alpha\rho + \beta^2\rho^2} + \rho\eta < 1$ for $0 < \rho < \frac{\alpha - \eta}{\beta^2 - \eta^2}$, $\rho\eta < 1$ and $\eta < \alpha$. Thus it follows that $0 < h < 1$. Hence from (4.7), it follows that the sequence $\{u_n\}$ is a

Cauchy sequence in H so that there exists $u \in H$ with $u_n \to u$ as $n \to \infty$. Also from (4.3), we have

$$\|y_{n+1} - y_n\| \leq M(V(u_{n+1}), V(u_n)) \leq \eta \|u_{n+1} - u_n\|,$$

from which it follows that the sequence $\{y_n\}$ is also a Cauchy sequence in H. Thus there exists $y \in H$ such that $y_n \to y$ as $n \to \infty$. Now by using the continuity of the operators T, V, P_K and Algorithm 4.1, we have

$$u = P_K[u - \rho(w + y)] \in K.$$

Now we shall show that $y \in V(u)$. In fact,

$$\begin{aligned} d(y, V(u)) &\leq \|y - y_n\| + d(y_n, V(u)) \leq \|y - y_n\| + M(V(u_n), V(u)) \\ &\leq \|y - y_n\| + \eta \|u_n - u\| \to 0 \text{ as } n \to \infty, \end{aligned}$$

where $d(y, V(u)) = \inf\{\|y - v\| : v \in V(u)\}$. Consequently, we have $d(y, V(u)) = 0$, from which it follows that $y \in V(u)$, since $V(u) \in C(H)$. By invoking Lemma 2.1, it follows that $u \in K$, $w \in T(u)$, $y \in V(u)$ are solutions of the variational inequality (2.1) and $u_n \to u$, $w_n \to w$, $y_n \to y$ strongly in H, which is the required result. □

5. Wiener-Hopf Equations Technique

In this section, we introduce a class of Wiener-Hopf equations and establish the equivalence between the multivalued Wiener-Hopf equation and the multivalued variational inequalities (2.1). To be more precise, let $Q_K = I - P_K$, where I is the identity operator and P_K is the projection of H into the closed convex set K in H. For given multivalued operators $T, V : H \to 2^H$, consider the problem of finding $z \in H, u \in K$, $w \in T(u)$, $y \in V(u)$ such that

$$w + \rho^{-1} Q_K z = -y, \tag{5.1}$$

where $\rho > 0$ is a constant. The equations of the type (5.1) are called the multivalued Wiener-Hopf equations.

If $T, V : H \to H$ are singlevalued operators, then the problem (5.1) is equivalent to finding $z \in H$ such that

$$T P_K z + \rho^{-1} Q_K z = -V z, \tag{5.2}$$

which are called strongly nonlinear Wiener-Hopf equations. One can easily establish the equivalence between the equations (5.2) and the variational inequalities (2.3). This equivalence can be used to suggest a number of iterative algorithms for solving strongly nonlinear variational inequalities. For the applications and formulation of the

Wiener-Hopf equations, see Speck [56], Noor-Noor and Rassias [44,45] and Robinson [52].

We now show that the multivalued Wiener-Hopf equations (5.1) are equivalent to the multivalued variational inequalities (2.1) essentially using Lemma 2.2 and the techniques of Shi [53,54] and Noor [38].

Theorem 5.1. The functions $u \in K$, $w \in T(u)$, $y \in V(u)$ are solutions of the multivalued variational inequalities (2.1) if and only if $z \in H$, $u \in K$, $w \in T(u)$, $y \in V(u)$ satisfy the multivalued Wiener-Hopf equations (5.1), where

$$u = P_K z \tag{5.3}$$

$$z = u - \rho(w+y), \tag{5.4}$$

and $\rho > 0$ is a constant.

Proof. Let $u \in K$, $w \in T(u)$, $y \in V(u)$ be a solution of the variational inequality (2.1). Then by Lemma 2.2, we have

$$u = P_K[u - \rho(w+y)]. \tag{5.5}$$

Now using the fact that $Q_K = I - P_K$ and the equation (5.5), we obtain

$$\begin{aligned} Q_K[u - \rho(w+y)] &= u - \rho(w+y) - P_K[u - \rho(w+y)] \\ &= -\rho(w+y), \end{aligned}$$

from which and (5.3) it follows that

$$w + \rho^{-1} Q_K z = -y.$$

Conversely, let $z \in H$, $u \in K$, $w \in T(u)$, $y \in V(u)$ be a solution of (5.1), then

$$\rho(w+y) = -Q_K z = P_K z - z. \tag{5.6}$$

Now from Lemma 2.1 and (5.6), for all $v \in K$, we have

$$0 \leq < P_K z - z, v - P_K z >= \rho < w + y, v - P_K z > .$$

Thus (u, w, y), where $u = P_K z$ is a solution of (2.1). □

Theorem 5.1 implies that the variational inequality (2.1) and the Wiener-Hopf equations (5.1) are equivalent. This equivalence is quite general and flexible. This equivalence is used to suggest a number of iterative algorithms for solving variational inequalities.

I. The equation (5.1) can be written as

$$Q_K z = -\rho(w+y)$$

from which it follows that

$$\begin{aligned} z &= P_K z - \rho(w+y) \\ &= u - \rho(w+y), \quad \text{using (5.3).} \end{aligned}$$

This fixed point formulation enables us to suggest the following iterative scheme.

Algorithm 5.1. For given $z_0 \in H, u_0 \in K, w_0 \in (Tu_0), y_0 \in V(u_0)$, compute the sequences $\{z_n\}, \{u_n\}, \{w_n\}$, and $\{y_n\}$ by the iterative schemes

$$\begin{aligned} &u_n = P_K z_n \\ &w_n \in (Tu_n) : \|w_{n+1} - w_n\| \leq M(T(u_{n+1}), T(u_n)) \\ &y_n \in V(u_n) : \|y_{n+1} - y_n\| \leq M(V(u_{n+1}), V(u_n)) \\ &z_{n+1} = (1-\lambda)u_n + \lambda\{u_n - \rho(w_n + y_n)\}, \quad n = 0, 1, 2, \cdots. \end{aligned}$$

where $0 < \lambda < 1$ is a parameter and $\rho > 0$ is a constant.

II. The equations (5.1) may be written as

$$Q_K z = (1 - \rho^{-1})Q_K z - (w + y),$$

which implies that

$$\begin{aligned} z &= P_K z - (w+y) + (1-\rho^{-1})Q_K z \\ &= u - (w+y) + (1-\rho^{-1})Q_K z, \quad \text{using (5.3).} \end{aligned}$$

Using this fixed point formulation, we can suggest the following iterative scheme.

Algorithm 5.2. For given $z_0 \in H, u_0 \in K, w_0 \in (Tu_0), y_0 \in V(u_0)$, compute the sequences $\{z_n\}, \{u_n\}, \{w_n\}$ and $\{y_n\}$ from the iterative schemes

$$\begin{aligned} &u_n = P_K z_n \\ &w_n \in (Tu_n) : \|w_{n+1} - w_n\| \leq M(T(u_{n+1}), T(u_n)) \\ &y_n \in V(u_n) : \|y_{n+1} - y_n\| \leq M(V(u_{n+1}), V(u_n)) \\ &z_{n+1} = (1-\lambda)u_n + \lambda\{u_n - (w_n) + y_n) + (1-\rho^{-1})Q_K z_n\}, \quad n = 0, 1, 2, \cdots \end{aligned}$$

where $0 < \lambda < 1$ is a parameter and $\rho > 0$ is a constant.

For suitable and appropriate choice of the operators T, V and the convex set K, one can obtain a number of iterative algorithms for solving various classes of variational inequalities and complementarity problems.

Using essentially the technique of Theorem 4.1, we can study the convergence analysis of the Algorithm 5.1 and Algorithm 5.2. See also Theorem 7.2.

6. Auxiliary Principle Technique

In this section, we consider a method for solving variational inequalities (2.1), which do not depend on the projection technique. This technique deals with an auxiliary variational inequality problem by proving that the solution of the auxiliary problem is the solution of the original problem. This technique is mainly due to Glowinksi, Lions and Tremolieres [18] and Noor [35,39,41-43]. This technique is quite general and flexible. By a suitable choice, one can consider an appropriate auxiliary problem related to the original problem. This approach has been used by Zhu and Marcotte [58] and Noor [46] to reformulate the nonsymmetric variational inequality problem into a variational problem. It has been used by Fukushima [17] and Zhu and Marcotte [58] to develop general descent frame for solving the standard variational inequality problem. We use this approach to study not only the existence of a solution of the problem (2.1) but also to suggest a general iterative algorithm for multivalued variational inequalities.

Theorem 6.1. Assume that K is a closed convex set in H. Let the multivalued operator, $T : H \to 2^H$, be a strongly monotone with constant $\alpha > 0$ and M-Lipschitz continuous with constant $\beta > 0$ and $V : H \to C(H)$ be a M-Lipschitz continuous with constant $\eta > 0$. If $\eta < \alpha$, then there exists a unique solution $u \in K$, $w \in T(u)$, $y \in V(u)$ such that the inequality (2.1) holds.

Proof. We now use the auxiliary principle technique to prove the existence of a solution of (2.1). For given $u \in K$, we consider the auxiliary problem of finding a unique $z \in K$, $w \in T(u)$, $y \in V(u)$ such that

$$< z, v - z > \geq < u, v - z > -\rho < w + y, v - z >, \quad \text{for all } v \in K\ , \tag{6.1}$$

where $\rho > 0$ is a constant.

It is clear that the relation (6.1) defines a mapping $u \to z$. In order to prove the existence of a solution of (2.1), it is enough to show that the mapping $u \to z$ defined by (6.1) has a fixed point belong to K satisfying (2.1). Let $z_1, z_2 \in K$, $z_1 \neq z_2$, be two solutions of (6.1) related to $u_1, u_2 \in H$ respectively. Taking $v = z_2$ (respectively z_1) in (6.1) related to u_1 (respectively u_2), we have

$$< z_1, z_2 - z_1 > \geq < u_1, z_2 - z_1 > -\rho < w_1 + y_1, z_2 - z_1 >$$

and

$$< z_2, z_1 - z_2 > \geq < u_2, z_1 - z_2 > -\rho < w_2 + y_2, z_1 - z_2 > .$$

Adding these inequalities, we have

$$\begin{aligned} < z_1 - z_2, z_1 - z_2 > &\leq < u_1 - u_2, z_1 - z_2 > -\rho < w_1 - w_2, z_1 - z_2 > \\ &\quad -\rho < y_1 - y_2, z_1 - z_2 >, \\ &= < u_1 - u_2 - \rho(w_1 - w_2), z_1 - z_2 > -\rho < y_1 - y_2, z_1 - z_2 >, \end{aligned}$$

from which, (3.4) and (3.7), we have

$$\begin{aligned}\|z_1 - z_2\|^2 &\leq \|u_1 - u_2 - \rho(w_1 - w_2)\|\|z_1 - z_2\| + \rho\|y_1 - y_2\|\|z_1 - z_2\| \\ &\leq \{\sqrt{1 - 2\rho\alpha + \rho^2\beta^2} + \rho\eta\}\|u_1 - u_2\|\|z_1 - z_2\|.\end{aligned}$$

Thus, it follows that

$$\begin{aligned}\|z_1 - z_2\| &\leq \{\sqrt{1 - 2\rho\alpha + \rho^2\beta^2} + \rho\eta\}\|u_1 - u_2\|. \\ &= \{t(\rho) + \rho\eta\}\|u_1 - u_2\| = \theta\|u_1 - u_2\|,\end{aligned}$$

where $\theta = t(\rho) + \rho\eta, \quad t(\rho) = \sqrt{1 - 2\rho\alpha + \rho^2\beta^2}$.

We have to show that $\theta < 1$. It is evident that $t(\rho)$ assumes its minimum value for $\bar{\rho} = \frac{\alpha}{\beta^2}$ with $t(\bar{\rho}) = \sqrt{1 - \frac{\alpha^2}{\beta^2}}$. For $\rho = \bar{\rho}$, $\rho\eta + t(\rho) < 1$ implies that $\rho\eta < 1$ and $\eta < \alpha$. Thus it follows that $\theta < 1$ for all ρ with $0 < \rho < 2\frac{\alpha - \eta}{\beta^2 - \eta^2}$,$\rho\eta < 1$ and $\eta < \alpha$. Since $\theta < 1$, so the mapping $u \rightarrow z$ defined by (6.1) has a fixed point belonging to K, which is the solution of (2.1). This completes the proof. □

Remark 6.1. We note that for a given $u \in K$, the solution $z \in K, w \in T(u), y \in V(u)$ satisfying the auxiliary variational inequality (6.1) is equivalent to finding the minimum of the auxiliary functional $F[z]$ on K in H, where

$$\begin{aligned}F[z] &= \frac{1}{2} < z - u, z - u > + \rho < w + y, z - u > \\ &= \|z - (u - (w + y))\|^2, \quad \text{for } \rho = 1,\end{aligned} \tag{6.2}$$

which is a differentiable quadratic convex functional. It is worth mentioning that the functional (6.1) reduces to the following functional

$$\begin{aligned}F_1[z] &= \frac{1}{2} < z - u, z - u > + \rho < Tu + V - u, z - u >, \\ &= \|z - (u - (Tu + Vu))\|^2, \quad \text{for } \rho = 1,\end{aligned} \tag{6.3}$$

for $T, V : H \rightarrow H$ single-valued operators. Note that for $V \equiv 0$, the functional $F_1[z]$ defined by (6.3) is exactly the auxiliary functional used by Fukushima [17] to suggest the descent framework for solving variational inequalities. It can be easily shown that the optimal solution of (6.2) is the projection of the point $(u - (w + y))$ onto the convex set K, that is,

$$z(u) = P_K[u - (w + y)]. \tag{6.4}$$

It follows from the relation (6.4) that the map $z(u)$ defined by (6.4) is the fixed point characterization of the variational inequality (2.1). In addition, we conclude that the auxiliary principle technique is equivalent to the projection method. Based on the above discussion and using the technique of Fukushima [17] and Zhu and Marcotte

[58], one can easily show that the variational inequality problem (2.1) is equivalent to finding the minimum of the functional $f[u]$ on K in H, where

$$f[u] = < w + y, u - z(u) > + \frac{1}{2} < z(u) - u, u - z(u) > . \tag{6.5}$$

The functional $f[u]$ defined by (6.5) is known as the gap (merit) function. For different choice of the gap (merit) functions, see [26] and the references therein.

Following the ideas of Cohen [7] and Noor [39,40], we propose and analyze a general auxiliary principle related with the multivalued variational inequality (2.1).

For given $u \in K$, $w \in T(u)$, $y \in V(u)$, we consider the general auxiliary problem of finding the minimum of the functional $I[z]$ on the convex set K in H, where

$$I[z] = E(z) - E(u) - < E'(u), z - u > + \rho < w + y, z - u > . \tag{6.6}$$

One can easily characterize the minimum of $I[z]$ on K by the following auxiliary variational inequality

$$< E'(z), v - z > \geq < E'(u), v - z > - \rho < w + y, v - z >, \quad \text{for all } v \in K. \tag{6.7}$$

We note that if $z = u$, then clearly z is a solution of the variational inequality (2.1). We like to point out the functional $I[z]$ defined by (6.6) is the variational principle associated with the multivalued variational inequality (2.1) in the terminology of Auchmuty [1] and Blum and Oettli [4]. Thus it is evident that for a suitable and appropriate choice of the auxiliary principle problem, one can suggest various gap (merit) functions for different classes of variational inequalities and complementarity problems. These gap functions have been used by many authors including Fukushima [17], Zhu and Marcotte [58] and Larsson and Patriksson [26] to suggest unified descent and Newton type algorithms for solving variational inequalities.

Continuing the above line of argument and observations, we may suggest a general auxiliary variational inequality problem which includes the above mentioned auxiliary problems as special cases.

Auxiliary Problem. For given $u \in K$, $w \in T(u)$, $y \in V(u)$, we consider the auxiliary multivalued variational inequality problem of finding $z \in K$ such that

$$< M(z), v - z > \geq < M(u), v - z) - \rho(w + y, v - z >, \quad \text{for all } v \in K . \tag{6.8}$$

where $M : H \rightarrow H$ is a nonlinear operator and $\rho > 0$ is a constant.

We note that if the operator M is a linear symmetric positive operator, then one can easily show that the auxiliary problem (6.8) is equivalent to finding the minimum of the auxiliary functional $N[z]$ on K in H, where

$$N[z] = \frac{1}{2} < M(u) - M(z), u - z > + \rho < w + y, u - z > .$$

We remark that if $z = u$, then z is a solution of the multivalued variational inequality (2.1). On the basis of the above observations and the discussions, we now suggest the following general iterative algorithm.

Algorithm 6.1.

(a) At $n = 0$, start with some initial z_0.

(b) At step n, solve the auxiliary problem (6.8) with $u = z_n$. Let z_{n+1} denote the solution of the problem (6.8).

(c) For given $\varepsilon > 0$, if $\|z_{n+1} - z_n\| \leq \varepsilon$, stop, otherwise repeat (b).

Using the technique of Noor [41], one can study the convergence analysis of Algorithm 6.1.

7. Quasi Variational Inequalities

We remark that if the convex set K in H depends on the solution u implicitly or explicitly, then the variational inequality is known as the quasi variational inequality. This type of problems was originally considered and studied by Bensoussan and Lions [3] in impulse control. Mosco [30] showed that many obstacle and unilateral problems can be studied in the general framework of quasi variational inequalities. For recent developments and applications, see [2,3,5,6,9,10,12,19,23,30,36,44,45].

Given a point-to-set mapping $K : u \to K(u)$, which associates a closed convex set $K(u)$ with any element u of H and multivalued operators $T : H \to 2^H$ and $V : H \to C(H)$, we consider the problem of finding $u \in K(u)$, $w \in T(u)$, $y \in V(u)$ such that

$$< w + y, v - u > \geq 0, \quad \text{for all } v \in K(u). \tag{7.1}$$

The problem (7.1) is known as the multivalued strongly nonlinear quasi variational inequality. Note that if $K(u) \equiv K$, that is, K is independent of the solution u, then the problem (7.1) is exactly the problem (2.1).

We remark that if $T, V : H \to H$ are single-valued operators and $K(u) \equiv K$, then problem (7.1) is equivalent to finding $u \in K$ such that

$$< T(u) + V(u), v - u > \geq 0, \quad \text{for all } v \in K,$$

which is known as the strongly nonlinear variational inequality problem. This problem is mainly due to Noor [32,33]. For the recent applications and generalizations, see [6,12,34-43].

Using Lemma 2.1, and the technique of Noor [36], we can easily show that the problem (7.1) is equivalent to the fixed point problem

$$u = P_{K(u)}[u - \rho(w + y)],$$

where $\rho > 0$ is a constant. This fixed point formulation can be used to suggest the following iterative scheme.

Algorithm 7.1. Assume that $K(u)$ is a closed convex set in H, $T : H \to 2^H$ and $V : H \to C(H)$. For given $u_0 \in K(u_0)$, $w_0 \in T(u_0)$, $y_0 \in V(u_0)$, compute the sequences $\{u_n\}$, $\{w_n\}$ and $\{y_n\}$ from the iterative schemes

$$\begin{aligned} &w_n \in T(u_n) : \|w_{n+1} - w_n\| \le M(T(u_{n+1}), T(u_n)) \\ &y_n \in V(u_n) : \|y_n - y_{n+1}\| \le M(V(u_n), V(u_{n+1})) \\ &u_{n+1} = (1-\lambda)u_n + \lambda P_{K(u_n)}[u_n - \rho(w_n + y_n)], \quad n = 0, 1, 2, \cdots \end{aligned}$$

where $0 < \lambda < 1$ is a parameter and $\rho > 0$ is a constant.

We note that if $K(u) = m(u) + K$, where m is a point-to-point mapping and K is a closed convex set in H, then Algorithm 7.1 reduces to:

Algorithm 7.2. Assume that $K(u) = m(u) + K$, $T : H \to 2^H$, and $V : H \to C(H)$. For given $u_0 \in K(u_0)$, $w_0 \in T(u_0)$, $y_0 \in V(u_0)$, find the approximate solutions u_n, w_n and y_n from the iterative schemes

$$\begin{aligned} &w_n \in T(u_n) : \|w_{n+1} - w_n\| \le M(T(u_{n+1}), T(u_n)) \\ &y_n \in V(u_n) : \|y_n - y_{n+1}\| \le M(V(u_n), V(u_{n+1})), \\ &u_{n+1} = (1-\lambda)u_n + \lambda\{m(u_n) + P_K[u_n - \rho(w_n + y_n) - m(u_n)], \quad n = 0, 1, 2, \cdots. \end{aligned}$$

Algorithm 7.3. Assume that $K(u) = m(u) + K$ and $T, V : H \to 2^H$. For given $u_0 \in K(u_0)$, $w_0 \in T(u_0)$, $y_0 \in V(u_0)$, compute the sequences $\{u_n$, $\{w_n\}$ and $\{y_n\}$ from the iterative schemes

$$\begin{aligned} &w_n \in T(u_n) : \|w_{n+1} - w_n\| \le M(T(u_{n+1}), T(u_n)) \\ &y_n \in V(u_n) : \|y_n - w_{n+1}\| \le M(V(u_n), V(u_{n+1})) \\ &u_{n+1} = (1-\lambda)u_n + \lambda\{m(u_n) + P_K[u_n - \rho(w_n + y_n) - m(u_n)]\}, \quad \text{for } n = 0, 1, 2, \cdots. \end{aligned}$$

Related to the multivalued quasi variational inequality problem (7.1), we can consider the Wiener-Hopf equations. Let $Q_{K(u)} \equiv I - P_{K(u)}$, where $P_{K(u)}$ is the projection of H onto $K(u)$ and I is the identity operator. For given multivalued operators $T : H \to 2^H$ and $V : H \to C(H)$, consider the problem of finding $z \in H$, $u \in K(u)$, $w \in T(u)$, $y \in V(u)$ such that

$$w + \rho^{-1} Q_{K(u)} z = -y, \tag{7.2}$$

where $\rho > 0$ is a constant. The equations (7.2) are called the multivalued strongly nonlinear implicit Wiener-Hopf equations. If $K(u) \equiv K$, then the equations (7.2) collapse to (5.1).

Using essentially the technique of Theorem 5.1, we can show that the multivalued quasi variational inequality (7.1) is equivalent to the multivalued implicit Wiener-Hopf equations (7.2). In fact, we have the following result in this direction.

Theorem 7.1. The multivalued quasi variational inequality (7.1) has a solution $u \in K(u)$, $w \in T(u)$, $y \in V(u)$ if and only if $z \in H$, $u \in K(u)$, $w \in T(u)$, $y \in V(u)$ satisfy the multivalued implicit Wiener-Hopf equations (7.2), where

$$u = P_{K(u)}z \tag{7.3}$$

$$z = u - \rho(w + y), \tag{7.4}$$

and $\rho > 0$ is a constant.

Proof. Its proof is similar to that of Theorem 5.1.

Theorem 7.1 establishes the equivalence between the multivalued quasi variational inequality (7.1) and the multivalued implicit Wiener-Hopf equations (7.2). Using this equivalence, we can suggest a number of new iterative algorithms for solving multivalued quasi variational inequalities (7.1). To convey an idea, we consider only the following.

The equations (7.2) can be written as

$$Q_{K(u)}z = -\rho(w + y),$$

which implies that

$$\begin{aligned} z &= P_{K(u)}z - \rho(w + y) \\ &= u - \rho(w + y), \text{using (7.3).} \end{aligned} \tag{7.5}$$

This fixed point formulation enables us to suggest:

Algorithm 7.4. For given $z_0 \in H$, $u_0 \in K(u_0)$, $w_0 \in T(u_0)$, $y_0 \in V(u_0)$, find the approximate sequences $\{z_n\}, \{u_n\}, \{w_n\}$ and $\{y_n\}$ by the iterative schemes

$$u_n = P_{K(u_n)}z_n \tag{7.6}$$

$$w_n \in T(u_n) : \|w_{n+1} - w_n\| \leq M(T(u_{n+1}), T(u_n)) \tag{7.7}$$

$$y_n \in V(u_n) : \|y_n - y_{n+1}\| \leq M(V(u_n), V(u_{n+1})), \tag{7.8}$$

$$z_{n+1} = u_n - \rho(w_n + y_n), \quad n = 0, 1, 2, \cdots. \tag{7.9}$$

where $\rho > 0$ is a constant.

We now study the convergence criteria of Algorithm 7.1. For this purpose, we need the following assumption, which plays a significant part in the convergence analysis of the Algorithm 7.4.

Assumption 7.1. For all $u, v, z \in H$, the operator $P_{K(u)}$ satisfies the relation

$$\|P_{K(u)}z - P_{K(v)}z\| \leq \gamma\|u - v\|,$$

where $\gamma > 0$ is a constant.

Remark 7.1. We remark that the assumption 7.1 is true for the special case, if

$$K(u) = m(u) + K, \tag{7.10}$$

where m is a point-to-point mapping and K is a closed convex set in H. Hence

$$P_{K(u)}v = P_{K+m(u)}v = m(u) + P_K[v - m(u)]. \tag{7.11}$$

If the point-to-point mapping m is Lipschitz continuous with constant $\mu > 0$, then from (7.10) and (7.11) we have

$$\begin{aligned} \|P_{K(u)}z - P_{K(v)}z\| &= \|m(u) - m(v) + P_K[z - m(u)] - P_K[z - m(v)]\| \\ &\leq \|m(u) - m(v)\| + \|P_K[z - m(u)] - P_K[z - m(v)]\| \\ &\leq 2\|m(u) - m(v)\| \leq 2\mu\|u - v\|, \end{aligned}$$

which shows that the assumption 7.1 is true for $\gamma = 2\mu$.

Theorem 7.2. Let the multivalued operator $T : H \to 2^H$ be strongly monotone with constant $\alpha > 0$ and M-Lipschitz continuous with constant $\beta > 0$. Let $V : H \to C(H)$ be M-Lipschitz continuous with constant $\eta > 0$ If the Assumption 7.1 holds and

$$\left|\rho - \frac{\alpha - \eta k}{\beta^2 - \eta^2}\right| < \frac{\sqrt{(\alpha - \eta k)^2 - (\beta^2 - \eta^2)(1 - k^2)}}{\beta^2 - \eta^2}, \tag{7.12}$$

$$\alpha > \eta k + \sqrt{(\beta^2 - \eta^2)(1 - k^2)}, \tag{7.13}$$

$$k = 1 - \gamma \neq 0, \tag{7.14}$$

then there exist $z \in H$, $u \in K(u)$, $w \in (Tu)$, $y \in V(u)$, which are the solutions of the multivalued implicit Wiener-Hopf equations (7.2) and the sequences $\{z_n\}, \{u_n\}, \{w_n\}$ and $\{y_n\}$ generated by Algorithm 7.4 converge to z, u, w and y strongly in H respectively.

Proof. From Algorithm 7.4, (3.4) and (3.7) we have

$$\begin{aligned} \|z_{n+1} - z_n\| &\leq \|u_n - u_{n-1} - \rho(w_n - w_{n-1})\| + \rho\|y_n - y_{n-1}\| \\ &\leq \{\sqrt{1 - 2\rho\alpha + \rho^2\beta^2} + \rho\eta\}\|u_n - u_{n-1}\|. \end{aligned} \tag{7.15}$$

From Assumption 7.1 and (7.6), we have

$$\begin{aligned} \|u_n - u_{n-1}\| &= \|P_{K(u_n)}z_n - P_{K(u_{n-1})}z_{n-1}\| \\ &\leq P_{K(u_n)}z_n - P_{K(u_{n-1})}z_n\| + \|P_{K(u_{n-1})}z_n - P_{K(u_{n-1})}z_n\| \\ &\leq \gamma\|u_n - u_{n-1}\| + \|z_n - z_{n-1}\|, \end{aligned}$$

which implies that

$$\|u_n - u_{n-1}\| \leq \frac{1}{1 - \gamma}\|z_n - z_{n-1}\| = \frac{1}{k}\|z_n - z_{n-1}\|, \tag{7.16}$$

where $k = 1 - \gamma \neq 0$.

Combining (7.15) and (7.16), we obtain

$$\begin{aligned} \|z_{n+1} - z_n\| &\leq \left\{ \frac{\sqrt{1 - 2\rho\alpha + \rho^2\beta^2} + \rho\eta}{k} \right\} \|z_n - z_{n-1}\| \\ &= \theta \|z_n - z_{n-1}\|, \end{aligned} \tag{7.17}$$

with $\theta = \dfrac{\sqrt{1 - 2\rho\alpha + \rho^2\beta^2} + \rho\eta}{k}$.

Now from (7.12), (7.13) and (7.14), it follows that $\theta < 1$. Hence from (7.17), we see that the sequence $\{z_n\}$ is a Caucy sequence in H, that is, $z_{n+1} \to z$ as $n \to \infty$. Similarly, the sequence $\{u_n\}$ is also a Cauchy sequences.

From (7.8), and the M-Lipschitz continuity of the operator $V : H \to C(H)$, we have

$$\|y_{n+1} - y_n\| \leq M(V(u_{n+1}), V(u_n)) \leq \eta \|u_{n+1} - u_n\|,$$

from which it follows that $\{y_n\}$ is a Cauchy sequence in H, that is, $y_n \to y$ as $n \to \infty$.

Using the continuity of the operators T, V, P_K and Theorem 7.1, we have

$$\begin{aligned} u &= P_{K(u)} z, \\ z &= u - \rho(w + y), \end{aligned}$$

that is,

$$z = u - \rho(w + y) \in H.$$

Now we show that $y \in V(u)$. In fact,

$$\begin{aligned} d(y, V(u)) &\leq \|y - y_n\| + d(y_n, V(u)) \leq \|y - y_n\| + M(V(u_n), V(u)) \\ &\leq \|y - y_n\| + \eta \|u_n - u\| \end{aligned}$$

where $d(y, V(u)) = \inf\{\|y - v\| : v \in V(u)\}$. As thesequences $\{y_n\}$ and $\{u_n\}$ are the Cauchy sequences, it follows from the above inequality that $d(y, V(u)) = 0$. This implies that $y \in V(u)$, since $V(u) \in C(H)$.

Using Theorem 7.1, we see that $z \in H$, $u \in K(u)$, $w \in (Tu)$, $y \in V(u)$ are solutions of the multivalued implicit Wiener- Hopf equations (7.2) and consequently $z_{n+1} \to z$, $u_{n+1} \to u$, $w_{n+1} \to w$ and $y_{n+1} \to y$ strongly in H. This completes the proof. □

Remark 7.2. In this section, we have shown that the projection method and the Wiener-Hopf equations techniques can be modified and extended to suggest a number of iterative type algorithms for the multivalued quasi variational inequalities. Extending the auxiliary principle technique considered in Section 6 for various classes of quasi variational inequalities and quasi complementarity problems is an open problem. This constitutes an interesting subject for future studies.

References

1. Auchmuty, G (1989) Variational principles for variational inequalities, Numer. Funct. Anal. Optim. 10, 863-874.

2. Baiocchi, C and Capelo, A (1984) Variational and Quasi- Variational Inequalities, J. Wiley and Sons, New York, London.

3. Bensoussan, A and Lions, J L (1978) Applications des Inequations Variationelles en Control et en Stochastiques, Dunod, Paris.

4. Blum, E and Oettli, W (1994) From optimization and variational inequalities to equilibrium problems, Math. Student 63, 123-145.

5. Chan, D and Pang, J S (1982) The generalized quasi variational inequality problems, Math. Oper. Research 7, 211-222.

6. Chang, S S and Huang, N J (1991) Generalized strongly nonlinear quasi complementarity problems in Hilbert spaces, J. Math. Anal. Appl. 158, 194-202.

7. Cohen, G (1988) Auxiliary problem principle extended to variational inequalities, J. Optim. Theory Appl. 59, 325- 333.

8. Cottle, R W, Pang, J S and Stone, R E (1992) The Linear Complementarity Problem, Academic Press, New York.

9. Cottle, R W, Giannessi, F and Lions, J L (1980) Variational Inequalities and Complementarity Problems: Theory and Applications, J. Wiley and Sons, New York.

10. Crank, J (1984) Free and Moving Boundary Problems, Clarendon Press, Oxford.

11. Dafermos, S (1990) Exchange price equilibrium and variational inequalities, Math. Program. 46, 391-402.

12. Ding, X P (1993) Generalized strongly nonlinear quasi variational inequalities, J. Math. Anal. Appl. 173, 577-587.

13. Duvaut, G and Lions, G (1972) Les Inequations en Mechanique et en Physique, Dunod, Paris.

14. Fang, S C and Peterson, E L (1982) Generalized variational inequalities, J. Optim Theory. Appl. 38, 363-383.

15. Fichera, G (1963-64) Problemi elastostatici con vincoli unilaterali: il problema di Signorini con ambigue condizioni al contorno, Atti Accad. Naz. Lincei Mem. Cl. Sci. Fis. Mat. Natur. Sez. IA (8) 7, 91-140.

16. Frèchet, M (1907) sur les ensembles de functions et les opertions lineaires, C.R. Acad. Sci., Paris, 1414-1416.

17. Fukushima, M (1992) Equivalent differentiable optimization problems and descent methods for assymetric variational inequality problems, Math. Program. 53, 99-110.

18. Glowinski, R, Lions J L and Tremolieres, R (1981) Numerical Analysis of Variational Inequalities, North-Holland, Amsterdam.

19. Harker, P T and Pang, J S (1990) Finite dimensional variational inequality and nonlinear complementarity problems: a survey of theory, algorithms and applications, Math. Programming 48, 161-220.

20. Hilbert, D (1900) Liber das Dirchletschen Princip., Jber Deutsch Math-Verrein, 8, 184-188.

21. Hilbert, D (1912) Grundzüge einer allegmeinen theorie der linearen integralgleichungen, Leipzig.

22. Hyers, D H, Isac, G and Rassias T M (1997) Topics in Nonlinear Analysis and Applications, World Scientific Publ. Co. Singapore.

23. Isac, G (1992) Complentarity problem, Lecture Notes in Mathematics, 1528, Springer Verlag, Berlin.

24. Kikuchi, N and Oden, J T (1988) Contact Problems in Elasticity, SIAM Publishing Co., Philadelphia.

25. Kinderlehrer, D and Stampacchia, G (1980) An Introduction to Variational Inequalities and Their Applications, Academic Press, London.

26. Larsson, T and Patriksson, M (1994) A class of gap functions for variational inequalities, Math. Program. 64, 53-79.

27. Lax, P D and Milgram, A N (1954) Parabolic equations, Annals Math. Study No. 33, Princeton, N.J., 167-190.

28. Lions, J L and Stampacchia, G (1967) Variational inequalities, Comm. Pure Appl. Math. 20, 493-519.

29. Minty, G (1962) Monotone (nonlinear) operators in Hilbert space, Duke Math. J., 29, 341-346.

30. Mosco, U (1976) Implicit variational problems and quasi variational inequalities, Lect. Notes Math. 543, Springer Verlag, Berlin, 83-126.

31. Von Neumann, J (1929) Allegmeine eigenwetheorie hermitscher functional operatoren, Math. Annal. 102, 49-131.

32. Aslam Noor, M (1971) Riesz-Frèchet Theorem and Monotonicity, M.Sc. Thesis, Queen's University, Kingston, Ontario, Canada.

33. Aslam Noor, M (1975) On Variational Inequalities, Ph.D. Thesis, Brunel University, London, U.K.

34. Aslam Noor, M (1982) Strongly nonlinear variational inequalities, C.R. Math. Rep. Acad. Sci. Canada 4, 213-218.

35. Aslam Noor, M (1987) General nonlinear variational inequalities, J. Math. Anal. Appl. 126, 78-84.

36. Aslam Noor, M (1985) An iterative scheme for a class of quasi variational inequalities, J. Math. Anal. Appl. 110, 463-468.

37. Aslam Noor, M (1991) An iterative algorithm for variational inequalities, J. Math. Anal. Appl. 158, 448-455.

38. Aslam Noor, M (1993) Wiener-Hopf equations and variational inequalities, J. Optim. Theory Appl. 79, 197-206.

39. Aslam Noor, M (1992) General algorithm for variational inequalities, J. Optim. Theory Appl. 73, 409-413.

40. Aslam Noor, M (1995) Equivalence of differentiable optimization problems for variational inequalities, J. Nat. Geometry, 8, 117-128.

41. Aslam Noor, M (1995) Some nonlinear variational inequalities, Tamkang J. Math. 26.

42. Aslam Noor, M (1994) General auxiliary principle for variational inequalities, PanAmer. Math. J. 4 (1), 27-44.

43. Aslam Noor, M (1995) On iterative algorithms for a class of nonlinear variational inequalities, Approx. Theory Appl., 11(3), 95-105.

44. Aslam Noor, M, Inayat Noor, K and Rassias, Th M (1993) Some aspects of variational inequalities, J. Comput. Appl. Math. 47 , 285-312.

45. Aslam Noor, M, Inayat Noor, K and Rassias, Th M (1993) Invitation to variational inequalities, in: Analysis, Geometry and Groups: A Riemann Legacy Volume (eds) H.M. Srivastava and Th.M. Rassias, Hadronic Press, U.S.A., 373-448.

46. Rassias, Th M (1989) Topics in Mathematical Analysis, World Scientific Publ. Co. Singapore.

47. Rassias, Th M (1986) Foundation of Global Nonlinear Analysis, Teubner, Leipsig.

48. Rassias, Th M (1987) (ed.), Nonlinear Analysis, World Scientific Publ. Co. Singapore.

49. Rassias, G M and Rassias, Th M (1985) (eds.), Differential Geometry, Calculus of Variations and their Applications, Lecture Notes in Pure and Applied Mathematics, 100, Marcel Dekker, New York.

50. Riesz, F (1907) Sur une espéce de geometrie analytique des systémes de fonctions sommables, C.R. Acad. Sci. Paris, 1409-1411.

51. Riesz, F (1934-1935) Zur theorie des Hilbertschen vaumes, Acta. Scient. 7, 34-38.

52. Robinson, SM (1992) Normal maps induced by linear transformations, Math. Opers. Research, 17, 691-714.

53. Shi, P (1991) Equivalence of variational inequalities with Wiener-Hopf equations, Proc. Amer. Math. Soc. 111, 339- 346.

54. Shi, P (1990) An iterative method for obstacle problems via Green's functions, Nonlinear Anaysis 15, 339-344.

55. Shih, M H and Tan, K K (1988) Browder-Hartman-Stampacchia variational inequalities for multivalued monotone operators, J. Math. Anal. Appl. 134, 431-440.

56. Speck, F O (1985) General Wiener-Hopf Factorization Methods, Pitman Advanced Publishing Program, London.

57. Stampacchia, G (1964) Formes bilineaires coercitives sur les ensembles convexes, C.R. Acad. Sci. Paris, 258, 4413-4416.

58. Zhu, D L and Marcotte, P (1994) An extended descent framework for variational inequalities, J. Optim. Theory Appl. 80, 329-366.

Mathematics Department,
College of Science
King Saud University
P.O. Box 2455, Riyadh 11451
Saudi Arabia

JEAN-PAUL PENOT

What is a subgradient in a space which is not an inner product space?

1 Introduction.

Linear forms on inner product spaces can conveniently be identified with vectors so that derivatives and gradients are not always distinguished. Similarly, transpose and adjoint mappings are often confused.

It is the purpose of the present paper to point out that a concept of adjoint mapping can be defined in general normed vector space (n.v.s.) and that a notion of subgradient of a non differentiable function can be introduced. When the function is differentiable one recovers the notion of gradient introduced in [11] and used in [8], [13], [21] under some restrictive assumptions for descent algorithms (see also [19]).

We show that the subgradient multimapping of a convex function satisfies an accretivity property. This property differs from the accretivity property considered classically [3], [6], [7], [10], [26] when the space is not a Hilbert space because in such a space the duality mapping is not linear [25].

Whereas monotonicity is a condition which does not depend on the norm on the space, conservativity (as accretivity) is a condition which does depend on the choice of the norm.

This notion shares several useful properties with monotonicity. We differ to [20] the study of some of these properties, in particular regularization properties.

We show in section 5 that the best approximation projection on a closed (not necessarily convex) subset of a normed vector space can be interpreted in terms of subgradients.

Using a deep theorem of D. Preiss [22], we deduce from this fact and from [17] existence and uniqueness results in reflexive Banach spaces which possess an equivalent smooth norm.

2 Adjoint multimappings

In the sequel we identify a multimapping (or set-valued mapping or relation or correspondence) with its graph : if X and Y are two sets, $F \subset X \times Y$ can be considered as the correspondence $x \to F(x) = \{y \in Y : (x, y) \in F\}$, and conversely, given

$F : X \to \mathcal{P}(Y)$ (the power set of Y) also denoted by $F : X \rightrightarrows Y$, its graph

$$G(F) = \{(x, y) \in X \times Y : y \in F(x)\}$$

characterizes F. The inverse of $F : X \rightrightarrows Y$ is the multimapping $F^{-1} : Y \rightrightarrows X$ given by

$$F^{-1} = \{(y, x) \in Y \times X : (x, y) \in F\}$$

so that $F^{-1}(y) = \{x \in X : y \in F(x)\}$ and similarly, for $B \subset Y$

$$F^{-1}(B) = \{x \in X : F(x) \cap B \neq \emptyset\}$$

is the image by F^{-1} of the subset B of Y : $F^{-1}(B) = \bigcup_{b \in B} F^{-1}(b)$.

The duality mapping $J_X : X \rightrightarrows X^*$ of a n.v.s. X with dual X^* is the multimapping given by

$$J_X(x) = \left\{x^* \in X^* : \langle x^*, x\rangle = \|x\|^2, \|x^*\| = \|x\|\right\}.$$

We also denote it by J if there is no risk of confusion.

It coincides with the subdifferential of the convex function $j : X \to \mathbb{R}$ given by $j(x) = \frac{1}{2}\|x\|^2$.

Definition 1 *The adjoint of a multimapping $F : X \rightrightarrows Y$ between two n.v.s. is the multimapping $F^* : Y \rightrightarrows X$ given by*

$$F^* = J_X^{-1} \circ F^T \circ J_Y$$

where F^T is the transpose mapping of F whose graph is $G\left(F^T\right) = G(F)^0$.

Here the polar G^0 of a subset G of $X \times Y$ is defined with respect to the coupling

$$\langle (y^*, x^*), (x, y)\rangle = \langle x^*, x\rangle - \langle y^*, y\rangle,$$

so that

$$G^0 = \{(y^*, x^*) \in Y^* \times X^* : \forall (x, y) \in G \;\; \langle x^*, x\rangle - \langle y^*, y\rangle \leq 1\}.$$

When F is positively homogeneous (i.e. $F(tx) = tF(x)$ for $t > 0$, $x \in X$) one has

$$F^T(y^*) = \{x^* \in X^* : \forall (x, y) \in F \;\; \langle x^*, x\rangle - \langle y^*, y\rangle \leq 0\}.$$

In particular when F is a linear mapping, one recovers the usual definition : $F^T(y^*) = y^* \circ F$ for $y^* \in Y^*$.

The transpose of a process (i.e. of a positively homogeneous mapping) is always a closed convex process i.e. a positively homogeneous multimapping with closed convex graph. Although the adjoint multimapping is positively homogeneous in such a case, it is no more convex in general.

The multivalued character of F^* (even when F is a mapping) is an inconvenience but it may also be an advantage, as the following lemma shows.

Lemma 1 *For any multimapping* $F : X \rightrightarrows Y$ *one has* $(F^{-1})^* = (F^*)^{-1}$.

Proof. It is easy to see that $(F^{-1})^T = \left(F^T\right)^{-1}$. Then the result follows from $(F^{-1})^* = J_Y^{-1} \circ \left(F^T\right)^{-1} \circ J_X = \left(J_X^{-1} \circ F^T \circ J_Y\right)^{-1}$. □

Proposition 2 *Let* $F : X \rightrightarrows Y, G : Y \rightrightarrows Z$ *be multimappings with closed convex graphs containing* 0. *Suppose* F *or* G *is positively homogeneous and*

$$\mathbb{R}_+ F(X) - \mathbb{R}_+ G^{-1}(Z) = Y.$$

Then

$$(G \circ F)^* \subset F^* \circ G^*.$$

If Y *is reflexive, smooth and strictly convex one has*

$$(G \circ F)^* = F^* \circ G^*.$$

Proof. The first relation is a consequence of the relation

$$(G \circ F)^T = F^T \circ G^T$$

established in [18] under our assumption (see also [5] for the case F and G are both positively homogeneous). It also uses the inclusion

$$I_Y \subset J_Y \circ J_Y^{-1}.$$

This inclusion is an equality when Y is reflexive, smooth and strictly convex as then J_Y is a bijection from Y onto Y^*. □

3 Conservative multimappings

In this section we introduce a generalization of the notion of monotone operator in spaces which are not Hilbert spaces.

Definition 2 *Given a n.v.s.* X, *a multimapping* $A : X \rightrightarrows X$ *is said to be conservative if there exists a monotone operator* $M : X \rightrightarrows X^*$ *such that* $A = J^{-1} \circ M$.

It follows immediately that the images $A(x)$, $x \in X$ are J-saturated, a subset S of X being said to be J-*satured* if for any $x \in S$ there exists $x^* \in J(x)$ such that for any $x' \in X$ with $x^* \in J(x')$ one has $x' \in S$. When the norm of X is smooth, a subset S is J-saturated iff $J^{-1}(J(x)) \subset S$ for each $x \in S$.

Proposition 3 *If $A : X \rightrightarrows X$ is conservative, then the images of A are J-saturated and for any $x_i \in X$, $y_i \in A(x_i)$ for $i = 1, 2$, there exists $y_i^* \in J(y_i)$ such that*

$$\langle y_1^* - y_2^*, x_1 - x_2 \rangle \geq 0.$$

If the norm of X is smooth, the converse holds.

Proof. For the first assertion, given a monotone operator M such that $A = J^{-1} \circ M$, and given $x_i \in X$, $y_i \in A(x_i)$ for $i = 1, 2$, it suffices to take $y_i^* \in J(y_i) \cap M(x_i)$ which is nonempty by contruction of A.

In order to check the second assertion, it suffices to set

$$M = \left\{ (x, y^*) \in X \times X^* : A(x) \cap J^{-1}(y^*) \neq \emptyset \right\}$$

so that $M(x) = J(A(x))$ for each $x \in X$. Then, as the values of J are singletons, the assumption ensures that M is monotone. Since the values of A are J-saturated, we have $J^{-1}(M(x)) = J^{-1}(J(A(x))) \subset A(x)$ and we get $A = J^{-1} \circ M$, the inclusion $A \subset J^{-1} \circ M$ being obvious. □

Remark 1. If A is conservative, $A = J_X^{-1} \circ M$, one has

$$A = J_{X^*} \circ M \cap (X \times X)$$

where $J_{X^*} : X^* \rightrightarrows X^{**}$ is the duality mapping of X^*. This relation follows from the equivalence

$$y \in X,\ y^* \in J_X(y) \iff y \in X,\ y \in J_{X^*}(y^*).$$

Definition 3 *The multimapping $A : X \rightrightarrows X$ is said to be m-conservative if there exists a maximal monotone operator $M : X \rightrightarrows X^*$ such that $A = J^{-1} \circ M$.*

Proposition 4 *Suppose X is reflexive with a smooth norm. Then a multimapping $A : X \rightrightarrows X$ is m-conservative iff A is maximal in the set of conservative multimappings (ordered by inclusion of graphs in $X \times X$).*

Proof. Let $B = J^{-1} \circ N$, with N monotone, be a conservative operator such that $B \supset A$, and let $M : X \rightrightarrows X^*$ be a maximal monotone operator such that $A = J^{-1} \circ M$. Since J is surjective, for each $(x, y^*) \in M$ we can find $y \in X$ such that $y^* \in J(y)$. Then $(x, y) \in A \subset B$ so that, $J(y)$ being the singleton $\{y^*\}$, we have $y^* \in N(x)$. As M is maximal monotone, we get $N = M$, and $B = A$.

Conversely, suppose A is maximal in the set of conservative operators and let $M : X \rightrightarrows X^*$ be a monotone operator such that $A = J^{-1} \circ M$. Let $N \supset M$ be a monotone operator. Set $B = J^{-1} \circ N$. Since B is conservative and $A \subset B$, we have $B = A$. Therefore, for each $(x, y^*) \in N$, and for each $y \in J^{-1}(y^*)$ we have $(x, y) \in B = A$, so that, $J(y)$ being a singleton and $J(y) \cap M(x)$ being nonempty, we have $y^* \in M(x) : M = N$ and M is monotone maximal. □

We will make use of the following important result due to Rockafellar [24] (see also [2], [3]).

Theorem 5 *Suppose X is a reflexive Banach space endowed with a strictly convex norm whose dual is also strictly convex. Then a monotone multimapping $M : X \rightrightarrows X^*$ is maximal monotone iff for each $\lambda > 0$ (equivalently some $\lambda > 0$) $J+\lambda M$ is surjective: $R(J+\lambda M) = X^*$.*

Theorem 6 *Suppose X is a reflexive Banach space endowed with a strictly convex norm whose dual is also strictly convex and let $A : X \rightrightarrows X$ be a m-conservative multimapping. Then for each $\lambda > 0$, $I + \lambda A$ is surjective.*

Proof. Since the dual norm of X is strictly convex, the norm of X is smooth. In fact J is a bijection of X onto X^* under our assumptions. Thus, if $M = J \circ A$, M is maximal monotone by Proposition 3.2. In view of the preceding theorem, for any $\lambda > 0$ and any given $y \in X$ the inclusion

$$x \in X : J(y - x) \in \lambda M(x)$$

is solvable since it can be equivalently written

$$0 \in J(x - y) + \lambda M(x)$$

or

$$0 \in J(u) + \lambda M_y(u)$$

for $M_y(u) = M(y+u)$, and since M_y is maximal monotone. Therefore the inclusion $y \in x + \lambda A(x)$, $x \in X$, is solvable. □

Several properties of conservative operators can be deduced from the corresponding properties of monotone operators. The following statement gives an example.

Proposition 7 *Any conservative operator $A : X \rightrightarrows X$ is locally bounded at each point of the interior of its domain $D(A)$.*

Proof. Let $M : X \rightrightarrows X^*$ be a monotone operator such that $A = J^{-1} \circ M$. Since $\|y\| = \|y^*\|$ for each $y \in J^{-1}(y^*)$ the result follows from the corresponding result for monotone operators. □

4 Subgradients of nondifferentiable functions

Let us start by recalling the notion of gradient mappings in general normed vector spaces. This notion is useful for the study of descent algorithms ([8], [11], [21], ...).

Definition 4 *Given a function $f : X \to \mathbb{R}$ on a n.v.s. X which is differentiable at some $x \in X$, the gradient set of f at x is the set*

$$\nabla f(x) = \{y \in X : f'(x) \in J(y)\} = J^{-1}(f'(x)).$$

This set may be empty when X is not reflexive ; it is always at most a singleton when X is strictly convex. When X is not reflexive, one could set $\nabla^{**} f(x) = J_{X^*}(f'(x)) \subset X^{**}$, but in such a case the benefit of using vectors instead of linear forms would be lost.

The preceding definition can be extended to the case of a nondifferentiable function. There exist abundant manners to introduce objects called subdifferentials, so that a list of properties are satisfied (see [12], [16], [14] and their references). Most of the known constructions have a unilateral (i.e. one -sided character). For instance the Fréchet subdifferential of f at x is the set $\partial^- f(x)$ of $x^* \in X^*$ such that there exists a function $\varepsilon : X \to \mathbb{R} \cup \{\infty\}$ with limit 0 at 0 such that

$$f(x+u) - f(x) - \langle x^*, u \rangle \geq -\varepsilon(u) \|u\| \quad \forall u \in X$$

and the Hadamard subdifferential of f at x is the set $\partial f(x)$ of $x^* \in X^*$ such that for each $u \in X$ there exists a function $\varepsilon : \mathbb{R}_+ \times X \to \mathbb{R} \cup \{\infty\}$ with limit 0 at $(0, u)$ such that

$$f(x+tv) - f(x) - \langle x^*, tv \rangle \geq -\varepsilon(t, v)\, t \quad \forall (t, v) \in \mathbb{R}_+ \times X.$$

Since the precise way in which the construction is performed is not important for our purposes, we simply assume that for $f : X \to \mathbb{R} \cup \{\infty\}$ finite at x, the subdifferential $\partial f(x)$ of f at x is a subset of X^*.

The duality multimapping J_X of X enables us to introduce the following definition.

Definition 5 *The subgradient set $\partial_X f(x)$ of f at x associated with the subdifferential ∂ is the set*

$$\partial_X f(x) = J_X^{-1}(\partial f(x)).$$

Therefore when f is differentiable at x and $\partial f(x)$ is the singleton $\{f'(x)\}$ (this is the case for the Fréchet subdifferential and the Hadamard subdifferential when differentiability of f at x is understood in the correponding sense) then $\partial_X f(x)$ coincides with the gradient set $\nabla f(x)$.

We observe that the lack of linearity of the duality multimapping J_X precludes calculus rules for sums, even in the case of gradients.

However one disposes of a chain rule.

Proposition 8 *Let $F : X \to Y$ be differentiable at x with derivative $A = F'(x)$ and let $g : Y \to \mathbb{R} \cup \{\infty\}$. Suppose the subdifferentials $\partial f(x)$ and $\partial g(y)$ of f and g at x and $y = F(x)$ respectively are such that $\partial f(x) \subset \partial g(y) \circ A$. Then, if Y is reflexive one has*

$$\partial_X f(x) \subset A^*(\partial_Y g(y)).$$

Equality holds when Y is reflexive, smooth and strictly convex and when $\partial f(x) = \partial g(y) \circ A$.

In particular this relation holds if ∂ is the Fréchet or the Hadamard subdifferential and F is strictly differentiable at x with a surjective derivative. It is also satisfied if g is convex, F is linear continuous and a qualification condition is satisfied (see [15] for instance).

The proof is obvious : one simply observes that since $I_{Y^*} \subset J_Y \circ J_Y^{-1}$ when Y is reflexive

$$\partial_X f(x) \subset J_X^{-1}\left(A^T(\partial g(y))\right) \subset J_X^{-1} \circ A^T \circ J_Y \circ J_Y^{-1}(\partial g(y)) = A^*(\partial_Y g(y))$$

with equality when J_Y is a bijection from Y into Y^*. □

The following statement is an immediate consequence of the definitions and of the maximal monotonicity of the subdifferential of a convex function.

Theorem 9 *Let $f : X \to \mathbb{R} \cup \{\infty\}$ be a closed proper convex function on a Banach space. Then its subgradient multimapping $\partial_X f$ is m-conservative.*

Example 1. Let $A : X \to Y$ be a continuous linear operator and let $j(y) = \frac{1}{2}\|y\|^2$. Then for $f = j \circ A$ one has

$$\partial_X f(x) = (A^* \circ A)(x).$$

In particular, if $X = Y$, $\partial_X j(x) = J_X^{-1} \circ J_X$ and for each $x \in X$, x is a subgradient of j.

Example 2. Let $f(x) = h(\|x\|)$ where $h : \mathbb{R} \to \mathbb{R}$ is differentiable. Then for $x \in X \backslash \{0\}$ one has

$$\partial_X f(x) = \frac{h'(\|x\|)}{\|x\|}\left(J_X^{-1} \circ J_X\right)(x).$$

In particular, if the norm is smooth (on $X\backslash\{0\}$) and strictly convex one has

$$\partial_X f(x) = \|x\|^{-1} h'(\|x\|) I_X.$$

5 Projections maps considered as subgradients

Given a nonempty closed subset C of a Banach space X, we set

$$d_C(w) = \inf\{\|w - x\| : x \in C\}.$$

In a striking paper [1] Asplund has shown that when X is a Hilbert space the projection mapping can be interpreted as a gradient. In a general Banach space one can extend this result in a slightlty different form, the duality mapping being nonlinear.

Proposition 10 *Let C be a nonempty closed subset of a Banach space X and let $w \in X \backslash C$. Suppose that d_C is subdifferentiable at w in the Hadamard sense and that the projection set*

$$P_C(w) := \{x \in C : \|x - w\| = d_C(w)\}$$

of w on C is nonempty. Then

$$w - P_C(w) \subset \partial_X \left(\frac{1}{2} d_C^2\right)(w).$$

Proof. In view of [17] Corollary 2.3, for each $x \in P_C(w)$ one has

$$\partial d_C(w) \subset S(w - x)$$

where $S(w - x) = \|w - x\|^{-1} J(w - x)$. Thus

$$\partial \left(\frac{1}{2} d_C^2\right)(w) \subset J(w - x).$$

By definition of the subgradient $\partial_X \left(\frac{1}{2} d_C^2\right)(w)$ one get that

$$w - x \in \partial_X \left(\frac{1}{2} d_C^2\right)(w)$$

when $\partial \left(\frac{1}{2} d_C^2\right)(w)$ is nonempty. □

Let us observe that the conclusion can be written

$$P_C \subset \partial_X j - \partial_X (\frac{1}{2} d_C^2).$$

Corollary 11 *If C is a nonempty closed convex subset of a Banach space X, then for each $w \in X \backslash C$ one has*

$$w - P_C(w) \subset \partial_X \left(\frac{1}{2} d_C^2\right)(w).$$

Proof. Since d_C is convex and continuous, it is everywhere subdifferentiable in the convex sense which coincides with the Hadamard sense. □

Corollary 12 *[17] Let C be a nonempty closed subset of a strictly convex Banach space X and let $w \in X \backslash C$ be such that d_C is Hadamard subdifferentiable at w. Then P_C is at most a singleton.*

Proof. Given $w^* \in \partial(\frac{1}{2} d_C^2)(w)$ and $x \in P_C(w)$ we have $w^* \in J(w - x)$. Since X is strictly convex, $J^{-1}(w^*)$ is a singleton (see [9] p 42 for instance). Therefore $P_C(w)$ is at most a singleton. □

Corollary 13 *Let C be a nonempty closed subset of a strictly convex Banach space X which has an equivalent norm which is Gâteaux-differentiable on $X \setminus \{0\}$. Then there is a dense subset D of $X \setminus C$ such that for each $w \in D$, $P_C(w)$ is at most a singleton.*

Proof. Preiss' theorem [22] Th. 2.4 ensures that the Lipschitzian function d_C is Gâteaux-differentiable on a dense subset D of $X \setminus C$. The result follows from the preceding corollary. □

In order to state existence results, let us recall the following definition.

Definition 6 *[17]*

The Banach space X is said to be metrically reflexive (in short M-reflexive) if for any $x^* \in X^* \setminus \{0\}$ any sequence (x_n) of X such that $\|x_n\| = 1$, $(\langle x^*, x_n \rangle) \to \|x^*\|$ has a converging subsequence.

Obviously, a M-reflexive space is reflexive by the James' theorem. On the other hand, any reflexive Banach space can be renormed in an equivalent way into a M-reflexive space. In fact X is M-reflexive whenever the dual norm is Fréchet differentiable on $X^* \setminus \{0\}$.

Theorem 14 *Let C be a nonempty closed subset of a metrically reflexive Banach space X. Suppose there exists an equivalent norm on X which is Fréchet differentiable on $X \setminus \{0\}$. Then there exists a dense subset D of $X \setminus C$ such that each $w \in D$ has a best approximation in C. If moreover the norm of X is strictly convex then each $w \in D$ has a unique best approximation in C.*

Proof. The first part result follows from Corollary 3.6 of [17] and from Theorem 2.4 of [22]. The last assertion is a consequence of Corollary 12. □

References

[1] E. Asplund, Differentiability of the metric projection in finite-dimensional Euclidean space, Proc. Amer. Math. Soc. 38 (1973), pp. 218-219.

[2] H. Attouch, *Variational convergence for functions and operators*, Pitman, Boston, 1984.

[3] V. Barbu, *Nonlinear semigroup and evolution equations in Banach spaces*, Noordhoff, Leyden, 1976.

[4] V. Barbu and A. Cellina, On the surjectivity of multivalued dissipative mappings, Boll. Un. Mat. Ital. 3 (1970), pp. 817-826.

[5] J.M. Borwein, Adjoint process duality, Math. Oper. Res. 8 (3) (1983),pp. 403-434.

[6] H. Brézis, Equations et inéquations non-linéaires dans les espaces vectoriels en dualité, Ann. Inst. Fourier, 18 (1968), 115-176.

[7] F.E. Browder, Existence theorems for nonlinear partial differential equations, Proc. Symposia in Pure Maths. Vol 16, American Math. Soc. (1970) pp. 1-60.

[8] R.H. Byrd and R.A. Tapia, An extension of Curry's theorem to steepest descent in normed linear spaces, Math. Programming 9 (1975), pp. 247-254.

[9] R. Deville, G. Godefroy and V. Zizler, *Smoothness and renormings in Banach spaces*, Pitman Monographs 64, Longman, Harlow, 1993.

[10] R. Dautray-J.-L. Lions, Analyse mathématique et calcul numérique pour les sciences et techniques, Vol. 8: Evolution, Masson, Paris (1984).

[11] M. Golomb and R.A. Tapia, The metric gradient in normed linear spaces, Numer. Math. 20 (1972), pp. 115-124.

[12] A.D. Ioffe, Codirectional compactness, metric regularity and subdifferential calculus, preprint Technion, Haifa, 1996.

[13] G.P. McCormick and R.A. Tapia, The gradient projection method under mild differentiability conditions, SIAM J. Control 10 (1972), pp. 93-98.

[14] J.-P. Penot, A mean value theorem with small subdifferentials, J. Opt. Th. Appl. 95 (2) (1997) , to appear.

[15] J.-P. Penot, Subdifferential calculus without qualification assumptions, J. of Convex Anal. 3 (2) (1996), 1-13.

[16] J.-P. Penot, Subdifferential calculus and subdifferential compactness, Proceedings of the Second Catalan Days on Applied Mathematics, November 1995, M. Sofonea and J.N. Corvellec, eds. Presses Universitaires de Perpignan, Perpignan, 1995, pp. 209-226.

[17] J.-P. Penot, Proximal mappings, preprint Univ. of Pau, May 1996

[18] J.-P. Penot, Transposition of relations, preprint Univ. of Pau, December 1996.

[19] J.-P. Penot, On the convergence of descent algorithms, preprint Univ. of Pau, 1996.

[20] J.-P. Penot and R. Ratsimahalo, Conservative mappings and their regularizations, in preparation.

[21] R. Phelps, Metric projection and the gradient projection method in Banach spaces, SIAM J. Control and Optim. 23 (1985), pp. 973-977.

[22] D. Preiss, Differentiability of Lipschitz functions on Banach spaces, J. Funct. Anal. 91 (1990), pp. 312-345.

[23] R. Ratsimahalo, *Etude de la projection métrique dans les espaces de Banach*, Ph. D. Thesis, Univ of Pau, June 1996.

[24] R.T. Rockafellar, Local boundedness of nonlinear monotone operators, Michigan Math. J. 16 (1969), 397-407.

[25] R.A. Tapia, The characterization of inner product spaces, Proc. Amer. Math. Soc. 41 (1973) pp. 569-574.

[26] E. Zeidler, *Nonlinear functional analysis and its applications, Vol. III, Variational methods and optimization*, Springer Verlag, New York, 1985.

Laboratoire de Mathématiques appliquées, CNRS UPRES A 5033 Faculté des Sciences, Avenue de l'Université 64000 PAU, France.

THEMISTOCLES M RASSIAS

Stability of the generalized orthogonality functional equation

Dedicated to Professor Donald H. Hyers

We study the generalized orthogonality equation and some of its characteristic stability properties.

Let $(E, \cdot)$ denote a real Hilbert space of dimension greater than one, and let $T : E \to E$ be a mapping which satisfies the functional equation

$$|T(x) \cdot T(y)| = |x \cdot y|, \tag{1}$$

known as *the generalized orthogonality equation*. We will follow the work of C. Alsina and J.L. Garcia-Roig (1991). It is easy to see that any solution of (1) will satisfy the following conditions for x, y in E and real μ:

(A) $\|T(x)\| = \|x\|$.

(B) $T(x) = 0$ if and only if $x = 0$.

(C) $T(x) \cdot T(y) = 0$ if and only if $x \cdot y = 0$.

(D) $|\cos A(x, y)| = |\cos A(T(x), T(y))|$, where $A(u, v)$ denotes the angle between the vectors u and v.

(E) $T(\mu x) = \pm \mu T(x)$.

Lemma 1. *If $T : E \to E$ satisfies* (1), *then T transforms each plane (i.e., a two dimensional linear manifold) in E into a plane.*

Proof. For a given plane P with an orthonormal basis e_1, e_2, we extend it to a complete orthonormal system $\{e_i\}_{i \in I}$ in E. By (1) and (A), $\{T(e_i)\}_{i \in I}$ is an orthonormal set which can be extended by means of some set $\{u_j\}_{j \in J}$ (if necessary) to a complete orthonormal system. Let v be any element of P, say $v = a_1 e_1 + a_2 e_2$. Then

$$T(v) = \sum_{i \in I} b_i T(e_i) + \sum_{j \in J} c_j u_j.$$

Note that, for $i = 1, 2$, we have $a_i = v \cdot e_i$ and $b_i = T(v) \cdot T(e_i)$, so that by (1), $|b_i| = |a_i|$ for $i = 1, 2$. Again, by (A), we find that

$$|a_1|^2 + |a_2|^2 = \|v\|^2 = \|Tv\|^2 = \sum_{i \in I} b_i^2 + \sum_{j \in J} c_j^2.$$

Hence, $c_j = 0$ for all $j \in J$ and $b_i = 0$ for $i \neq 1, 2$, and thus

$$T(v) = b_1 T(e_1) + b_2 T(e_2).$$

Q.E.D.

Corollary 2. *If $v = \lambda e_1 + \mu e_2$ then $Tv = \pm\lambda Te_1 \pm \mu Te_2$.*

Notation. With $w = \lambda Te_1 + \mu Te_2$ we write $\overline{w} = \lambda Te_1 - \mu Te_2$. Similarly, with $v = \lambda e_1 + \mu e_2$ we put $\overline{v} = \lambda e_1 - \mu e_2$. With this notation,

$$Tv \in \{\pm w, \pm \overline{w}\}. \tag{2}$$

Lemma 3. *Any mapping $T : \mathbf{R}^2 \to \mathbf{R}^2$ which satisfies the functional equation* (1) *is of the form*

$$T(x) = \sigma(x)U(x), \tag{3}$$

where U is an orthogonal transformation and σ is a mapping from $\mathbf{R}^2$ into the set $\{\pm 1\}$.

Proof. Clearly, a function of the form (3) satisfies (1). By using a suitable rotation, we may suppose that the vector $e_1 = (1, 0)$ is fixed under a transformation T which satisfies (1). By (2), we have $Tv \in \{\pm v, \pm \overline{v}\}$.

Case 1. All points in $\mathbf{R}^2$ are eigenvectors for T. Their eigenvalues are ± 1, so (3) follows.

Case 2. There exists a non-eigenvector u of T, which by (2) must satisfy $T(u) = \pm \overline{u}$ where $\overline{u}$ is not a scalar multiple of u. That is, we have $u = ae_1 + be_2$ with $ab \neq 0$. In this case, we shall show that for all $v \in \mathbf{R}^2$

$$T(v) = \pm \overline{v}. \tag{4}$$

Now (4) is trivially true for the line generated by e_1 or that generated by e_2, and (4) is also true for the line generated by u. Suppose that there exists a v not in any of these lines and not satisfying (4), so that $Tv = \pm v$. Then we would have

$$|\cos A(\overline{u}, v)| = \frac{|\overline{u} \cdot v|}{\|\overline{u}\| \|v\|} = \frac{|T(u) \cdot T(v)|}{\|T(u)\| \|T(v)\|} = |\cos A(Tu, Tv)| = |\cos A(u, v)|$$

which would mean that v is in one of the lines generated by e_1 or e_2, and that is a contradiction.

Q.E.D.

Theorem 4. *Let E be a real Hilbert space of dimension > 1 and suppose that $T : E \to E$ satisfies the generalized orthogonality functional equation* (1)*. Then $T(x) = \sigma(x)U(x)$, where σ maps E into $\{\pm 1\}$ and U is a linear isometry of E onto a closed subspace of E.*

This theorem follows from the Lemma by the use of direct sums.

A Stability Problem

The stability problem for the functional equation (1) has been studied by J. Chmieliński (1994a) and (1994b), using two different definitions of approximate solutions. We give details here for his paper (1994a) only. Here the approximating mapping $T : E \to E$ is defined as a solution of the inequality

$$||T(x) \cdot T(y)| - |x \cdot y|| < \varepsilon \quad \text{for all } x, y \in E \text{ and a fixed } \varepsilon > 0. \tag{5}$$

Lemma 5. *Suppose that x, y, z in E are all different from 0 and satisfy either of the conditions:*

$$\text{(A)} \quad \begin{cases} \cos A(x,y) > \dfrac{\sqrt{2}}{2}, \\ \cos A(x,z) > \dfrac{\sqrt{2}}{2} \end{cases} \quad \text{or} \quad \text{(B)} \quad \begin{cases} \cos A(x,y) < -\dfrac{\sqrt{2}}{2}, \\ \cos A(x,z) < -\dfrac{\sqrt{2}}{2}. \end{cases}$$

Then $y \cdot z > 0$.

Proof. If x, y, z satisfy (B), then $-x, y, z$ satisfy (A), so it is sufficient to prove this lemma for the case (A) only. Put $W = (\operatorname{lin} x)^\perp$ (that is the orthogonal complement of the line generated by x). Let w_1 and w_2 in W be chosen so that $y = \lambda x + w_1$, $z = \mu x + w_2$. Then we have

$$\frac{\sqrt{2}}{2} < \cos A(x,y) = \frac{x \cdot (\lambda x + w_1)}{\|x\|\sqrt{(\lambda x + w_1) \cdot (\lambda x + w_1)}} = \frac{\lambda\|x\|}{\sqrt{\lambda^2\|x\|^2 + \|w_1\|^2}}.$$

Hence, $\lambda > 0$ and $\|w_1\| < \lambda\|x\|$. Similarly, we have $\mu > 0$ and $\|w_2\| < \mu\|x\|$, so that $w_1 \cdot w_2 > -\|w_1\|\|w_2\| > -\lambda\mu\|x\|^2$. Hence, $y \cdot z = (\lambda x + w_1) \cdot (\mu x + w_2) = \lambda\mu\|x\|^2 + w_1 \cdot w_2 > 0$. Q.E.D.

Lemma 6. *Let $T : E \to E$ satisfy the inequality* (5) *and let $\{T_n\}_0^\infty$ be the sequence defined by*

$$T_n(x) = 2^{-n}T(2^n x), \quad x \text{ in } E. \tag{6}$$

Then for each x in E and for $m, n = 0, 1, 2, \ldots$, we have

(I) $\|x\|^2 - \varepsilon \le \|T(x)\|^2 \le \|x\|^2 + \varepsilon.$

(II) $\|x\|^2 - 4^{-n}\varepsilon \le \|T_n(x)\|^2 \le \|x\|^2 + 4^{-n}\varepsilon.$

(III) $\|x\|^2 - 2^{-(m+n)}\varepsilon \le |T_m(x) \cdot T_n(x)| \le \|x\|^2 + 2^{-(m+n)}\varepsilon.$

(IV) $\|x\|^2 - 2^{-n}\varepsilon \le |T(x) \cdot T_n(x)| \le \|x\|^2 + 2^{-n}\varepsilon.$

(V) $T_n(x) = 0$ *implies that* $x = 0$.

Proof. Inequality (5) may be written as

$$|x \cdot y| - \varepsilon \leq |T(x) \cdot T(y)| \leq |x \cdot y| + \varepsilon.$$

Replace x by $2^m x$, y by $2^n x$, divide the result by 2^{m+n} and use (6) to obtain (III). Inequality (IV) follows from (III) by putting $m = 0$, since $T_0 = T$. With $m = n$ in (III), we have (II), and (I) follows from (II) when we put $n = 0$. In order to demonstrate (V), suppose that $T(x) = 0$. Then by (5), we have $|x \cdot y| \leq \varepsilon$ for all y in E. Put $y = kx$, where $k \in \mathbf{N}$, to see that $k\|x\|^2 \leq \varepsilon$ for all $k \in \mathbf{N}$. Hence, $x = 0$. But, by (6), $T_n(x) = 0$ implies that $T(2^n x) = 0$, and thus $x = 0$. Q.E.D.

We might hope that the sequence $\{T_n(x)\}$ defined by (6) would converge when x belongs to E. Unfortunately, this is not the case, as is shown by Chmieliński's

Example. Choose $x_0 \neq 0$ in E and define $s : E \to \mathbf{R}$ by

$$s(x) = \begin{cases} 1 & \text{if } x \neq 2^n x_0 \\ (-1)^n & \text{if } x = 2^n x_0, \quad n = 0, 1, 2, \ldots. \end{cases}$$

Let $T : E \to E$ be given by $T(x) = s(x)x$. Then T satisfies (5) for all $\varepsilon \geq 0$. But $T_n(x_0) = 2^{-n}T(2^n x_0) = 2^{-n}(-1)^n 2^n x_0 = (-1)^n x_0$ does not converge.

Proposition 7. *Let $T : E \to E$ satisfy the inequality* (5). *Given any x in E, there is an increasing sequence $\{j(n)\}_{n=1}^{\infty}$ of positive integers and a constant $\sigma \in \{-1, 1\}$ such that the sequence $\left\{\sigma T_{j(n)}\right\}$ converges and*

$$\sigma T(x) \cdot T_{j(n)}(x) \geq 0 \quad \textit{for all } n \in \mathbf{N}. \tag{7}$$

Proof. When $x = 0$, we have

$$T(0) \cdot T_n(0) = 2^{-n}\|T(0)\|^2 \geq 0,$$

so (7) is true for $x = 0$ and with $j(n) = n$. Now take any x in E with $x \neq 0$. By (V) of Lemma 6, we see that $T_n(x) \neq 0$ for all $n = 0, 1, 2, \ldots$. Choose a positive integer p such that

$$\frac{-\|x\|^2 + 4^{-p}\varepsilon}{\|x\|^2 + 4^{-p}\varepsilon} < -\frac{\sqrt{2}}{2} \tag{8}$$

and

$$\frac{\|x\|^2 - 4^{-p}\varepsilon}{\|x\|^2 + 4^{-p}\varepsilon} > \frac{\sqrt{2}}{2}. \tag{9}$$

Our first objective is to show that there exists an integer $q > p$ such that the set $\mathcal{A} = \{m \geq q : \ T_q(x) \cdot T_m(x) > 0\}$ is infinite.

If the set $\{m \geq p : \ T_p(x) \cdot T_m(x) \geq 0\}$ is infinite, we take $q = p$. If this is not so, there exists $q > p$ such that, for all $m \geq q$, $T_m(x) \cdot T_p(x) < 0$. By Lemma 6 (III), we find that for $m \geq q$,

$$\|x\|^2 - 4^{-p}\varepsilon < \|x\|^2 - 2^{-m}2^{-p}\varepsilon \leq |T_m(x) \cdot T_p(x)| = -T_m(x) \cdot T_p(x)$$

$$\leq -\|T_m(x)\| \cdot \|T_p(x)\| \cos A(T_m(x), T_p(x)),$$

while, by Lemma 6 (II), we have $\|T_m(x)\| \leq \sqrt{\|x\|^2 + 4^{-m}\varepsilon} \leq \sqrt{\|x\|^2 + 4^{-p}\varepsilon}$, and $\|T_p(x)\| \leq \sqrt{\|x\|^2 + 4^{-p}\varepsilon}$, so that $\|T_m(x)\| \cdot \|T_p(x)\| \leq \|x\|^2 + 4^{-p}\varepsilon$. Therefore, when $m \geq q$, we have $\|x\|^2 - 4^{-p}\varepsilon \leq -(\|x\|^2 + 4^{-p}\varepsilon) \cos A(T_m(x), T_p(x))$, that is

$$\cos A(T_m(x), T_p(x)) \leq \frac{-\|x\|^2 + 4^{-p}\varepsilon}{\|x\|^2 + 4^{-p}\varepsilon} < -\frac{\sqrt{2}}{2},$$

according to (8). Since in particular $\cos A(T_q(x), T_p(x)) < -\frac{\sqrt{2}}{2}$, the last two inequalities imply by Lemma 5 that

$$T_m(x) \cdot T_q(x) > 0 \quad \text{for all } m \geq q,$$

so the set $\mathcal{A}$ is infinite.

By using (III) and (II) of Lemma 6, as in the previous argument, we find that for all m in $\mathcal{A}$

$$\frac{\|x\|^2 - 4^{-q}\varepsilon}{\|x\|^2 + 4^{-q}\varepsilon} < \cos A(T_q(x), T_m(x)).$$

Thus, from (9), when $m \in \mathcal{A}$ and $n \in \mathcal{A}$, it follows that $\cos A(T_q(x), T_m(x)) > \frac{\sqrt{2}}{2}$ and $\cos A(T_q(x), T_n(x)) > \frac{\sqrt{2}}{2}$. Hence, by Lemma 5, we find that $T_m(x) \cdot T_n(x) > 0$, and thus, by Lemma 6 (III), that

$$T_m(x) \cdot T_n(x) \geq \|x\|^2 - 2^{-(m+n)}\varepsilon. \tag{10}$$

We now define a sequence $\{k(n)\}_{n=1}^{\infty}$ by arranging the elements of the infinite set $\mathcal{A}$ in increasing order, and then proceed to show that $T_{k(n)}(x)$ is a Cauchy sequence. Indeed, for two given positive integers n and m, we have

$$\|T_{k(n)}(x) - T_{k(m)}(x)\|^2 = \|T_{k(n)}(x)\|^2 + \|T_{k(m)}(x)\|^2 - 2T_{k(n)}(x) \cdot T_{k(m)}(x).$$

Thus, by Lemma 6 (II) and inequality (10),

$$\|T_{k(n)}(x) - T_{k(m)}(x)\|^2 \leq \|x\|^2 + 4^{-k(n)}\varepsilon + \|x\|^2 + 4^{-k(m)}\varepsilon - 2\left(\|x\|^2 - \varepsilon 2^{-k(n)-k(m)}\right)$$

$$= 4^{-k(n)}\varepsilon + 4^{-k(m)}\varepsilon + 2\varepsilon 2^{-k(n)-k(m)}.$$

Wken $k(n)$ and $k(m)$ are both greater than some k in $\mathbf{N}$, we find, for all x in E, that

$$\|T_{k(n)}(x) - T_{k(m)}(x)\| < \frac{\sqrt{\varepsilon}}{2^{k-1}}.$$

Hence, $\left\{T_{k(n)}(x)\right\}_{n=1}^{\infty}$ is a Cauchy sequence and so is convergent.

In order to prove (7), we consider a set $\mathcal{B} = \{m \in \mathcal{A} : \ \sigma T(x) \cdot T_m(x) \geq 0\}$, where $\sigma \in \{-1, 1\}$ is chosen so that $\mathcal{B}$ is infinite. This is possible because of the fact that for q fixed, $T_q(x) \cdot T_m(x) \geq 0$ for all $m \geq q$ as was shown above. A sequence $\{j(n)\}_{n=1}^{\infty}$ consisting of the elements of $\mathcal{B}$ is a subsequence of $\{k(n)\}_{n=1}^{\infty}$ so $\left\{\sigma T_{j(n)}(x)\right\}_{n=1}^{\infty}$ converges and satisfies (7). Q.E.D.

We are ready to prove the stability theorem:

Theorem 8 (Chmieliński). *Given a real Hilbert space E of dimension ≥ 2, suppose that the mapping $T : E \to E$ satisfies the inequality* (5) *for some $\varepsilon \geq 0$ and all x, y in E. Then there exists a mapping $V : E \to E$ which is a solution of equation* (1) *and also satisfies the inequality*

$$\|V(x) - T(x)\| \leq \sqrt{\varepsilon} \quad \textit{for all } x \textit{ in } E. \tag{11}$$

Moreover, any two solutions of (1) *which also satisfy* (11) *are either equal at a given point x of E or differ there only in sign.*

Proof. By Proposition 7, for each x in E, we may choose $\sigma(x) \in \{-1, 1\}$ and a sequence $\{j(n, x)\}_{n=1}^{\infty}$ such that the limit

$$V(x) = \lim_{n\to\infty} \sigma(x) T_{j(n,x)}(x)$$

exists. Similarly, $V(y) = \lim_{n\to\infty} \sigma(y) T_{j(n,y)}(y)$. By (6) and (5), we have

$$\left| |\sigma(x)T_{j(n,x)}(x) \cdot \sigma(y)T_{j(n,y)}(y)| - |x \cdot y| \right| = \left| |T_{j(n,x)}(x) \cdot T_{j(n,y)}(y)| - |x \cdot y| \right|$$

$$= 2^{-j(n,x)-j(n,y)} \left| |T(2^{j(n,x)}x) \cdot T(2^{j(n,y)}y)| - |2^{j(n,x)}x \cdot 2^{j(n,y)}y| \right| \leq 2^{-(j(n,x)+j(n,y))}\varepsilon.$$

As $n \to \infty$, we obtain $|V(x) \cdot V(y)| - |x \cdot y| = 0$, so that $V : E \to E$ is a solution of (1).

Given any x in E, we may write

$$\|\sigma(x)T_{j(n,x)}(x) - T(x)\|^2 = \|T_{j(n,x)}(x)\|^2 + \|T(x)\|^2 - 2\sigma(x)T_{j(n,x)}(x) \cdot T(x).$$

From Proposition 7, we know that $\sigma(x)T(x) \cdot T_{j(n,x)}(x) \geq 0$. Now we use Lemma 6, (I), (II) and (IV) to see that $\|T(x)\|^2 \leq \|x\|^2 + \varepsilon$, $\|T_{j(n,x)}(x)\|^2 \leq \|x\|^2 + 4^{-j(n,x)}\varepsilon$ and $\sigma(x)T_{j(n,x)}(x) \cdot T(x) \geq \|x\|^2 - 2^{-j(n,x)}\varepsilon$. Therefore,

$$\|\sigma(x)T_{j(n,x)}(x) - T(x)\|^2 \leq \varepsilon + 4^{-j(n,x)}\varepsilon + 2\left(2^{-j(n,x)}\right)\varepsilon.$$

Taking the limit as $n \to \infty$, we have $\|V(x) - T(x)\| \leq \varepsilon$, so that (11) holds.

It remains to prove the "uniqueness up to a change of sign" property of V. Suppose that V and V' are two solutions of the generalized orthogonality functional equation (1) which also satisfy the inequality (11) for all x in E. Then

$$\|V(x) - V'(x)\| \leq 2\sqrt{\varepsilon} \quad \text{for all } x \text{ in } E. \tag{12}$$

By Theorem 4 above, there are linear isometries U and U' from E into E and functions $\sigma, \sigma' : E \to \{-1, 1\}$ such that $V(x) = \sigma(x)U(x)$ and $V'(x) = \sigma'(x)U'(x)$ for all x in E. By (12), we have $\|\sigma(x)U(x) - \sigma'(x)U'(x)\| \leq 2\sqrt{\varepsilon}$ for x in E. Let A, B denote the subsets of E defined by

$$A = \{x \in E : \sigma(x) = \sigma'(x)\}$$

and

$$B = \{x \in E : \sigma(x) = -\sigma'(x)\} = E \setminus A.$$

Then, by (12),

$$\|U(x) - U'(x)\| \le 2\sqrt{\varepsilon} \quad \text{for } x \text{ in } A, \tag{13}$$

$$\|U(x) + U'(x)\| \le 2\sqrt{\varepsilon} \quad \text{for } x \text{ in } B. \tag{14}$$

Take x_0 in A with $x_0 \neq 0$.

Suppose that $(U - U')(x_0) = 0$. Then, since $U - U'$ is linear, we would have $(U - U')(x) = 0$ for all $x \in \text{lin}\,(x_0) = \{u \in E : u = \alpha x_0, \ -\infty < \alpha < \infty\}$.

On the other hand, suppose that $(U - U')(x_0) \neq 0$. If there exists a real number $\beta > \frac{2\sqrt{\varepsilon}}{\|(U-U')(x_0)\|}$ such that βx_0 lies in A, then we would have $\|(U - U')(\beta x_0)\| = \beta\|(U - U')(x_0)\| > 2\sqrt{\varepsilon}$, which contradicts (13). Thus, we must have $\beta x_0 \in B$ for all $\beta > \frac{2\sqrt{\varepsilon}}{\|(U-U')(x_0)\|}$. Choose a number β which satisfies this last inequality, so that $\beta x_0 \in B$. If $(U + U')(\beta x_0) \neq 0$, then, if we take $\gamma > \max\left\{1, \frac{2\sqrt{\varepsilon}}{\|(U+U')(\beta x_0)\|}\right\}$, we would have $\gamma\beta x_0 \in B$ and $\|(U + U')(\gamma\beta x_0)\| = \gamma\|(U + U')(\beta x_0)\| > 2\sqrt{\varepsilon}$, which contradicts (14). Hence, $(U + U')(\beta x_0) = 0$, and it follows that $(U + U')(x) = 0$ for all $x \in \text{lin}\,(x_0)$. We conclude that, when $0 \neq x_0 \in A$, then $U = U'$ on $\text{lin}\,(x_0)$ or else $U = -U'$ on $\text{lin}\,(x_0)$. Similarly, we see that for $0 \neq x_0 \in B$, either $U = U'$ on $\text{lin}\,(x_0)$ or $U = -U'$ on $\text{lin}\,(x_0)$. For each x in E, then we must have either $U(x) = U'(x)$ or $U(x) = -U'(x)$. Thus $V(x) = \pm V'(x)$ at each point of E. Q.E.D.

Finally, Chmieliński (1994a) shows that a mapping satisfying (1) and (11) need not be unique, and also that the estimate of $\sqrt{\varepsilon}$ in (11) is the best possible, by constructing the following

Example. Let l^2 denote the real classical Hilbert space of sequences and let k be a functional defined on l^2 whose range lies in the closed interval $-\sqrt{\varepsilon} \le t \le \sqrt{\varepsilon}$. Define $T : l^2 \to l^2$ by $T(x) = T(x_1, x_2, \ldots) = (k(x), x_1, x_2, \ldots)$. Then $T(x) \cdot T(y) = x \cdot y + k(x)k(y)$, so that

$$\big|\,|T(x) \cdot T(y)| - |x \cdot y|\,\big| \le |T(x) \cdot T(y) - x \cdot y| = |k(x)k(y)| \le \varepsilon.$$

Thus, T satisfies (5). Clearly the function $V : l^2 \to l^2$ defined by

$$V(x) = V(x_1, x_2, \ldots) = (0, x_1, x_2, \ldots)$$

satisfies (1) and (11).

In particular, take a functional k as above, but such that $k\left(\frac{\sqrt{\varepsilon}}{4}, 0, 0, \ldots\right) = \frac{\sqrt{\varepsilon}}{2}$, and define a dunction $\sigma : l^2 \to \{-1, 1\}$ as follows:

$$\sigma\left(\frac{\sqrt{\varepsilon}}{4}, 0, 0, \ldots\right) = -1, \quad \text{while} \quad \sigma(x) = 1 \ \text{ for all other } x \text{ in } l^2.$$

The mapping $V' : l^2 \to l^2$ defined by $V'(x) = \sigma(x)V(x)$ also satisfies (1), but $V' \neq V$. Also, we have $\|V'(x) - T(x)\| \leq \sqrt{\varepsilon}$ for $x \neq \left(\frac{\sqrt{\varepsilon}}{4}, 0, 0, \ldots\right)$ and for $x = \left(\frac{\sqrt{\varepsilon}}{4}, 0, 0, \ldots\right)$,

$$\left\| V'\left(\frac{\sqrt{\varepsilon}}{4}, 0, 0, \ldots\right) - T\left(\frac{\sqrt{\varepsilon}}{4}, 0, 0, \ldots\right) \right\| = \left\| \left(0, -\frac{\sqrt{\varepsilon}}{4}, 0, 0, \ldots\right) - \left(\frac{\sqrt{\varepsilon}}{2}, \frac{\sqrt{\varepsilon}}{4}, 0, 0, \ldots\right) \right\|$$

$$= \left\| \left(-\frac{\sqrt{\varepsilon}}{2}, -\frac{\sqrt{\varepsilon}}{2}, 0, 0, \ldots\right) \right\| = \sqrt{\frac{\varepsilon}{2}} \leq \sqrt{\varepsilon}.$$

Thus, V' also satisfies (11).

Now take a constant function $k(x) = \sqrt{\varepsilon}$, i.e., $T(x_1, x_2, \ldots) = (\sqrt{\varepsilon}, x_1, x_2, \ldots)$ for x in l^2. If V is a solution of (1) then $V(0, 0, \ldots) = (0, 0, \ldots)$ and we have $V(0, 0, \ldots) - T(0, 0, \ldots) = (\sqrt{\varepsilon}, 0, \ldots) = \sqrt{\varepsilon}$. Hence, $\sqrt{\varepsilon}$ is the best possible estimate in (11).

Comments

Chmieliński (1994a) pointed out that he had already solved a simpler stability problem, namely that, if $|T(x) \cdot T(y) - x \cdot y| < \varepsilon$ for x, y in E, then there exists a unique function $V : E \to E$ satisfying the orthogonality equation $V(x) \cdot V(y) = x \cdot y$ together with the inequality $\|V(x) - T(x)\| \leq \sqrt{\varepsilon}$, in his Ph. D. dissertation.

Moreover, Chmieliński (1994c) proves that in the case $E = \mathbf{R}^n$ superstability holds, i.e., each function $T : \mathbf{R}^n \to \mathbf{R}^n$ satisfying (5) is a solution of the generalized orthogonality functional equation (1) (the same is true in the $\mathbf{R}^n \to \mathbf{R}^n$ case for the orthogonality equation).

In another paper, Chmieliński (1994b) adopted a different class of solutions, defined by the inequalities

$$(1 + \varepsilon)^{-1}|x \cdot y| \leq |T(x) \cdot T(y)| \leq (1 + \varepsilon)|x \cdot y|, \tag{15}$$

where T is a mapping of $\mathbf{R}^n$ into itself. This scheme of approximation was suggested by J. Tabor's treatment of the Cauchy equation in Tabor (1988), (1990). The main theorem of Chmieliński (1994b) is:

Theorem 9. *Let $T : \mathbf{R}^n \to \mathbf{R}^n$, with $n > 1$, satisfy the inequalities* (15) *for all x, y in $\mathbf{R}^n$. Then there exists a mapping $V : \mathbf{R}^n \to \mathbf{R}^n$ which is a solution of the equation*

$$|V(x) \cdot V(y)| = |x \cdot y| \quad \textit{for all } x, y \textit{ in } \mathbf{R}^n$$

and satisfies the inequalities

$$(1 + \varepsilon)^{-\frac{3}{2}}\|T(\xi)\| \leq \|V(\xi)\| \leq (1 + \varepsilon)^{\frac{3}{2}}\|T(\xi)\| \quad \textit{for all } \xi \textit{ in } \mathbf{R}^n.$$

Remarks on the complex case of equation (1)

So far in this section we have been concerned with the case of equation (1) in which the Hilbert space E is real. When E is complex the analog of Theorem 4 above, is the well known (at least to physicists) theorem of E.P. Wigner (see Wigner (1931), p. 251, Wigner (1939), and the book Wigner (1959)). This basic theorem justifies the restriction of group representations in quantum mechanics to unitary and antiunitary operators. Among the various discussions and proofs of Wigner's theorem, we shall cite only three. The first is the treatment of J.S. Lomont and P. Mendelson (1963), with the following set up.

Definitions. A mapping T of a complex Hilbert space E into itself is called *antilinear* if $T(\alpha x + \beta y) = \overline{\alpha} Tx + \overline{\beta} Ty$ for all x, y in E and all α, β in $\mathbf{C}$ where the bar over the scalar denotes the complex conjugate. If $T : E \to E$ is bijective and antilinear, then T is called *antiunitary* if $(Tx, Ty) = (y, x)$ for all x, y in E, where $(\cdot, \cdot)$ denotes the inner product on E. Two maps $S : E \to E$ and $T : E \to E$ are called *phase-equivalent* if there exists a functional $\omega : E \to \mathbf{C}$ such that $|\omega(x)| = 1$ for all x in E and $Tx = \omega(x) Sx$ for all x in E.

The following version of Wigner's theorem was demonstrated by Lomont and Mendelson (1963).

Theorem 10. *Let E be a Hilbert space, of dimension greater than one, over F, where F is either the field of real or complex numbers. Suppose that $T : E \to E$ is a bijective mapping such that*

$$|(Tx, Ty)| = |(x, y)| \quad \textit{for all } x, y \textit{ in } E. \tag{16}$$

Then, (1) if F is the real field, T is phase-equivalent to a unitary operator; (2) if F is the complex field, T is phase-equivalent either to a unitary or to an antiunitary operator (operator = a mapping of E into itself).

We note that conclusion (1) of this theorem was demonstrated above in Theorem 4 by Alsina and Garcia-Roig (1991) without the hypothesis of bijectivity.

Another proof of Wigner's famous theorem was given by C.S. Sharma and D.F. Almeida (1990). Their version reads as follows.

Theorem 11. *Let T be an operator on a complex Hilbert space E of dimension > 1 which satisfies* (16)*. Then there exists a linear isometry U or an antilinear isometry V on E such that, at each point x in E, Tx and Ux or else Tx and Vx are phase-equivalent. If in addition to the above conditions, T is surjective, then U is unitary and V is antiunitary.*

Another proof of Wigner's theorem was given by J. Rätz (1996).

Recently J. Chmieliński (to appear) proved the stability of the Wigner equation, i.e., generalized Theorem 8 to the case of a complex Hilbert space. The result reads as follows

Theorem (Chmieliński). *Let E_1 be a complex inner product space and let E_2 be a complex Hilbert space. Then, for $T : E_1 \to E_2$ being a solution of*

$$\big|\,|(Tx, Ty)| - |(x, y)|\,\big| \le \varepsilon \quad \textit{for } x, y \in E_1$$

there exists $T_*: E_1 \to E_2$ – a solution of

$$|(T_*x, T_*y)| = |(x,y)| \quad \textit{for } x, y \in E_1$$

such that

$$\|Tx - T_*x\| \leq \sqrt{\varepsilon} \quad \textit{for } x \in E_1.$$

Moreover, two functions T_* and T_*' satisfying the above assertion are phase-equivalent.

A similar result (with much simpler proof) concerning the stability of the orthogonality equation can be found in Chmieliński (this volume).

A generalization of Wigner's theorem

In section 6 of the paper by Th. M. Rassias and C.S. Sharma (1993), the theory of operators on a "quaternionic Hilbert space" is briefly presented, as follows.

Let S denote the skew field of quaternions. S forms a normed associative division algebra over the real field $\mathbf{R}$. Any γ in S has a unique representation in terms of the basis $1, i, j, k$, where $i^2 = j^2 = k^2 = -1$, $ij = -ji = k$, $jk = -kj = i$ and $ki = -ik = j$. Conjugation is defined by $\overline{1} = 1$, $\overline{i} = -i$, $\overline{j} = -j$, $\overline{k} = -k$. Thus,

$$\gamma = \gamma_0 + \gamma_1 i + \gamma_2 j + \gamma_3 k, \quad \text{where } \gamma_n \in \mathbf{R}\ (n = 0,1,2,3); \quad \overline{\gamma} = \gamma_0 + \gamma_1\overline{i} + \gamma_2\overline{j} + \gamma_3\overline{k}.$$

The term "modulus", instead of norm, will be used for the quantity

$$\|\gamma\| = \left(\sum_{n=0}^{3} \gamma_n^2\right)^{\frac{1}{2}} = (\gamma\overline{\gamma})^{\frac{1}{2}}.$$

This modulus has the property that for two elements β and γ of S we have $\|\beta\gamma\| = \|\beta\|\|\gamma\|$.

For a quaternionic Hilbert space $\mathcal{H}$ over S we define a positive definite Hermitian form $\langle\cdot,\cdot\rangle$ which for u, v in $\mathcal{H}$ and p, q in S satisfies the following conditions:

(a) $\langle pu, qv\rangle = p\langle u, v\rangle\overline{q}$,

(b) $\langle u + v, w\rangle = \langle u, w\rangle + \langle v, w\rangle$,

(c) $\overline{\langle u, v\rangle} = \langle v, u\rangle$,

(d) $\langle u, u\rangle = 0$ if and only if $u = 0$.

The authors point out that quaternionic vector spaces including quaternionic Hilbert spaces are used in various branches of physics such as celestial mechanics and relativity as well as quantum mechanics. Wigner's theorem for quaternionic Hilbert spaces was studied by V. Bargmann (1964), whose statement of the theorem was essentially that of Theorem 12 below, which corresponds to theorem 6.14 in the paper by Th. M. Rassias and C.S. Sharma (1993). These authors use the term "Wigner-equivalent" instead of phase-equivalent. Thus, two operators T and U on $\mathcal{H}$ are said to be *Wigner-equivalent* if there is a quaternion p of unit modulus such that $U = pT$.

Theorem 12. *Let T be an operator on a quaternionic Hilbert space $\mathcal{H}$ which is not two-dimensional. If T satisfies*

$$\|\langle Tx, Ty\rangle\| = \|\langle x, y\rangle\| \quad \textit{for all } x, y \textit{ in } \mathcal{H},$$

then T is Wigner-equivalent to a linear isometry A. If in addition T is surjective, then A is unitary.

As indicated by Rassias and Sharma (1993), their proof of this theorem is a generalization of the proof for the complex case given by Sharma and Almeida (1990).

An unsolved problem

It appears that the problem of generalization of Chmieliński's stability theorems for the functional equation $|(Tx, Ty)| = |(x, y)|$ to the case of a quaternionic Hilbert space remains open.

Orthogonally additive mappings

Recently several authors have considered functional congruences which apply only to orthogonal pairs of independent variables, with orthogonality defined in some sense.

A number of definitions of orthogonality in vector spaces, in addition to the usual one for inner product spaces, have appeared in the literature during the past half century. Many of these are mentioned in an article by H. Drljevic (1989), pp. 232-234. Perhaps the best known of these is the "Birkhoff-James" orthogonality (see R.C. James (1947)) for real normed vector spaces, where x orthogonal to y means that $\|x + \beta y\| \geq \|x\|$ for all β in $\mathbf{R}$.

In giving his axiomatic definition of orthogonality, J. Rätz (1985) modified the definition given on pages 427-428 of S. Gudder and D. Strawther (1975) and arrived at the following:

Definition (0). Let X be a real vector space of dimension ≥ 2 and $\perp$ a binary relation on X with the following properties.

(01) $x \perp 0$ and $0 \perp x$ for all x in X.

(02) If $x, y \in X \setminus \{0\}$ and $x \perp y$, then x and y are linearly independent.

(03) If $x, y \in X$ and $x \perp y$, then $\alpha x \perp \beta y$ for all $\alpha, \beta \in \mathbf{R}$.

(04) If P is a two dimensional subspace of X and $x \in P$, $\lambda \geq 0$, then there exists $y \in P$ such that $x \perp y$ and $x + y \perp \lambda x - y$.

Then $(X, \perp)$ is called an *orthogonality space.*

Rätz points out that this definition is more restrictive than that given by S. Gudder and D. Strawther (1975), but he showed that his definition includes the following basic examples.

Example I. "Trivial" orthogonality, defined by (01) and $x, y \in X \setminus \{0\}$ implies $x \perp y$ if and only if x and y are linearly independent.

Example II. The usual orthogonality of vectors defined on an inner product space.

Example III. The Birkhoff-James orthogonality on a normed space, defined above.

Following J. Rätz (1985) we give a brief account of his principal results.

Lemma 13. *Let $(X, \perp)$ be an orthogonality space, let x be different from 0 and let P be a 2-dimensional subspace of X with $x \in P$. Then there exists a $y \in P$ with the property that $x \perp y$ and* $\operatorname{lin}\{x, y\} = P$. *(Here* $\operatorname{lin}\{\ldots\}$ *is the linear hull of the set indicated).*

Proof. Put $\lambda = 1$ in (04) to get $y \in P$ with $x \perp y$ and $x + y \perp x - y$. Then $y \neq 0$, for otherwise we would have $x \perp x$, which is impossible by (02). Thus $x + y$ and $x - y$ are linearly independent, so that $\operatorname{lin}\{x, y\} = P$. Q.E.D.

Definition. If $(X, \perp)$ is an orthogonality space and $(Y, +)$ is an Abelian group, then a mapping $f : X \to Y$ is called *orthogonally additive* if

$$f(x_1 + x_2) = f(x_1) + f(x_2) \quad \text{for all } x_1, x_2 \text{ in } X \text{ with } x_1 \perp x_2. \tag{17}$$

The set of all solutions of (17) is denoted by $\operatorname{Hom}_\perp(X, Y)$.

Note. The relation $\perp$ is not symmetric in general. However, for $f \in \operatorname{Hom}_\perp(X, Y)$ we do have

$$f(x_1 + x_2) = f(x_1) + f(x_2) \quad \text{for all } x_1, x_2 \text{ in } X \text{ with } x_2 \perp x_1, \tag{18}$$

since X and Y are Abelian groups.

The next two lemmas follow directly from the definitions.

Lemma 14. *(a) If $f \in \operatorname{Hom}_\perp(X, Y)$, then $f(0) = 0$. (b) If $f, g \in \operatorname{Hom}_\perp(X, Y)$, then $f + g \in \operatorname{Hom}_\perp(X, Y)$. (c) If $f \in \operatorname{Hom}_\perp(X, Y)$ and $h(x) = f(-x)$ for all $x \in X$, then $h \in \operatorname{Hom}_\perp(X, Y)$.*

Lemma 15. *Let $(Y, +)$ be uniquely 2-divisible and let $f \in \mathrm{Hom}_\perp(X, Y)$. Put $g(x) = \dfrac{f(x) + f(-x)}{2}$, $h(x) = \dfrac{f(x) - f(-x)}{2}$ for all $x \in X$. Then $g, h \in \mathrm{Hom}_\perp(X, Y)$, where g is even, h is odd and $f = g + h$.*

Lemma 16. *Let u, v be in X with $u \perp v$. Suppose that $f \in \mathrm{Hom}_\perp(X, Y)$ is additive on $\mathrm{lin}\,\{u\}$ and on $\mathrm{lin}\,\{v\}$. Then f is additive on $\mathrm{lin}\,\{u, v\}$.*

Proof. Take any x_1, x_2 in $\mathrm{lin}\,\{u, v\}$ with $x_1 = \alpha_1 u + \beta_1 v$, $x_2 = \alpha_2 u + \beta_2 v$. Then $x_1 + x_2 = (\alpha_1 + \alpha_2)u + (\beta_1 + \beta_2)v$, where $(\alpha_1 + \alpha_2)u \perp (\beta_1 + \beta_2)v$ by (03). By hypothesis we have

$$f(x_1 + x_2) = f\left((\alpha_1 + \alpha_2)u\right) + f\left((\beta_1 + \beta_2)v\right) = f(\alpha_1 u) + f(\alpha_2 u) + f(\beta_1 v) + f(\beta_2 v)$$

$$= f(\alpha_1 u + \beta_1 v) + f(\alpha_2 u + \beta_2 v) = f(x_1) + f(x_2).$$

Q.E.D.

Theorem 17 (J. Rätz (1980)). *Let $(X, \perp)$ be an orthogonality space and let Y be an Abelian group. Then $h : X \to Y$ is an odd solution of (17) if and only if h is additive.*

Proof. The "if" part is obvious. To prove the "only if" part, let $h \in \mathrm{Hom}_\perp(X, Y)$ be an odd function. We shall proceed in steps and show that

(i) if $x \in X$ and $\lambda > 0$ then $h(x + \lambda x) = h(x) + h(\lambda x)$,

(ii) if x_1, x_2 are linearly dependent then $h(x_1 + x_2) = h(x_1) + h(x_2)$.

To prove step (i), let P be a 2-dimensional subspace of X containing x. By (04) there exists a $y \in P$ such that $x \perp y$ and $x + y \perp \lambda x - y$. Hence by (03), $\lambda x \perp (-y)$. Thus

$$h(x + \lambda x) = h(x + y + \lambda x - y) = h(x + y) + h(\lambda x - y)$$

$$= h(x) + h(y) + h(\lambda x) + h(-y) = h(x) + h(\lambda x).$$

To prove step (ii), we see that the assertion is true if $x_1 = 0$, by Lemma 14 (a). Take $x_1 \neq 0$. Then there exists $\mu \in \mathbf{R}$ with $x_2 = \mu x_1$.

Case 1. $\mu \geq 0$. Then by (i),

$$h(x_1 + x_2) = h(x_1 + \mu x_1) = h(x_1) + h(\mu x_1) = h(x_1) + h(x_2).$$

Case 2. $-1 < \mu < 0$. Put $y = (1 + \mu)x_1$, and $\lambda = -\dfrac{\mu}{\mu + 1}$ and get $\lambda > 0$, $\lambda y = -\mu x_1$. Hence by (i), $h(y + \lambda y) = h(y) + h(\lambda y)$, so that

$$h(x_1) = h\left((1 + \mu)x_1 - \mu x_1\right) = h\left((1 + \mu)x_1\right) + h(-\mu x_1)$$

$$= h(x_1 + x_2) + h(-x_2) = h(x_1 + x_2) - h(x_2),$$

i.e., $h(x_1 + x_2) = h(x_1) + h(x_2)$.

Case 3. $\mu \leq -1$. Here we put $\lambda = -1-\mu$, $y = -x_1$, so that $\lambda \geq 0$, $\lambda y = (1+\mu)x_1 = x_1 + x_2$. Hence by (i), $h(y + \lambda y) = h(y) + h(\lambda y)$, that is,

$$h(-x_1+x_1+x_2) = h(-x_1)+h(x_1+x_2), \text{ or } h(x_1+x_2) = h(x_2)-h(-x_1) = h(x_1)+h(x_2)$$

and (ii) has been demonstrated.

Now, take an arbitrary pair x_1, x_2 in X and let P be a 2-dimensional subspace containing x_1 and x_2. According to Lemma 13 there exist x, y in P such that $x \perp y$ and $\operatorname{lin}\{x, y\} = P$. By (ii), h is additive on $\operatorname{lin}\{x\}$ and on $\operatorname{lin}\{y\}$. Therefore, by Lemma 16, it is also additive on P and we have

$$h(x_1 + x_2) = h(x_1) + h(x_2) \quad \text{for each pair } x_1, x_2 \text{ in } X.$$

Q.E.D.

Theorem 18 (J. Rätz (1981)). *If $(X, \perp)$ is an orthogonality space, $(Y, +)$ an Abelian group and $g : X \to Y$ an even solution of (17), then g is quadratic, i.e., for all x_1, x_2 in X,*

$$g(x_1 + x_2) + g(x_1 - x_2) = 2g(x_1) + 2g(x_2). \tag{19}$$

Proof. Given an even solution g of (17), we shall proceed in steps, as follows:

(iii) $u, v \in X$, $u + v \perp u - v$ implies that $g(u) = g(v)$,

(iv) $g(2u) = 4g(u)$ for all u in X,

(v) $x \in X$, $\lambda > 0$ implies that $g(x + \lambda x) + g(x - \lambda x) = 2g(x) + 2g(\lambda x)$,

(vi) $x \in X$, $\alpha, \beta \in \mathbf{R}$ implies that $g(\alpha x + \beta x) + g(\alpha x - \beta x) = 2g(\alpha x) + 2g(\beta x)$.

Step (iii). By (03), $\dfrac{u+v}{2} \perp \pm \dfrac{u-v}{2}$, and we have

$$g(u) = g\left(\frac{u+v}{2} + \frac{u-v}{2}\right) = g\left(\frac{u+v}{2}\right) + g\left(\frac{u-v}{2}\right) = g\left(\frac{u+v}{2}\right) + g\left(\frac{v-u}{2}\right)$$

$$= g\left(\frac{u+v}{2} + \frac{v-u}{2}\right) = g(v).$$

Step (iv). By (04), there exists v in X with $u \perp v$ and $u + v \perp u - v$, and by (03), also $u \perp (-v)$. Using the evenness of g and step (iii) we get

$$g(2u) = g(u + v + u - v) = g(u + v) + g(u - v) = g(u) + g(v) + g(u) + g(-v)$$

$$= 2g(u) + g(v) + g(v) = 2g(u) + 2g(v) = 4g(u).$$

Step (v). By (04), there exists y in X such that $x \perp y$, $x + y \perp \lambda x - y$, where $\lambda > 0$. Then

$$g(x+\lambda x) + g(x - \lambda x) + g(2y) = g(x + y + \lambda x - y) + g(x - \lambda x) + g(2y)$$

$$= g(x + y + \lambda x - y) + g(x - \lambda x + 2y) = g(x + y + \lambda x - y) + g(x + y - \lambda x + y)$$

$$= g(x + y) + g(\lambda x - y) + g(x + y) + g(-\lambda x + y) = 2g(x + y) + 2g(\lambda x - y)$$

$$= 2g(x)+2g(y)+2g(\lambda x)+2g(-y) = 2g(x)+2g(\lambda x)+4g(y) = 2g(x)+2g(\lambda x)+g(2y),$$

by Step (iv). Hence $g(x + \lambda x) + g(x - \lambda x) = 2g(x) + 2g(\lambda x)$.

Step (vi). Since $g(0) = 0$ by Lemma 14 (a), the assertion is clear when $\alpha = 0$. Take $\alpha \neq 0$, and put $\lambda = \left|\frac{\beta}{\alpha}\right|$ and $z = \alpha x$.

Case 1. $\frac{\beta}{\alpha} \geq 0$, so that $\lambda = \frac{\beta}{\alpha}$ and $\lambda z = \beta x$. Then by Step (v) we have

$$g(\alpha x+\beta x)+g(\alpha x-\beta x) = g(z+\lambda z)+g(z-\lambda z) = 2g(z)+2g(\lambda z) = 2g(\alpha x)+2g(\beta x).$$

Case 2. $\frac{\beta}{\alpha} < 0$, so that $\lambda = -\frac{\beta}{\alpha}$, $\lambda z = (-\beta)x$, $z = \alpha x$. Then

$$g(\alpha x + \beta x) + g(\alpha x - \beta x) = g(z - \lambda z) + g(z + \lambda z) = 2g(z) + 2g(\lambda z)$$

$$= 2g(\alpha x) + 2g(-\beta x) = 2g(\alpha x) + 2g(\beta x),$$

since g is even.

Now take any two points x_1, x_2 in X. If $x_1 = 0$, then (19) holds by Lemma 14 (a). When $x_1 \neq 0$, let P be a 2-dimensional subspace of X containing x_1 and x_2. By Lemma 13 there exists a y in P such that $x_1 \perp y$ and $\operatorname{lin}\{x_1, y\} = P$. Thus $x_2 = \alpha x_1 + \beta y$ for some pair α, β in $\mathbf{R}$. By (03) we see that $(1+\alpha)x_1 \perp \beta y$ and $(1-\alpha)x_1 \perp (-\beta y)$. Then, on using Step (vi) we obtain

$$g(x_1 + x_2) + g(x_1 - x_2) = g\left((1+\alpha)x_1 + \beta y\right) + g\left((1-\alpha)x_1 - \beta y\right)$$

$$= g\left((1+\alpha)x_1\right)+g(\beta y)+g\left((1-\alpha)x_1\right)+g(-\beta y) = g(x_1+\alpha x_1)+g(x_1-\alpha x_1)+2g(\beta y)$$

$$= 2g(x_1) + 2g(\alpha x_1) + 2g(\beta y) = 2g(x_1) + 2g(\alpha x_1 + \beta y) = 2g(x_1) + 2g(x_2).$$

Q.E.D.

Corollary 19. *If $(Y, +)$ is uniquely 2-divisible, then every solution f of (19) has the form*

$$f = g + h,$$

where g is quadratic and h is additive.

This corollary follows immediately from Lemma 15 and Theorems 17 and 18.

As we might expect, more precise results are true in the case of inner product spaces.

Theorem 20 (J. Rätz (1980)). *If $(X, \perp)$ is an inner product space and $(Y, +)$ is an Abelian group, then $g : X \to Y$ is an even solution of (17) if and only if there exists an additive mapping $a : \mathbf{R} \to Y$ such that $g(x) = a(\|x\|^2)$ for all x in X.*

Proof. If the additive mapping a exists, then $g(x) = a(\|x\|^2)$ is an even function and by the theorem of Pythagoras g is a solution of (17). Now suppose that g is an even solution of (17). If $u, v \in X$ and $\|u\| = \|v\|$, then $u + v \perp u - v$ and by Theorem 18 (iii), $g(u) = g(v)$. Thus g is constant on each sphere centered at 0, so that the function $\tilde{a}$ is well defined on $\mathbf{R}_+$ by $\tilde{a}(\|x\|^2) = g(x)$ for all $x \in X$. When $\lambda > 0$ and $\mu > 0$, there exist $x, y \in X$ such that $x \perp y$ and $\|x\|^2 = \lambda$, $\|y\|^2 = \mu$. Thus $\|x+y\|^2 = \|x\|^2 + \|y\|^2 = \lambda + \mu$, and we have

$$\tilde{a}(\lambda+\mu) = \tilde{a}\left(\|x+y\|^2\right) = g(x+y) = g(x)+g(y) = \tilde{a}\left(\|x\|^2\right) + \tilde{a}\left(\|y\|^2\right) = \tilde{a}(\lambda)+\tilde{a}(\mu),$$

so $\tilde{a}$ is additive on $\mathbf{R}_+$. As an additive mapping on $\mathbf{R}_+$, $\tilde{a}$ has an additive extension $a : \mathbf{R} \to Y$ and so $g(x) = a(\|x\|^2)$ for all $x \in X$. Q.E.D.

The next result appeared in the article by J. Rätz (1980) and was also found independently by R. Ger and Gy. Szabó.

Corollary 21. *Let $(X, \perp)$ be an inner product space and let $(Y, +)$ be a uniquely 2-divisible Abelian group. Then $f : X \to Y$ is a solution of (17) if and only if there exist additive mappings $a : \mathbf{R} \to Y$ and $h : X \to Y$ such that*

$$f(x) = a\left(\|x\|^2\right) + h(x) \quad \textit{for all } x \textit{ in } X.$$

Proof. The "if" statement follows from Theorem 20, Theorem 17 and Lemma 14 (b). The "only if" statement is established by the use of Lemma 15 and the same two theorems. Q.E.D.

Corollary 22. *Let $(X, \perp)$ be an inner product space and let $(Y, +)$ a real Hausdorff topological vector space. If $f : X \to Y$ is a continuous solution of (17), then there exist a $c \in Y$ and a continuous linear function $h : X \to Y$ such that $f(x) = \|x\|^2 c + h(x)$ for all x in X.*

Proof. Since f is continuous, so are its even and odd parts, $x \mapsto a(\|x\|^2)$ and h, respectively. Take $\lambda > 0$ and rational numbers $\alpha_n > 0$ such that $\alpha_n \to \lambda$ as $n \to \infty$. Also let $x \in X$ be such that $\|x\|^2 = \lambda$ and let $x_n = \left(\frac{\alpha_n}{\lambda}\right)^{\frac{1}{2}} x$ for all $n \in \mathbf{N}$. Then $a(\|x_n\|^2) \to a(\|x\|^2) = a(\lambda)$ as $n \to \infty$. On the other hand, $a(\|x_n\|^2) = a\left(\left(\frac{\alpha_n}{\lambda}\right)\|x\|^2\right) = a(\alpha_n) = \alpha_n a(1) \to \lambda a(1)$ as $n \to \infty$. Hence since Y is Hausdorff, $a(\lambda) = \lambda a(1)$ for $\lambda \geq 0$. Since a is an odd function we have $a(\lambda) = \lambda a(1)$ for all real λ. Putting $c = a(1)$ we obtain $a(\|x\|^2) = \|x\|^2 c$.

Similarly, for any real number λ, let $\{\beta_n\}$ be a sequence of rationals with $\lambda = \lim_{n\to\infty} \beta_n$. Then since h is additive we have $h(\beta_n x) = \beta_n h(x)$. Since h is continuous we have

$$h(\lambda x) = \lim_{n\to\infty} h(\beta_n x) = \lim_{n\to\infty} \beta_n h(x) = \lambda h(x),$$

so h is linear. Q.E.D.

Stability of Orthogonal Additivity

Following R. Ger and J. Sikorska (1995) we shall consider solutions of the inequality

$$\|f(x+y)-f(x)-f(y)\|\leq\varepsilon \tag{20}$$

for orthogonal pairs x, y.

Theorem 23. *Let $(X,\perp)$ be an orthogonality space in the sense of Definition (0), above. Given a real Banach space Y and a number $\varepsilon\geq 0$, let $f: X\to Y$ be a mapping such that for x, y in X,*

$$x\perp y \quad \textit{implies that} \quad \|f(x+y)-f(x)-f(y)\|\leq\varepsilon. \tag{21}$$

Then there exists a unique orthogonally additive mapping $g: X\to Y$ satisfying the inequality $\|f(x)-g(x)\|\leq\dfrac{16\varepsilon}{3}$ for all x in X.

The following two propositions will be needed in the proof of this theorem.

Proposition 24. *Under the asumptions of the theorem, if $f: X\to Y$ is an odd solution of (21) then there exists a unique additive mapping $a: X\to Y$ such that $\|f(x)-a(x)\|\leq 3\varepsilon$ for all x in X.*

Proof. According to (04) of Definition (0), for a given $x\in X$ there exists a $y\in X$ such that $x\perp y$ and also $(x+y)\perp(x-y)$. Then, by (03), $x\perp -y$, so by the oddness of f and (21) we have

$$\|f(x-y)-f(x)+f(y)\|\leq\varepsilon \quad\text{and}\quad \|f(2x)-f(x+y)-f(x-y)\|\leq\varepsilon.$$

Thus,

$$\begin{aligned}
&\|f(2x)-2f(x)\|\\
&\leq\|f(2x)-f(x+y)-f(x-y)+f(x+y)-f(x)-f(y)\|+\|f(x-y)-f(x)+f(y)\|\\
&\leq\|f(2x)-f(x+y)-f(x-y)\|+\|f(x+y)-f(x)-f(y)\|+\|f(x-y)-f(x)+f(y)\|\leq 3\varepsilon.
\end{aligned}$$

It follows that

$$\|f(x)-2^{-n}f(2^n x)\|\leq(1-2^{-n})3\varepsilon \quad\text{for all } x\in X,\ n\in\mathbf{N}, \tag{22}$$

and that $\{2^{-n}f(2^n x)\}$ is a Cauchy sequence. Put

$$b(x)=\lim_{n\to\infty}2^{-n}f(2^n x) \quad\text{for all } x \text{ in } X.$$

Then from (22) we see that $\|f(x)-b(x)\|\leq 3\varepsilon$ for all x in X.

Given any orthogonal pair x, y we have $2^n x \perp 2^n y$ for $n \in \mathbf{N}$, so that

$$\|2^{-n} f(2^n(x+y)) - 2^{-n} f(2^n x) - 2^{-n} f(2^n y)\| \leq 2^{-n}\varepsilon.$$

As $n \to \infty$ we get $b(x+y) - b(x) - b(y) = 0$ when $x \perp y$, so b is orthogonally additive. Thus by Corollary (19), $b : X \to Y$ admits a representation of the form $b = a + q$, where $a : X \to Y$ is additive and $q : X \to Y$ is quadratic. Hence for each x in X we have

$$\|f(x) - a(x) - q(x)\| \leq 3\varepsilon,$$

and

$$\| - f(x) + a(x) - q(x)\| = \|f(-x) - a(-x) - q(-x)\| \leq 3\varepsilon.$$

It follows that

$$\|2q(x)\| = \|f(x) - a(x) - q(x) - f(x) + a(x) - q(x)\| \leq 6\varepsilon, \quad \text{or} \quad \|q(x)\| \leq 3\varepsilon$$

for all x in X. But since q is quadratic we have

$$\|q(x)\| = n^{-2}\|q(nx)\| \leq 3\varepsilon n^{-2} \quad \text{for each } n \text{ in } \mathbf{N}.$$

Hence $q = 0$ and $\|f(x) - a(x)\| \leq 3\varepsilon$ for x in X.

To prove the uniqueness, suppose that $a_1 : X \to Y$ is another additive mapping such that $\|f(x) - a_1(x)\| \leq 3\varepsilon$ for all x in X. Then for each $x \in X$ and $n \in \mathbf{N}$ we have

$$\|a(x) - a_1(x)\| = n^{-1}\|a(nx) - a_1(nx)\| \leq n^{-1}\|a(nx) - f(nx) - a_1(nx) + f(nx)\| \leq n^{-1}6\varepsilon,$$

so that $a = a_1$. Q.E.D.

Proposition 25. *With the assumptions of Theorem 23, if $f : X \to Y$ is an even solution of* (21)*, then there exists a unique quadratic mapping $q : X \to Y$ which is orthogonally additive and such that*

$$\|f(x) - q(x)\| \leq \frac{7\varepsilon}{3} \quad \textit{for all } x \in X.$$

Proof. As before, given $x \in X$ we may choose a $y \in X$ such that $x \perp y$ and $(x+y) \perp (x-y)$, and we also have $x \perp -y$. By (21) and the fact that f is even we get

$$\|f(2x) - f(x+y) - f(x-y)\| \leq \varepsilon \quad \text{and} \quad \|f(x-y) - f(x) - f(y)\| \leq \varepsilon.$$

Since $\frac{x+y}{2} \perp \frac{x-y}{2}$ and $\frac{x+y}{2} \perp \frac{y-x}{2}$, we may use (21) again to obtain

$$\left\|f(x) - f\left(\frac{x+y}{2}\right) - f\left(\frac{x-y}{2}\right)\right\| \leq \varepsilon,$$

$$\left\|f(y) - f\left(\frac{x+y}{2}\right) - f\left(\frac{y-x}{2}\right)\right\| \leq \varepsilon.$$

The last two inequalities together with the evennes of f imply that

$$\|f(x) - f(y)\| \le 2\varepsilon.$$

Hence

$$\|f(2x) - 4f(x)\| \le \|f(2x) - f(x+y) - f(x-y)\| + \|f(x+y) + f(x-y) - 4f(x)\|$$

$$\le \varepsilon + \|f(x+y) - f(x) - f(y)\| + \|f(x-y) - f(x) - f(y)\| + 2\|f(y) - f(x)\|$$

$$\le \varepsilon + \varepsilon + \varepsilon + 4\varepsilon = 7\varepsilon, \quad \text{for all } x,$$

and therefore

$$\|f(x) - 4^{-1}f(2x)\| \le \frac{7\varepsilon}{4} = \frac{(1-4^{-1})7\varepsilon}{3}.$$

By an induction it follows that for $x \in X$ and $n \in \mathbf{N}$,

$$\|f(x) - 4^{-n}f(2^n x)\| \le \frac{(1-4^{-n})7\varepsilon}{3}. \tag{23}$$

It is easily shown that $\{4^{-n}f(2^n x)\}$ is a Cauchy sequence. Define $q : X \to Y$ by

$$q(x) = \lim_{n\to\infty} 4^{-n}f(2^n x).$$

By (23) we obtain

$$\|f(x) - q(x)\| \le \frac{7\varepsilon}{3} \quad \text{for all } x \in X.$$

Take x, y in X with $x \perp y$. Then $2^n x \perp 2^n y$ for each $n \in \mathbf{N}$, and we have

$$\|4^{-n}f\left(2^n(x+y)\right) - 4^{-n}f(2^n x) - 4^{-n}f(2^n y)\| \le 4^{-n}\varepsilon.$$

Let $n \to \infty$ to obtain $q(x+y) - q(x) - q(y) = 0$ when $x \perp y$, so q is orthogonally additive. Hence, again by Corollary 19, there exists an additive mapping $A : X \to Y$ and a quadratic mapping $Q : X \to Y$ such that $q = A + Q$. Thus for each $x \in X$ we have

$$\|f(x) - A(x) - Q(x)\| \le \frac{7\varepsilon}{3},$$

as well as

$$\|f(x) + A(x) - Q(x)\| = \|f(-x) - A(-x) - Q(-x)\| \le \frac{7\varepsilon}{3}.$$

It follows that $\|A(x)\| \le \frac{7\varepsilon}{3}$ for all $x \in X$. But the only additive mapping bounded on the whole space X is the zero mapping, so $A = 0$, $q = Q$ is a quadratic mapping and the existence is proved. As to the uniqueness, let q_1 be another quadratic orthogonally additive mapping from X into Y such that $\|f(x) - q_1(x)\| \le \frac{7\varepsilon}{3}$ for x in X. Then for $n \in \mathbf{N}$ we have

$$\|q(x)-q_1(x)\| = n^{-2}\|q(nx)-q_1(nx)\| = n^{-2}\|q(nx)-f(nx)+f(nx)-q_1(nx)\| \le \frac{n^{-2}14\varepsilon}{3}.$$

Hence $q_1 = q$. Q.E.D.

Proof of Theorem 23. Let f_o and f_e denote the odd and even parts of a solution $f : X \to Y$ of (21), respectively. By definition,

$$f_o(x) = 2^{-1}\left(f(x) - f(-x)\right), \quad f_e(x) = 2^{-1}\left(f(x) + f(-x)\right).$$

Since $x \perp y$ implies $-x \perp -y$, it is easily seen that both f_o and f_e are solutions of (21). From Propositions 24 and 25 we have the existence of an additive function $a : X \to Y$ and an orthogonally additive quadratic function $q : X \to Y$ such that for all $x \in X$,

$$\|f_o(x) - a(x)\| \le 3\varepsilon \quad \text{and} \quad \|f_e(x) - q(x)\| \le \frac{7\varepsilon}{3}.$$

Thus the function $g = a + q$ is orthogonally additive and we have for all $x \in X$,

$$\|f(x) - g(x)\| = \|f_o(x) - a(x) + f_e(x) - q(x)\| < \frac{16\varepsilon}{3}.$$

To prove the uniqueness, let $g_1 : X \to Y$ be another orthogonally additive function such that $\|f(x) - g_1(x)\| \le \frac{16\varepsilon}{3}$ for $x \in X$. Then $g_0 = g - g_1$ is orthogonally additive and $\|g_0(x)\| \le \frac{32\varepsilon}{3} =: \eta$ for all $x \in X$. According to Corollary 19 we have the representation $g_0 = a_0 + q_0$, where a_0 is additive and q_0 is quadratic. Then for all $x \in X$, $n \in \mathbf{N}$, we have

$$\eta \ge \|g_0(nx)\| = \|na_0(x) + n^2 q_0(x)\|,$$

that is,

$$\|q_0(x) + n^{-1}a_0(x)\| \le n^{-2}\eta,$$

which implies that $q_0 = 0$. Then $g_0 = a_0$ is an additive function which is bounded on all of X and hence vanishes, so $g_1 = g$. Q.E.D.

Acknowledgments
Special thanks to J. Chmieliński and J. Garcia-Roig.

References:

ALSINA, C. and GARCIA-ROIG, J. , *On the functional equation* $Tx \cdot Ty = x \cdot y$, in: Constantin Catathéodory: An International Tribute, (Th. M. Rassias, ed.), World Sci. Publ. Co., Singapore, New Jersey, London, Hong Kong, 1991, pp. 47-52.

BARGMANN, V. , *Note on Wigner's theorem on symmetry operations*, J. Math. Phys. 5(1964), 862-868.

CHMIELIŃSKI, J. , *On the Hyers–Ulam stability of the generalized orthogonality equation in real Hilbert spaces*, in: *Stability of Mappings of Hyers-Ulam Type* (Th. M. Rassias and J. Tabor eds.), Hadronic Press, Palm Harbor, Florida, 1994a, pp. 31-41.

CHMIELIŃSKI, J. , *On the stability of the generalized orthogonality equation*, in: *Stability of Mappings of Hyers-Ulam Type* (Th. M. Rassias and J. Tabor eds.), Hadronic Press, Palm Harbor, Florida, 1994b, pp. 43-57.

CHMIELIŃSKI, J. , *On the superstability of the generalized orthogonality equation in Euclidean space*, Ann. Math. Sil. 8(1994c), 127-140.

CHMIELIŃSKI, J. , *On the stability of isometric operators for Hilbert spaces*, 1997, this volume.

CHMIELIŃSKI, J. , *The stability of the Wigner equation for complex Hilbert spaces*, to appear.

DRLJEVIC, H. , *On the representation of functionals and the stability of mappings in Hilbert and Banach spaces*, in: Topics in Math. Anal., (Th. M. Rassias, ed.), World Sci. Publ. Co., Singapore, New Jersey, London, Hong Kong, 1989, pp. 231-245.

GER, R. and SIKORSKA, J. , *Stability of the orthogonal additivity*, Bull. Polish Acad. of Sciences Math. 43(1995), 143-151.

GUDDER, S. and STRAWTHER, D. , *Orthogonally additive and orthogonally increasing functions in vector spaces*, Pacific J. Math. 58(1975), 427-436.

JAMES, R. C. , *Orthogonality and linear functionals in normed linear spaces*, Trans. Amer. Math. Soc. 61(1947), 265-292.

LOMONT, J. S. and MENDELSON, P. , *The Wigner unitarity - antiunitarity theorem*, Annals of Math. 78(1963), 548-559.

RASSIAS, TH. M. and SHARMA C. S. , *Properties of Isometries*, J. Natural Geometry **3** (1993), 1-38.

RÄTZ, J. , *On orthogonally additive mappings*, in: Proc. 18th Intern. Symp. on Functional Equations, Univ. of Waterloo (1980), 22-23.

RÄTZ, J. , *On orthogonally additive mappings II*, in Proc. 19th Intern. Symp. on Functional Equations, Univ. of Waterloo (1981), 38-39.

RÄTZ, J. , *On orthogonally additive mappings*, Aeq. Math. 28(1985), 35-49.

RÄTZ, J. , *On Wigner's theorem: Remarks, complements, comments, and corollaries*, Aeq. Math. 52(1996), 1-9.

SHARMA, C. S. and ALMEIDA, D. F. , *A direct proof of Wigner's theorem on maps wchich preserve probabilities between pure states of quantum systems*, Ann. Phys. 197(1990), 300-309.

TABOR, J. , *On functions behaving like additive functions*, Aeq. Math. 35(1988), 164-185.

TABOR, J. , *Quasi-additive functions*, Aeq. Math. 39(1990), 179-197.

WIGNER, E. P. , *Gruppentheorie und ihre Anwendungen auf die Quantenmechanik der Atomspektren.*, Friedr. Vieweg und Sohn Akt.-Ges., 1931.

WIGNER, E. P. , *On unitary representations of the inhomogeneous Lorentz group*, Ann. Math. 40(1939), 149-204.

WIGNER, E. P. , *Group Theory*, Academic Press, New York, 1959.

NATIONAL TECHNICAL UNIVERSITY OF ATHENS
DEPARTMENT OF MATHEMATICS, ZOGRAFOU CAMPUS
157 80 ATHENS, GREECE

PARFENY P SAWOROTNOW

Hilbert algebras and Hilbert modules

1. Introduction. This is an expository paper. It deals with two loosely related topics, each of which is centered around the notion of $H^\star$-algebra, an algebraic system that was introduced originally by W. Ambrose in his celebrated and well written paper [Ambrose, 1945].

First topic deals with various generalizations of the concept of $H^\star$-algebra, the second part of the paper deals essentially with modules over an $H^\star$-algebra.

In view of the fact that Hilbert modules and their application were discussed in another author's monograph [Saworotnow, 1983], we shall concentrate mostly on the first topic, in particular on those aspects of it with which the author is most familiar and which so far could be found in the original papers only.

2. $H^\star$-algebras. Concept of an $H^\star$-algebra was introduced by W. Ambrose in [Ambrose, 1945]. It could be used to characterize Hilbert-Schmidt operators on a Hilbert space (see, for example [Stone, 1966, Ch. II, §3]. The original definition in [Ambrose, 1945] allowed for some pathology. It was stated in effect as follows:

Let A be a complex Hilbert algebra. This means that A is a Banach algebra with a scalar product (,) which is related to the norm $\|\ \|$ of A by the identity "$(x, x) = \| x \|^2$, $x \in A$".

Definition 2.1. A Hilbert algebra A is an $H^\star$-algebra if for each $x \in A$ there exists some $x^\star \in A$ ($x^\star$ does not have to be unique) such that $(xy, z) = (y, x^\star z)$ and $(yx, z) = (y, zx^\star)$ for all $y, z \in A$.

A pathological example of an $H^\star$-algebra is a Hilbert space H with trivially defined multiplication : $xy = 0$ for all $x, y \in H$. For obvious reasons Ambrose refers to this algebra as "trivial algebra". To avoid this kind of pathology, Ambrose talks about <u>proper</u> $H^\star$-algebras.

Definition 2.2. An $H^\star$-algebra A is said to be proper if $x^\star$ is unique for each $x \in A$.

This means that a proper $H^\star$-algebra A has an involution $x \to x^\star$ such that $R_x^\star = R_{x^\star}$ and $L_x^\star = L_{x^\star}$ for all $x \in A$ (Here $R_x : y \to yx$ and $L_x : y \to yx$ are the operators of multiplying with x on the right and on the left respectively).

Examples of proper $H^\star$-algebras fall in the following three categories:

Example 2.1. Let G be a compact group and let $A = L^2(G)$ be the class of all square summable complex valued measurable functions $x(t)$ on G such that

$$\int |x(t)|^2 \, dt < \infty$$

(we integrate with respect to the Haar measure here). Then A is an $H^\star$-algebra with respect to the convolution $f \star g(t) = \int_G f(s)g(s^{-1}t)dt$ (we assume that the Haar measure here is normalized in a proper way).

For the next example we shall use notation on pages 65–70 of [Stone, 1966] (see also Appendix 1). Stone does not talk about Hilbert-Schmidt operators but they are exactly those operators T in [Stone, 1966] for which $N(T) < \infty$. (Reader can use the definition: a bounded operator T is a Hilbert-Schmidt operator if and only if it is of finite norm, $N(T) < \infty$ (Def. 2.15 of [Stone, 1966]).

Example 2.2. Let H be a Hilbert space and let $HS(H)$ be the class of all Hilbert-Schmidt operators on H. Then $HS(H)$ is an $H^\star$-algebra with respect to the usual operations of linear operator and with respect to the scalar product $(T_1, T_2) = Q(T_1, T_2)$ (it is easy to verify that $Q(T_1T_2, T_3) = Q(T_2, T_1^\star T_3) = Q(T_1, T_3T_2^\star)$ for all $T_1, T_2, T_3 \in HS(H)$.

Example 2.3. A direct sum

$$A = \sum_{\alpha\in\Gamma} \oplus A_\alpha$$

of any family $\{A_\alpha\}_{\alpha\in\Gamma}$ of proper $H^\star$-algebras is also a proper $H^\star$-algebra. (Direct sums of Hilbert spaces are discussed for example in §5 of [Naimark, 1968]. Multiplication on A is defined coordinatewise).

Non-proper $H^\star$-algebras are of not much interest. Fortunately, there is a theorem, originally observed by Ambrose [Ambrose, 1945] which allows us to eliminate this pathology:

Theorem 2.1. Each $H^\star$-algebra A is a direct sum, $A = A_p \oplus A_t$ of a proper $H^\star$-algebra A_p and a trivial algebra A_t ($ab = 0$ for all $a, b \in A_t$), each of which is a two-sided ideal in A. Moreover : $A_tA = AA_t = (0)$.

Here are the essential facts about proper $H^\star$-algebras:

Theorem 2.2. Each proper $H^\star$-algebra is semi-simple.

This theorem could be established rather easily without first proving existence of idempotents in each ideal. However, once Theorem 2.2 is established, the next theorem can be then derived using Jacobson's theory of radicals [Jacobson, 1945]. Both the proof of Theorem 2.2 and Theorem 2.3 are discussed in Appendix 2. It provides an alternative way of establishing existence of self-adjoint idempotents (not widely known fact in mathematical community).

Theorem 2.3. Every closed ideal in a proper $H^\star$-algebra A contains a self-adjoint idempotent e ($e^2 = e^\star = e$).

In the sequel we shall refer to e as a projection (if $e = e^\star = e^2$). Consequence of this theorem is an existence of a maximal family of minimal projections $\{e_\alpha\}$ such that

$$A = \sum_\alpha \oplus e_\alpha A = \sum_\alpha \oplus A e_\alpha$$

(a projection is minimal if it cannot be written as a sum $e = e_1 + e_2$ of non-zero projections e_1, e_2 such that $e_1 e_2 = e_2 e_1 = 0$). Another consequence of this theorem is existence of an approximate identity e_Δ such that

$$\lim_\Delta \| x e_\Delta - x \| = 0 \quad \text{and} \quad \lim_\Delta \| e_\Delta x - x \| = 0.$$

Define

$$e_\Delta = \sum_{i=1}^{n} e_i$$

where $\Delta = \{e_1 \dots e_n\}$ is any finite subset of $\{e_\alpha\}_{\alpha \in \Gamma}$.

Most important results of [Ambrose, 1945] are the following structure theorems:

Theorem 2.4. Every proper $H^\star$-algebra A is a direct sum of simple (proper) $H^\star$-algebra, each of which is a minimal two-sided ideal in A. This means that there exists a family $\{A_\alpha\}_{\alpha \in \Gamma}$ of simple $H^\star$-algebras such that

$$A = \sum_{\alpha \in \Gamma} \oplus A_\alpha$$

and each A_α is a minimal two-sided ideal in A. If A is commutative then each A_α is isomorphic to the complex field. In this case A is a direct sum of complex fields, i.e. $A = L^2(\Gamma)$.

Theorem 2.5. For each simple proper $H^\star$-algebra A there exists a Hilbert space H such that A is isomorphic and isometric to the algebra $HS(H)$ of Hilbert-Schmidt operators on H.

In the original paper [Ambrose, 1945] this result was formulated more or less as follows (as paraphrased by the author of the present paper):

Theorem 2.6. For each simple (complex) $H^\star$-algebra there exists a set J such that A is isomorphic and isometric to the algebra $L^2(J \times J)$ of all complex valued functions $a(i, j)$ on $J \times J$ such that

$$\sum_{i,j} | a(i, j) |^2 < \infty.$$

Algebraic operations on $L^2(J \times J)$ are the usual operations of matrices and the scalar product is defined by

$$(a, b) = \sum_{i,j} a_{ij} \bar{b}_{ij}$$

where $a, b \in L^2(J \times J)$.

Now assume that each of the operators $R_a : x \to xa$ is completely continuous. Then one can show that each minimal ideal A_α is finite dimensional (as it was established in [Ambrose, 1945]), hence each A_α is isomorphic to the algebra of all $n \times n$ matrices for some n. This is exactly the case for the algebra $L^2(G)$, where G is a compact group. This implies the following result (established in [Ambrose, 1945]).

Theorem 2.7. For each compact group g there exists a family $\{J_\alpha\}_{\alpha \in \Gamma}$ of finite set J_α, such that $L^2(G)$ is isomorphic and isometric to the algebra

$$\sum_{\alpha \in \Gamma} \oplus L^2(J_\alpha \times J_\alpha)$$

where $L^2(J_\alpha)$ denote the class of all $n_\alpha \times n_\alpha$ matrices $a(i, j)$ with the norm

$$\| a \| = \left(\sum_{i=1}^{n_\alpha} \sum_{j=1}^{n_\alpha} | a(i, j) |^2 \right)^{1/2}$$

where n_α denotes the number of members of J_α.

3. Right, left and two-sided $H^\star$-algebras. The concept of right $H^\star$-algebra was introduced by Smiley [Smiley, 1953] as a Hilbert algebra A with an involution $x \to a^\star$ such that $R_{a^\star} = R_a^\star$ (but not necessarily $L_{a^\star} = L_a^\star$). (We shall use a^r instead of $a^\star$ below). Every proper $H^\star$-algebra is a right $H^\star$-algebra but there are right $H^\star$-algebras, for which $L_{a^\star} \neq L_a^\star$, in general (see Example 3.1 below). M. F.

Smiley showed in [Smiley, 1953] that a right $H^\star$-algebra admits structure theorems similar to those for Ambrose's $H^\star$-algebras (see Theorems 3.1 and 3.2 below).

Definition 3.1. A Hilbert algebra A is a proper right $H^\star$-algebra if it has an involution $x \to x^r$ such that $(yx, z) = (y, zx^r)$ for all $x, y, z \in A$.

Definition 3.2. A proper right $H^\star$-algebra A is a two-sided $H^\star$-algebra if it is also a left $H^\star$-algebra, i.e. it has a second involution $x \to x^\ell$ ($x^\ell \neq x^r$, in general) such that "$(xy, z) = (y, x^\ell z)$" is valid.

Below is an example of a right $H^\star$-algebra which is not an $H^\star$-algebra (unless the operator α below is a positive multiple of the identity operator). It was originally given in [Smiley, 1953] in terms of (infinite) matrices. We present it in terms of operators on a Hilbert space.

Example 3.1. Let H be a Hilbert space and let α be a norm-increasing bounded positive (hence self-adjoint) operator acting on H. Let $HS(H)$ be the class of all Hilbert-Schmidt (linear) operators a on H with the inner product (,) as in Example 2.2 above (we use different notation, e.g. a, b instead of T_1, T_2, etc.). Consider the system A consisting of members of $HS(H)$ together with usual algebraic operation by with the scalar product $[a, b] = (\alpha a, b)$. Then A is a right $H^\star$-algebra with respect to this scalar product, for which the condition "$[ab, c] = [b, a^\star c]$" does not need to be true. The fact that α is norm-increasing is used to verify completeness of A with respect to the norm $[a, a]^{1/2}$ as well as a continuity of multiplications.

Smiley proved that each simple (proper) right $H^\star$-algebra is of the type described in Example 3.1 (see [Smiley, 1953]):

Theorem 3.1. For each simple proper right $H^\star$-algebra A there exists a Hilbert space H and a norm-increasing bounded positive operator α on H such that A is isomorphic to the algebra $HS(H)$ of Hilbert-Schmidt operators on H and the inner product of A corresponds to the inner product $[a, b] = (\alpha a, b)$ where a and b are operators in $HS(H)$ corresponding via isomorphism to appropriate members of A.

Also first Structure Theorem is valid for right $H^\star$-algebras

Theorem 3.2. Each proper right $H^\star$-algebra is a direct sum of proper right $H^\star$-algebras each of which is a two-sided ideal in the original algebra. An interesting fact about a right $H^\star$-algebra is the assertion that it is also a left (hence a two-sided) $H^\star$-algebra.

Theorem 3.3. For each member a of a proper right $H^\star$-algebra A there exists $a^\ell \in A$ such that $(ax, y) = (x, a^\ell y)$ for all $x, y \in A$ ($L_a^\star = L_{a^\star}$).

Similar theorems also hold for more general algebras (see section 4).

Here is a more or less symmetric example of a two-sided $H^\star$-algebra:

Example 3.2. Let H_1 and H_2 be Hilbert spaces and let $T : H_2 \to H_1$ be a norm-decreasing linear transformation (note that $\| T \| \leq 1$) having the bounded inverse $S : H_1 \to H_2$. Let A be the class of all linear Hilbert-Schmidt transformations from H_1 into H_2. Define multiplication on A by setting $a \star b = aTb$. Then A is a proper two-sided $H^\star$-algebra with respect to this multiplication and the standard scalar product $(a, b) = Q(a, b)$ $(a, b \in A)$ [Stone, 1966, p. 67] (see also Appendix 1). Note that the right and left adjoint of $a \in A$ are defined by $a^r = Sa^\star T^\star$ and $a^\ell = T^\star a^\star S$.

It was shown in [Saworotnow, 1957, p. 54] that Example 3.2 is a canonical example of two-sided proper $H^\star$-algebras:

Theorem 3.4. For each simple two-sided $H^\star$-algebra A there are complex Hilbert spaces H_1 and H_2 and a norm-decreasing bounded linear transformation $T : H_2 \to H_1$ having a bounded inverse S such that A is isomorphic and isometric to the class of all Hilbert-Schmidt linear transformations from H_1 into H_2 and multiplication corresponds to $a \star b = aTb$.

4. Weak $H^\star$- and complemented algebras. This section deals with a larger class of algebras, which of course includes those considered in previous sections. In the present case involution does not have to be defined everywhere. We assume that it is defined on a dense subset.

Definition 4.1. A Hilbert algebra A is said to be a proper weak right $H^\star$-algebra [Saworotnow, 1988] if it has an involution $x \to x^r$ defined on dense subset D_r of A such that $(yx, z) = (y, zx^r)$ for all $y, x \in A$ (and $x^r \in D_r$).

Weak left and two-sided $H^\star$-algebras are defined in a similar fashion.

Example 4.1. Let α be a norm-increasing, positive (self-adjoint), densely defined, linear operator on a Hilbert space H (α does not have to be continuous) and let $(\, , \,)$ denote (as in Example 3.2 above) the scalar product on the space $HS(H)$ of all Hilbert-Schmidt operators on H. Let $N(\ )$ be the corresponding norm (as in [Stone, 1966, p. 66]). Let A be the class of all members a of $HS(H)$ such that $N(\alpha^{1/2}a) < \infty$. Define the inner product $[\, , \,]$ on A by setting $[a, b] = (\alpha a, b)$.

As in Example 3.1, it is easy to see that α has a bounded inverse, and this fact could be used to establish completeness of A (whether α is bounded or not) and the fact that the multiplication on A (the restriction of the usual multiplication of members of $HS(H)$ to A) is continuous. It is also not difficult to see that there is a dense subset D_r of A such that $a^\star \in A$ for each $a \in D_r$ ($a^\star$ denotes the usual adjoint of the Hilbert-Schmidt operator a), and D_r cannot be entire A unless α is bounded.

It was established in [Saworotnow, 1988] that each proper right weak $H^\star$-algebra has also left involution:

Theorem 4.1. Every proper weak right $H^\star$-algebra is a proper weak two-sided $H^\star$-algebra.

Other characterizations of these algebras are established in the next theorems. We shall need the concept of complemented algebras.

Definition 4.2. A Hilbert algebra A is said to be right complemented if the orthogonal complement of every right ideal is again a right ideal. Left complemented algebra is defined in the obvious way. Complemented algebra is a Hilbert algebra which is both right and left complemented (we do not need adjective "two-sided" here).

Proposition 4.1. Every weak right $H^\star$-algebra is a right complemented algebra. Every proper weak right $H^\star$-algebra is semi-simple.

First part of this proposition is almost immediate. Second part is proven in [Saworotnow, 1992].

The converse is also true:

Theorem 4.2. Every semi-simple right complemented algebra is a weak two-sided $H^\star$-algebra.

This theorem was originally established in author's dissertation [Saworotnow, 1957] by showing first that each semi-simple right complemented algebra A has a dense subset D_ℓ of A such that each $x \in D_\ell$ has a left adjoint x^ℓ (satisfying "$(xy, z) = (y, x^\ell z)$").

Also there is generalization of Theorem 3.1 for complemented algebra:

Theorem 4.3. Each semi-simple right complemented algebra is isomorphic and isometric to the algebra A in Example 4.1.

Other representations of a weak $H^\star$-algebra are discussed below.

5. Characterization of $H^\star$-algebras in terms of complementation. The title refers mostly to algebras in section 3 above (right, left, two-sided $H^\star$-algebras). We shall need the following definition:

Definition 5.1. A semi-simple right complemented algebra A is said to be well-complemented if every proper closed right ideal in A has a non-zero left annihilator and every proper closed left ideal has a non-zero right annihilator.

{A right annihilator of a set S is the set $r(S) = \{x \in A : Sx = (0)\}$, left annihilator $\ell(S)$ is defined in a similar fashion.} Concept of annihilators was used originally by Kaplansky in [Kaplansky, 1948] in his generalization of $H^\star$-algebras.

Theorem 5.1. Every right well-complemented algebra is a two-sided $H^\star$-algebra (both involutions are defined everywhere).

This theorem was established by the author in his dissertation [Saworotnow, 1957, p. 53]. An important consequence of this theorem is the fact that each simple well-complemented algebra is isomorphic and isometric to the algebra described in Example 3.1 above (it should be noted here that originally the author did establish Theorem 5.1 as a consequence of this last fact). There is also a characterization of $H^\star$-algebras in terms of complementation:

Theorem 5.2. A right complemented algebra is an $H^\star$-algebra in the sense of Ambrose [Ambrose, 1945] if and only if every left self-adjoint idempotent in A is also right self-adjoint (if $e^2 = e$ and $(ex, y) = (x, ey)$ then $(xe, y) = (x, ye)$ for all $x, y \in A$).

6. Discrete complemented algebras. This section deals with a certain subclass of complemented algebras that admit certain representation which is symmetric with respect to both involutions (see Theorem 6.1 below).

Definition 6.1. Let A be a complemented algebra. A closed right ideal R in A is said to be invariant if $\ell(R^\perp) = \ell(R)^\perp$ (the left annihilator of the orthogonal complement $R^\perp$ of R coincides with the orthogonal complement of the left annihilator of R).

If the involutions on A are defined everywhere (if A is a proper two-sided $H^\star$-algebra) then one can use the following equivalent invariant definition: R is invariant if $R^r = R^\ell$ i.e. the set $\{x^r : x \in R\}$ is identical with the set $\{x^\ell : x \in R\}$ of left adjoints of R (note that there are certain difficulties with defining R^r in a general case).

Definition 6.2. A complemented algebra is said to be discrete if each closed right invariant ideal in it contains a two-sided projection (an idempotent e which is both right and left self-adjoint).

It was shown in [Saworotnow, 1966] that each finite dimensional complemented algebra is discrete.

Example 6.1. Let J be any set (it does not have to be countable) and let $\{\alpha(j) : j \in J\}$ be a real valued function on J such that $1 \leq \alpha(j)$ for all $j \in J$. Let $L^2(J, \alpha)$ be the space of all complex-valued matrices $a(i, j)$ on $J \times J$ such that

$$\sum_{i,j} \alpha(i) \, |a_{ij}|^2 \, \alpha(j) < \infty.$$

Then $L^2(J, \alpha)$ is a discrete complemented algebra. If α is bounded (from above) then A is a discrete two-sided $H^\star$-algebra.

Theorem 6.1. For each semi-simple discrete complemented algebra A there exists a set J and a real valued function $\alpha(j)$ on J such that $1 \leq \alpha(j)$ for all $j \in J$ and A is isomorphic and isometric to the algebra $L^2(J, \alpha)$ described in Example 6.1 above. If A is a two-sided $H^\star$-algebra then there exists constant k such that $\alpha(j) \leq k$ for all $j \in J$.

This theorem was essentially established in [Saworotnow, 1966, Proc. of A.M.S.] and [Saworotnow, 1966, Pac. J.].

7. Examples, involving measure theory. There is an interesting application of Gelfand theory to complemented algebras that allows for representation theorems involving measure. More specifically it was established in [Saworotnow, 1964] and [Saworotnow, 1970, Ill. J. of Math.] that each simple (and semi-simple) complemented algebra is essentially as in the next example.

Example 7.1. Let (S, μ) be a measure space and let $K(s)$ be a finite almost everywhere real valued measurable function on S such that $1 \leq K(s)$ for all $s \in S$. In particular μ can be a Borel measure on a locally compact space S and $K(s)$ could be a continuous function on S, finite locally almost everywhere (as defined in §6 of [Naimark, 1968]).

Let $L^2(S \times S, \mu, K)$ be the class of all complex valued members x of $L^2(S \times S, \mu \times \mu)$ such that $\int\int | \, x(t, s) \, |^2 \, K(s) d\mu(t) \, d\mu(s) < \infty$. Then $L^2(S \times S, \mu, K)$ is a complemented algebra with respect to the scalar product

$$(x, x') = \int x(t, s)\overline{x'}(t, s)K(s)d\mu(t)\, d\mu(s)$$

and the multiplication $(x \times x')(t, s) = \int x(t, r)x'(r, s)d\mu(r)$. If $K(s)$ is bounded (also from above) then $L^2(S \times S, \mu, K)$ is well-complemented (i.e. it is a two-sided $H^\star$-algebra).

It turns out that the algebra in this example is of a canonical nature:

Theorem 7.1. For each simple and semi-simple complemented algebra A there exists a locally compact Hansdorff space S, a Borel measure μ on S and a continuous real valued function $K(s)$, finite locally almost everywhere such that A is isometric and isomorphic to $L^2(S \times S, \mu, K)$.

If A is well-complemented then $K(s)$ is bounded on S.

It would be interesting to see if each simple complemented algebra would be of the form described in the next example.

Example 7.2 Let S, μ and $K(s)$ be as above. Let A be the algebra of all functions $x(s, t)$ such that $\int\int K(t) \mid x(t, s) \mid^2 K(s)d\mu(t)\, d\mu(s) < \infty$. Then A is a weak two-sided $H^\star$-algebra with respect to the multiplication as in Example 7.1 above and with respect to the scalar product $(x, x') = \int\int K(t)x(t, s)\overline{x'}(t, s)K(s)d\mu(t)\, d\mu(s)$ (note that $x^\ell(t, s) = K(t)^{-1}\overline{x}(s, t)K(s)$, $x^r(t, s) = K(t)\overline{x}(s, t)K(s)^{-1}$).

8. Systems with partially defined multiplication. In this section we shall consider briefly Hilbert spaces with a densely defined multiplication. In particular it could be an associative algebra, with an inner product, that is not complete with respect to the norm associated with this inner product. Most of the authors listed below assume existence of an involution $x \to x^\star$ such that "$(xy, z) = (y, x^\star z) = (x, zy^\star)$" is valid. M. F. Smith in [Smith and Saworotnow, 1961] uses a weaker assumption that the orthogonal complement of a right ideal is a right ideal (cf section 4.1 above).

W. Ambrose was also the first one to study such systems (as in the case of $H^\star$-algebras). In [Ambrose, 1949] he considered the Hilbert space $L^2(G)$, where G is some unimodular locally compact group. Unless G is compact, the convolution $x \star g(s) = \int x(ts^{-1})g(s)\, ds$ is not defined everywhere on $L^2(G)$; but is defined if $x \in L'(G)$ (and $y \in L^2(G)$).

Systems of this kind were considered by Dixmier, Nakano, Phillips, Riefel, Rohlin, Smith, Takesaki, Tomita, Yood and others. They are discussed in Chapter 14 of Istrătescu, 1987]. Interested readers may also want to read the manuscript of Takesaki [Takesaki, 1970]. Here he considers so-called modular and generalized Hilbert algebras (these two systems are equivalent in a certain sense). He shows (Chapter 12) that for each Von Neumann algebra M there exists a modular Hilbert algebra $\mathcal{A}$ such that M is equivalent to the Von Neumann algebra $\mathcal{L}(\mathcal{A})$ generated by operators $x \to ax$, $a \in \mathcal{A}$.

9. Non-associative Hilbert algebras. These algebras are more or less outside of the author's field of interest, so we shall be brief.

J. R. Schue was the first one to consider such systems. In [Schue, 1960, 1961] he talks about $L^\star$-algebras. An $L^\star$-algebra is both a Hilbert space and a Lie algebra with an involution $x \to x^\star$ such that "$([x, y], z) = (y, [x^\star, z])$" is valid (here $[\ ,\ ]$ denotes the multiplication on A). { Example: take any $H^\star$-algebra A and consider the (non-associative) multiplication $[x, y] = xy - yx \quad (x, y \in A)$.}

The field seems to be expanding rapidly. Interested readers may want to read the recently published work by M. Cabrera and co-authors [Cabrera, Marrakchi, Martinez and Rodrigues, 1994]. This manuscript gives a rather good overview of the field and contains an extensive bibliography. Section 17.13 of [Istrătescu, 1987] also deals with non-associative Hilbert algebras.

10. Trace-class of an $H^\star$-algebra. These algebras are not Hilbert algebras per se but they play an important role in the theory of Hilbert modules. Let A be an $H^\star$-algebra. For simplicity we assume that A is proper (no loss of generality). The trace-class of A is the algebra $\tau A = \{xy : x, y \in A\}$. It is a Banach algebra with respect to a certain norm, denoted by $\tau(\ )$ such that $\tau(a^\star a) = \| a \|^2, \quad a \in A$, and has a trace tr (a positive linear functional) such that $tr(xy) = tr(yx)$, and $tr(xy^\star) = (x, y)$ for all $x, y \in A$ and $tr(a) = \tau(a)$ if $a \geq 0$ ("$\geq$" here is defined in the obvious manner: "$a \geq b$" if "$(ax, x) \geq (bx, x)$ for each $x \in A$"). The trace tr has norm 1 i.e. $| tr(a) | \leq \tau(a)$ for all $a \in A$.

If A is proper and finite dimensional then $\tau A = A$. This fact is relevant in comparing author's theory of Hilbert modules with the work of Goldstine and Horwitz.

11. Hilbert modules. This section deals with a class of modules that could be considered as a generalization of Hilbert spaces. Many properties of Hilbert spaces are valid for the members of this class. This includes Schwarz inequality, Reisz theorem about general form of a (bounded) linear functional, Spectral Theorem and its consequences, and others.

This class of modules was first considered (for the case, when the algebra of "scalars" was finite dimensional) by Goldstine and Horwitz [Goldstine and Horwitz, 1966]. General case was studied by the author [Saworotnow, 1968].

The idea of generalization is simple: in a definition of Hilbert space we replace the scalars by an $H^\star$-algebra and adjust the axioms accordingly. As a result we obtain a system, which we call "Hilbert module."

More specifically, let A be a proper $H^\star$-algebra and let τA be its trace-class.

Definition 11.1. A right module H over A is called Hilbert A-module (or Hilbert module, if A is fixed) if there is a τA-valued function $[\ ,\ ]$ defined on $H \times H$ with following properties:

(i) $[\ ,\ ]$ is additive in both of its factors i.e. $[f+g, h] = [f, h] + [g, h]$ and $[f, h+g] = [f, h] + [f, g]$ for all $f, g \in H$.

(ii) $[\ ,\ ]$ is skew symmetric in the sense that $[f, g] = [g, f]^\star$ for all $f, g \in H$.

(iii) $[\ ,\ ]$ is A-homogeneous in the second factor i.e. $[f, ga] = [f, g]a$ for all $f, g \in H$ and $a \in A$ (note that this implies "$[fa, g] = a^\star[f, g]$" because of (ii)).

(iv) $[\ ,\ ]$ is strictly positive definite i.e. $(f, f) \geq 0$ for all $f \in H$ and $[f, f]$ is the "zero member" of A if and only if $f = \vec{0}$ (f is the "zero member" of H).

(v) H is complete with respect to the metric $\rho(f, g) = \sqrt{\tau[f-g, f-g]}$ (it is not difficult to show that $\rho(\ ,\ )$ is indeed a metric).

Remark 1. Originally the author postulated also the weak Schwarz inequality

$$|tr[f, g]|^2 \leq \tau[f, f]\tau[g, g] \quad \text{for all} \quad f, g \in H$$

but it was pointed out to him by Hugo D. Junghenn that this inequality is a consequence of other axioms.

Remark 2. If A is finite dimensional and proper then $\tau A = A$ and so one can state axioms of a Hilbert module the way it was done in [Goldstine and Horwitz, 1966].

In fact, in this case A has an identity and so it is easy to prove existence of a linear structure on the Hilbert module, something that is more difficult to establish in an infinite dimensional case. It is stated formally in the next theorem.

Theorem 11.1. For each Hilbert A-module H there is an operation on $H \times C$ into $H : < f, \lambda > \to \lambda f$ of multiplying a member f of H by a complex number λ, such that

$$\lambda(fa) = (\lambda f)a = f(\lambda a), \quad a \in A$$

holds, with respect to which H becomes an ordinary Hilbert space, if we consider $(f, g) = tr[g, f]$ as the inner product (also, it is easy to verify that (f, g) is indeed an inner product).

We shall denote the norm on H by $\| \ \|$:

$$\| f \|^2 = \tau[f, f] = tr[f, f] = (f, f), \quad f \in H$$

The metric ρ in the above definition is related to this norm by "$\rho(f, g) = \| f - g \|$".

Theorem 11.2 (strong Schwarz inequality). "$\tau[f, g] \leq \| f \| \cdot \| g \|$" is valid for any two members f and g of a Hilbert module.

Proof of both above theorems can be found in [Saworotnow, 1968].

As an example of a Hilbert module one can take a direct sum

$$H = \sum_{\alpha \in \Gamma} \oplus R_\alpha$$

of any family $\{R_\alpha\}_{\alpha \in \Gamma}$ of closed right ideals in a proper $H^\star$-algebra A: each member $x = \{x(\alpha)\}_{\alpha \in \Gamma}$ of H is a function $x(\ )$ such that $x(\alpha) \in R_\alpha$ for each $\alpha \in \Gamma$ and

$$\sum_{\alpha \in \Gamma} \|x(\alpha)\|^2 < \infty.$$

Also the algebra A in Example 4.1 can be considered an example of a Hilbert module over the $H^\star$-algebra $HS(H)$ considered there.

It was shown in [Saworotnow, 1968] that each Hilbert module is of the form described in the next example.

Example 11.1. Let A be a proper $H^\star$-algebra and let $\{R_\alpha\}_{\alpha \in \Gamma}$ be a family of closed right ideals in A. For each α let $a_\alpha \in A$ be such that $\| a_\alpha \| \leq 1$. Take

$$H = \sum_{\alpha \in \Gamma} \oplus R_\alpha$$

and define the vector inner product

$$[x, y] = \sum_{\alpha\in\Gamma} x(\alpha)^\star a_\alpha^\star a_\alpha y(\alpha) \quad \text{for each} \quad x, y \in H.$$

$$(x = \{x(\alpha)\}_{\alpha\in\Gamma}, \quad y = \{y(\alpha)\}_{\alpha\in\Gamma}).$$

Then H is a Hilbert module with respect to this inner product (algebraic operations are defined coordinatewise).

We have a generalization of Reisz Theorem for Hilbert modules but we need to make precise the generalization of a bounded linear functional. Let H be a Hilbert A-module over a proper $H^\star$-algebra A.

Definition 11.2. A map $\ell : H \to \tau A$ is said to be an A-linear functional if $\ell(f+g) = \ell(f) + \ell(g)$ and $\ell(fa) = \ell(f)a$ for all $f, g \in H$ and $a \in A$. It is bounded if there exists $\mathcal{M} \geq 0$ such that $\tau(\ell(f)) \leq \mathcal{M} \parallel f \parallel$ for all $f \in H$.

If $g \in H$ then the mapping $\ell_g : f \to [g, f]$ $(\ell_g(f) = [g, f])$ is a bounded A-linear functional.

For each bounded linear functional ℓ there exists a member g of H such that $\ell = \ell_g$ $(\ell(f) = [g, f]$ for all $f \in H)$ and

$$\parallel g \parallel = \inf\{\mathcal{M} : \tau(\ell(f)) \leq \mathcal{M} \parallel f \parallel, \quad f \in H\},$$

which is equivalent the fact that $\parallel g \parallel = \parallel \ell_g \parallel$.

Definition 11.3. A mapping $T : H \to H$ is said to be A-linear if $T(f+g) = Tf+Tg$ and $T(fa) = (Tf)a$ for all $f, g \in H$ and $a \in A$. It is said to be bounded if there exists $\mathcal{M} \geq 0$ such that $\parallel Tf \parallel \leq \mathcal{M} \parallel f \parallel$ for all $f \in H$.

Theorem 11.4. Each bounded A-linear mapping $T : H \to H$ is a bounded linear operator on H (with respect to the Hilbert space structure of H). Its adjoint $T^\star$ (in the sense that $(Tf, g) = (f, T^\star g)$) is also A-linear and has the property that $[T^\star f, g] = [f, Tg]$ for all $f, g \in H$

Corollary. The class of all bounded A-linear transformations on H is a $B^\star$-algebra.

All above theorems are established in [Saworotnow, 1968].

Spectral Theorem is also valid for a self-adjoint A-linear operator on a Hilbert module. It is a simple consequence of a classical statement. The statement and the proof can be found in [Saworotnow, 1968]. The statement differs from the classical formulation [e.g. Theorem IV in subsection 4 of §17 in [Naimark, 1968] only in the fact that we assert that the projections in questions are A-linear (the classical formulation states that for each self-adjoint operator T there exist a family $P(\lambda)$ of projections, with certain properties, such that $T = \int_{-\infty}^{\infty} P(\lambda)d\lambda$. Our statement asserts also that each $P(\lambda)$ is A-linear (for the case when T is).

Application of Hilbert modules to stationary processes and harmonic analysis on a group are discussed in [Saworotnow, 1983].

There are also some diagonalization theorems for self-adjoint A-linear operators. For example it was shown in [Saworotnow, 1985] that for each bounded A-linear operator T acting on a Hilbert module H there is a locally compact Hansdorff space X, a τA-valued regular measure μ, defined on certain Borel subsets of X and a bounded continuous real valued function h on X such that H is isomorphic to $L^2(\mu)\otimes A$ and T corresponds to multiplying members of $L^2(\mu)$ with h (T takes $\sum f_i \otimes a_i$ into $\sum(f_i h) \otimes a_i$). Some explanations: $L^2(\mu)$ is the set of all complex valued $tr\,\mu$-measurable functions such that $\int |f|^2 \, d\,tr\,\mu < \infty$. The space $L^2(\mu) \otimes A$ is defined as the closure of the set of all finite expressions $\xi = \sum_{i=1}^{n} f_i \otimes a_i$: where $f_i \in L^2(\mu)$, $a_i \in A$, the closure being taken with respect to the norm

$$\| \xi \| = \left(\sum_{i,j} tr\,(a_i^\star \int \overline{f}_i f_i \, d\mu a_j) \right)^{1/2} .$$

The reader may have noticed that there is room for improvement in the theorem. Similar theorems hold also for certain unbounded operators and for a commuting family of operators [Saworotnow, 1991].

12. Modules over other algebras. Study of these systems was originated by I. Kaplansky in [Kaplansky, 1953], where he considered modules over $C^\star$-algebras A with an inner product with values in A. This topic is discussed in Chapter 12 of [Istrătescu, 1987]. Interested readers are advised to read this portion of Istrătescu's book.

Appendix 1. To assist those who are not familiar with a theory of Hilbert Schmidt operators the author decided to give a short outline of the theory included in [Stone, 1966]. It is done in a more general set up, essentially because of Example 3.2 above.

We are discussing the class of Hilbert Schmidt transformations from one Hilbert space into another, which may or may not be identical with the first one (if they are identical then we have the case considered in [Stone, 1966].

The reader is expected to be familiar with the concept of an orthonormal base of a Hilbert space (a maximal family of mutually orthogonal vectors of norm one).

Let H_1 and H_2 be Hilbert spaces and let $T : H_1 \to H_2$ be a bounded linear transformation.

Definition. T is said to be a Hilbert Schmidt transformation (Stone uses term "T is ... of finite norm" in Definition 2.15 of [Stone, 1966]) if

$$N(T;\, \phi,\, \psi) = (\sum_{\alpha,\beta} |(T\phi_\alpha,\, \psi_\beta)|^2)^{1/2}$$

if finite for some orthonormal bases $\phi = \{\phi_\alpha\}_{\alpha\in\Gamma_1}$, and $\psi = \{\psi_\beta\}_{\beta\in\Gamma_2}$ of H_1 and H_2 respectively. The identities

$$\text{“}N(T;\, \phi,\, \psi)^2 = \sum_{\alpha,\beta}(T\phi_\alpha,\, \psi_\beta)(\psi_\beta,\, T\phi_\alpha) = \sum_\alpha (T\phi_\alpha, \sum_\beta (T\phi_\alpha,\, \psi_\beta))$$

$$= \sum_\alpha (T\phi_\alpha,\, T\phi_\alpha) = \sum_\alpha \| T\phi_\alpha \|^2\text{”}$$

and

$$\text{“}N(T;\, \phi,\, \psi)^2 = \sum_\beta \| T^\star \psi_\beta \|^2\text{”}$$

show that $N(T) = N(T;\, \phi,\, \psi)$ does not depend on a particfular choice of either ϕ or ψ (at least for the case when $N(T,\, \phi,\, \psi) < \infty$ for some ϕ and ψ).

It is not difficult to see that T_1, T_2 are of Hilbert Schmidt type $Q(T_1,\, T_2) = Q(T_1,\, T_2,\, \phi,\, \psi) = \sum_{\alpha,\beta}(T_1\phi_\alpha,\, \psi_\beta)(\psi_\beta,\, T_2\phi_\alpha)$ is finite and is independent of both ϕ and ψ. Also Q satisfies all the postulates of an inner product and is related to the function N by the identity "$N(T)^2 = Q(T,\, T)$".

In fact the class $HS(H_1,\, H_2)$ of all Hilbert Schmidt transformations from H_1 into H_2 is a Hilbert space. If $H_2 = H_1$, then the class $HS(H_1) = HS(H_1,\, H_1)$ of all Hilbert Schmidt operator on a Hilbert space H_1 (e.g. Example 2.2 above).

Appendix 2. In this appendix we give proofs of Theorem 2.2 and Theorem 2.3 above. We hope that the reader will be able to see how these proofs can be modified for more general situation (like proving second part of Proposition 4.1 above or proving existence of a left self-adjoint idempotent in a right complemented algebra (see the remark after statement of Theorem 4.2 above)).

Proof of Theorem 2.2 (Every proper $H^\star$-algebra A is semi-simple). According to [Jacobson, 1945] it is sufficient to show that each closed ideal in A contains a vector which is not a generalized nilpotent. It was established in 11B of [Loomis, 1953] that each linear operator T on a Hilbert space satisfies the condition $\| T^\star T \| = \| T \|^2$. In particular, we have

$$\|R_{a^\star a}\| = \|R_a^\star R_a\| = \|R_a\|^2$$

for any mapping

$$R_a : x \to xa.$$

If $a \in A$ is self-adjoint and

$$\| R_a \| = 1,$$

then

$$\| R_{a^2} \| = \| R_a \|^2 = 1$$

also. In this case we have

$$\| a^{2^n} \| \geq \| R_{a^{2^n}} \| = 1$$

for any n, and this means that a could not be a generalized nilpotent.

This implies that, in order to show semi-simplicity of A, it is sufficient to show that each closed ideal in A contains a non-zero self-adjoint vector a with above property. But this can always be done, since A is proper and

$$\| R_{\lambda a} \| = \lambda \| R_a \|$$

if $\lambda > 0$ (note also that $R_a \neq 0$ if $a \neq 0$).

Proof of Theorem 2.3 (Every closed ideal I in a proper $H^\star$-algebra A contains a projection). Assume that I is a left ideal, then use one of the characterizations of the radical [Jacobson, 1945] as the join of all left ideals L such that every member of L has a right quasi inverse (this characterization is not difficult to establish — see, for example, proof of Lemma 4 on page 52, in [Saworotnow, 1957]). Since A is semi-simple, I must contain a vector a with no right quasi-inverse (it is not possible to find

$b \in A$ such that $a+b+ab=0$). This means that the right ideal $R=\{ax+x : x \in A\}$ could not contain $-a$. Also it is not difficult to show that $\overline{R}$ is a proper ideal in A. Write $-a = e+u$ with $e \in \overline{R}^{\perp}$ and $u \in \overline{R}$. Then it is easy to show that e is a projection ($e^2 = e = e^\star$) and that $e(-a) = e$ (which implies that $e \in L$).

REFERENCES

[1] W. Ambrose (1945). "Structure Theorems for a Special Class of Banach Algebras." *Trans. Amer. Math. Soc.*, 57, 364–386.

[2] W. Ambrose (1949). "The L^2 System of a Unimodular Group." *Trans. Amer. Math. Soc.*, 65, 27–48.

[3] M. Cabrera, A. El Marrakchi, J. Martinez and A. Rodriguez (1994). "An Allison-Kantor-Koecher-Tits Construction for Lie $H^\star$-Algebras." *Journal of Algebra*, 164, 361–408.

[4] J. Dixmier (1952). "Algebres quasi-unitaires." *Comment. Math. Helvet.*, 26, 275–322.

[5] George R. Giellis (1971). "Trace-class for a Hilbert Module." *Proc. Amer. Math. Soc.*, 29, 63–68.

[6] ________, (1972). "A Characterization of Hilbert Module." *Proc. Amer. Math. Soc.*, 36, 440–442.

[7] H. H. Goldstine and L. P. Horwitz (1966). "Hilbert Space with Non-associative Scalars, II." *Math. Ann.*, 164, 291–316.

[8] V. I. Istrătescu (1987). *Inner Product Spaces.* D. Reidel Publ. Co.

[9] N. Jacobson (1945). "The Radical and Semi-simplicity for Arbitrary Rings." *Amer. J. Math.*, 67, 300–320.

[10] N. Jacobson (1962). *Lie Algebras.* Interscience Publishers, John Wiley.

[11] I. Kaplansky (1948). "Dual Rings." *Annals of Math.*, 49, 689–701.

[12] ________, (1953). "Modules over Operator Algebras." *Amer. J. Math.*, 75, 839–858.

[13] E. R. Keown (1955). "Reflexive Banach Algebras." *Proc. Amer. Math. Soc.*, 6, 252–259.

[14] ________, (1967). "Some New Hilbert Algebras." *Trans. Amer. Math. Soc.*, 128, 71–87.

[15] L. H. Loomis (1953). *An Introduction to Abstract Harmonic Analysis.* D. Van Nostrand Co.

[16] M. A. Naimark (1968). *Normed Rings.* English translation of an earlier edition by L. F. Baron, Groningen, 1964.

[17] H. Nakano (1950). "Hilbert Algebras." *Tohuku Math. J.*, 2(2), 4–23.

[18] John Phillips (1973). "Positive Integrable Elements Relative to a Left Hilbert Algebra." *J. of Functional Analysis*, 13, 390–409.

[19] M. Riefel (1969). "Square Integrable Representation of Hilbert Algebras." *J. Functional Analysis*, 3, 265–300.

[20] V. Rohlin (1948). "Unitary Rings." *Doklady Akad. Nauk USSR (N.S.)*, 59, 643–646.

[21] P. P. Saworotnow (1957). "On a Generalization of the notion of $H^\star$-algebra." *Proc. Amer. Math. Soc.*, 8, 49–55.

[22] ________, (1957). "On the Imbedding of a Right Complemented Algebra into Ambrose's $H^\star$-algebra." *Proc. Amer. Math. Soc.*, 8, 56–62.

[23] ________, (1964). "On a Realization of a Complemented Algebra." *Proc. Amer. Math. Soc.*, 15, 964–966.

[24] ________, (1964). "On Continuity of Multiplication in a Complemented Algebra." *Pacific J. of Math.*, 14, 1399–1403.

[25] ________, (1966). "On Two-sided $H^\star$-algebras." *Pacific J. of Math.*, 16, 365–370.

[26] ________, (1966). "Discrete Complemented Algebra." *Proc. Amer. Math. Soc.*, 17, 499–501.

[27] ________, (1968). "A Generalized Hilbert Space." *Duke Math. Journal*, 35, 191–198.

[28] P. P. Saworotnow and John C. Friedell (1970). "Trace-class for an Arbitrary $H^\star$-algebra." *Proc. Amer. Math. Soc.*, 26, 95–100.

[29] P. P. Saworotnow (1970). "Trace-class and Centralizers of an $H^\star$-algebra." *Proc. Amer. Math. Soc.*, 26, 101–104.

[30] ________, (1970). "Representation of a Complemented Algebra on a Locally Compact Space." *Ill. J. Math.*, 14, 674–677.

[31] ________, (1976). "A Characterization of Hilbert Modules." *Bull. Calcutta Math. Soc.*, 68, 39–41.

[32] ________, (1980). "Characterization of the Trace-class." *Proc. Amer. Math. Soc.*, 78, 545–547.

[33] ________, (1980). "Linear Spaces with an $H^\star$-algebra-valued Inner Product." *Trans. Amer. Math. Soc.*, 262, 543–549.

[34] ________, (1981). "A Simple Characterization of the Trace-class of Operators." *International J. Math. & Math. Sci.*, 4, 411–412.

[35] ________, (1983). "Generalized Positive Definite Functions and Stationary Processes." *Prediction Theory and Harmonic Analysis, The Pesi Masani Volume*, edited by V. Mandrekar and H. Salehi, North-Holland.

[36] ________, (1985). "Diagonalization of a Self-adjoint Operator Acting on a Hilbert Module." *International J. Math. & Math. Sci.*, 8, 669–675.

[37] ________, (1988). "Relation between Right and Left Involutions of a Hilbert Algebra." *Proc. Amer. Math. Soc.*, 102, 57–58.

[38] ________, (1992). "Semi-simplicity of a Proper Weak $H^\star$-algebra." *International J. Math. & Math. Sci.*, 15, 409–411.

[39] ________, (1991). "Diagonalization of Unbounded Self-adjoint Operators Acting on a Hilbert Module." *Glasnik Matematicki*, 26 (46), 89–95.

[40] J. R. Schue (1960). "Hilbert Space Method in the Theory of Lie Algebras." *Trans. Amer. Math. Soc.*, 95, 69–80.

[41] ________, (1961). "Cartan Decomposition for $L^\star$-algebras." *Trans. Amer. Math. Soc.*, 98, 334–349.

[42] M. F. Smiley (1953). "Right $H^\star$-algebra." *Proc. Amer. Math. Soc.*, 4, 1–4.

[43] J. F. Smith and P. P. Saworotnow (1961). "On Some classes of Scalar-product Algebras." *Pacific J. of Math.*, 11, 739–750.

[44] James F. Smith (1974). "The Structure of Hilbert Modules." *J. London Math. Soc.*, 8(2), 741–749.

[45] M. H. Stone (1966). "Linear Transformations in Hilbert Space and Their Applications." *A.M.S. Colloquium Publication, Vol. XV.*

[46] M. Takesaki (1970). "Theory of Modular Hilbert Algebras." *Lecture Notes in Mathematics*, 128, Springer-Verlag.

[47] B. Yood (1969). "On Algebras Which Are Pre-Hilbert Algebras." *Duke Math. J.*, 36, 261–272.

Mathematics Department
Catholic University of America
Washington, DC 20064-0001, USA
Internet e-mail address: saworotnow@cua.edu

K SUNDARESAN

Geometry of spaces of quadratic forms

INTRODUCTION

The spaces of homogeneous forms on Banach spaces gained considerable importance in the past decade mainly because of the fundamental role they play in function theory, differential analysis and in the theory of approximation of C^k–functions on Banach spaces by polynomials on these spaces, see Aron and Globevenik [1], Dineen [3], Llavona [4], and Sundaresan and Swaminathan [7]. Convexity and smoothness properties of Banach space of n–homogeneous forms on $L_p(\ell_p)$, $1 \leq p \leq \infty$ have been investigated by the author recently, see Sundaresan [8]. In this paper it is proposed to obtain a complete characterization of the extreme points and smooth points of the unit ball of the normed linear space of quadratic forms (2–homogeneous forms) on the finite dimensional Euclidean spaces. It is stressed that part of the motivation for this investigation arises from the importance of extremal structure and smoothness of convex sets in optimization, see Phelps [6].

The paper is divided into two sections. The necessary terminology and notation are stated in the first section. Further in this section several basic facts on linear transformations on finite dimensional vector spaces and matrices are recalled without proof for the convenience of reference. The main results are presented in section 2.

1. TERMINOLOGY, NOTATION AND BASIC FACTS

Let $(E, \| \ \|)$ be an Euclidean space of dimension $n \geq 2$ on the field of real numbers R with the norm, $\| \ \|$, determined by the inner product $(\ , \ )$, and $\widetilde{e} = \{e_i\}_{i=1}^n$ be a fixed ordered orthonormal basis of E. Linear transformations on $E \to E$ are represented by matrices canonically with respect to this basis.

Several well known basic facts on matrices and linear transformation on finite dimensional vector spaces, see for instance Mirsky [5], are useful here. Some of these are recalled here for convenience of reference without proof. If A is a matrix its transpose is noted at A^T. The vector space of symmetric $n \times n$ matrices is denoted by S_n. It is of dimension $\frac{1}{2}n(n+1)$. If $A \in S_n$, it has all real eigenvalues and whenever it is enumerated for instance as $\{\lambda_i\}_{i=1}^n$, it is understood that each appears with respective multiplicity. If $A \in S_n$, it is diagonalizable i.e. there is an orthogonal matrix Q such that $Q^T A Q$ is a diagonal matrix $D = (d_{ij})$ with the diagonal entries $\{d_{ii}\}_{i=1}^n$ same as the eigenvalues of A in some order. The operator norm of $A \in S_n$, defined by $\|A\|_\infty = \sup_{\|x\| \leq 1} \|Ax\|$, is also given by either of the numbers

1991 *Mathematics Subject Classification.* Primary 46B20, Secondary 46Co5.

$\sup_{\|x\|\leq 1} |(Ax, x)|$ or $\max_{1\leq i\leq n} |\lambda_i|$, where $\{\lambda_i\}_{i=1}^n$ are the eigenvalues of A. The normed linear space $(S_n, \|\ \|_\infty)$ is noted as X in the sequel. If Q is an orthogonal matrix, the map $B \to Q^T BQ$ is a linear isometry on X onto X.

The normed linear space $(P_2(E), |||\ |||)$ of concern here is the vector space $P_2(E)$ of 2–homogeneous forms on E, equipped with the norm,

$$|||P||| = \sup_{\|x\|\leq 1} |P(x)|,$$

for all $P \in P_2(E)$. It is well known that the map $\sigma\colon P_2(E) \to S_n$, defined by $\sigma(P) =$ the matrix $A = (a_{ij})$, where $a_{ij} = \frac{1}{2}[P(e_i + e_j) - P(e_i) - P(e_j)]$, $1 \leq i, j \leq n$, is a linear isomorphism on $P_2(E)$ onto S_n. The symmetric matrix $\sigma(P) = A$, is referred here as the matrix associated with P. It is verified that $P(x) = (Ax, x)$ for all $x \in E$. The Lemma 1 below is an easy consequence of the formula for $\|A\|_\infty$, for $A \in S_n$, stated in the preceding paragraph.

Lemma 1. *The map* $\sigma : (P_2(E), |||\ |||) \to (S_n, \|\ \|_\infty)$ *is a linear isometric surjection.*

The trace of a square matrix A is noted in the sequel as $\operatorname{tr} A$.

Remark 1. Since $\operatorname{tr} A$ is invariant under similarities, Frobenius equations for a matrix $A \in S_n$, $A = (a_{ij})$ stating that

$$\text{(1)} \qquad \operatorname{tr}(A^2) = \operatorname{tr}(A^T A) = \sum_{i=1}^n \sum_{j=1}^n a_{ij}^2 = \sum_{i=1}^n \lambda_i^2$$

are verified, if $\{\lambda_i\}_{i=1}^n$ are the eigenvalues of A. Further, in passing, it is noted here that if $A = (a_{ij})$ is an $n \times n$ matrix, $\|A\|_\infty \leq 1$, then $|a_{ij}| \leq 1$, $1 \leq i, j \leq n$, since for the choice of A, $\|Ae_i\| \leq 1$ for $1 \leq i \leq n$.

If $A \in S_n$, and $\{\mu_i\}_{i=1}^n$ are the eigenvalues of A, the trace norm of A, $\|A\|_\tau$ is by definition $\sum_{i=1}^n |\mu_i|$. The normed space $(S_n, \|\ \|_\tau)$ is congruent to the space X^*, the dual of the space X. The next lemma which makes the congruence very precise is stated here without proof.

Lemma 2. *The normed linear space* X^**, the dual of the space* X *is isometrically isomorphic with* $(S_n, \|\ \|_\tau)$ *under the mapping* $\sigma'\colon X^* \to (S_n, \|\ \|_\tau)$*, where* σ' *is such that if* $\sigma'(\ell) = C$*, then* $\ell(A) = \operatorname{tr}(AC)$ *for all* $A \in S_n$.

The lemma follows by arguing as in the proof of Theorem 1 on page 46 in Schatten [7] for the more general assertion which arises when X is replaced by the Banach space of compact operators with the supremum norm on a separable Hilbert space.

2. MAIN RESULTS

In this section a complete characterization of extreme points and smooth points of the unit ball of $P_2(E)$ are provided. For an account of extreme and smooth points, see Day [2]. For convenience the set of extreme points of a set M is noted as Ext M. The unit ball of $(P_2(E), ||| \, |||)$ is noted as U.

Theorem 1. *A 2–homogeneous polynomial P is an extreme point of U if and only if the matrix $\sigma(P)$ associated with P has unimodular eigenvalues.*

Proof. Since σ is a linear isometry as stated in Lemma 2, it follows that P is an extreme point of U if and only if the matrix $\sigma(P) = A$, associated with P is an extreme point of the unit ball U_n of X. As already noted in section 1, there is a linear isometry F on X onto X such that $F(A) = D$ is a diagonal matrix (d_{ij}) with the diagonal entries $\{d_{ii}\}_{i=1}^n$ same as the eigenvalues $\{\lambda_i\}_{i=1}^n$ of A in some order. Since F is a linear isometry, $A \in \text{Ext } U_n$ if and only if $D \in \text{Ext } U_n$.

Thus to prove the theorem it is enough to show that a diagonal matrix $D \in \text{Ext } U_n$ if and only if the diagonal entries of D are unimodular. Let $D = (d_{ij})$ be a $n \times n$ diagonal matrix, with $|d_{ii}| = 1$, $1 \leq i \leq n$. Let $D = \frac{P+Q}{2}$, where $P, Q \in U_n$. Since $\|D\|_\infty = 1$, $\|P\|_\infty = \|Q\|_\infty = 1$. Let $P = (p_{ij})$, $Q = (q_{ij})$, $1 \leq i, j \leq n$. Now $D = \frac{P+Q}{2}$ implies

$$n = \sum_{i=1}^n d_{ii}^2 = \sum_1^n \left(\frac{p_{ii} + q_{ii}}{2}\right)^2 \leq \frac{1}{2}\left(\sum_1^n p_{ii}^2 + \sum_1^n q_{ii}^2\right) \leq \frac{1}{2}(n+n) = n,$$

noting that $|p_{ii}| \leq 1$, $|q_{ii}| \leq 1$, $1 \leq i \leq n$ since $\|P\|_\infty \leq 1$ and $\|Q\|_\infty \leq 1$, see Remark 1. Thus $\sum_1^n \left(\frac{p_{ii} + q_{ii}}{2}\right)^2 = \frac{1}{2}\left(\sum_1^n p_{ii}^2 + \sum_1^n q_{ii}^2\right)$, which implies that $p_{ii} = q_{ii}$, $1 \leq i \leq n$. Hence $d_{ii} = p_{ii} = q_{ii}$, $1 \leq i \leq n$. Now it is claimed that the nondiagonal entries p_{ij}, q_{ij} in the matrices P, Q are all equal to 0. To verify the claim let the eigenvalues of P be $\{\mu_i\}_{i=1}^n$. Thus by the Frobenius equation

$$(*) \qquad \sum_{i=1}^n \sum_{j=1}^n p_{ij}^2 = \sum_{i=1}^n \mu_i^2 \leq n,$$

since

$$\|P\|_\infty = \sup_{1 \leq i \leq n} |\mu_i| \leq 1.$$

Now since $p_{ii} = d_{ii}$, $1 \leq i \leq n$ are unimolular, it follows from $(*)$ that

$$\sum_{i=1}^n \sum_{\substack{j=1 \\ i \neq j}}^n p_{ij}^2 + n \leq n.$$

Hence the nondiagonal entries in P vanish. Hence $P = D = Q$, and $D \in \text{Ext } U_n$.

Conversely if D is a diagonal matrix $= (d_{ij})$, $D \in \text{Ext } U_n$, then it is verified that the diagonal entries in D are unimodular. Let, if possible, $|d_{ii}| < 1$ for some $i = j$, say. Thus there is a $\delta > 0$, such that $|d_{jj} \pm \delta| \leq 1$. Now choose two diagonal $n \times n$ matrices D_1, D_2, say (α_{ij}), (β_{ij}) respectively such that $\alpha_{ii} = \beta_{ii} = d_{ii}$ if $i \neq j$, $\alpha_{jj} = d_{jj} + \delta$, and $\beta_{jj} = d_{jj} - \delta$. It is verified that $D_1 \neq D_2$, $D = \frac{D_1 + D_2}{2}$, and D_1, D_2 are in U_n. Thus $D \notin \text{Ext } U_n$, a contradiction completing the proof of the theorem.

Before stating the characterization of smooth points, a vector $\{a_i\}_{i=1}^n \in R^n$, is called a *peak vector* if $\max_{1 \leq i \leq n} |a_i| = 1$, and $|a_i| = 1$ for exactly one integer i.

Theorem 2. *A 2-homogeneous polynomial P is a smooth point of U if and only if the vector of eigenvalues $\{\lambda_i\}_{i=1}^n$ of the matrix $\sigma(P)$ associated with P, is a peak vector.*

Proof. Arguing as in the proof of the preceding theorem with isometries, it is at once inferred that it suffices to prove that a $n \times n$ diagonal matrix D is a smooth point of U_n if and only if the vector of diagonal entries in D is a peak vector.

Let $D = (d_{ij})$ be a $n \times n$ diagonal matrix, be a smooth point of U_n. Thus $\|D\|_\infty = 1$. Hence $\max_{1 \leq i \leq n} |d_{ii}| = 1$. If possible, let there be integers t, j, $t \neq j$ such that $|d_{tt}| = |d_{jj}| = 1$. It is now verified that there are two distinct hyperplanes of support for the ball U_n at D as follows. Consider the $n \times n$ diagonal matrices C_1, C_2. $C_1 = (\alpha_{ij})$, $C_2 = (\beta_{ij})$, where $\alpha_{tt} = \text{sign } d_{tt}$, $\alpha_{ii} = 0$ if $i \neq t$, $\beta_{jj} = \text{sign } d_{jj}$, $\beta_{ii} = 0$ if $i \neq j$. It is at once verified that $\|C_1\|_\tau = \|C_2\|_\tau = 1$, $\text{tr}(DC_1) = \text{tr}(DC_2) = 1$, and $C_1 \neq C_2$. Thus the linear functionals corresponding to C_1 and C_2 as described in Lemma 2, are distinct supporting functionals of the unit ball U_n at D, contrary to the hypothesis. Hence the diagonal entries in D form a peak vector as desired.

Conversely it will be shown that if $D = (d_{ij})$ is a $n \times n$ diagonal matrix with diagonal entries forming a peak n-vector, then D is a smooth point of U_n. Let $D = (d_{ij})$, with $|d_{jj}| = 1$, $|d_{ii}| < 1$ if $i \neq j$, $1 \leq i \leq n$. Let $\ell \in X^*$ supporting U_n at D i.e. $\|\ell\| = 1 = \ell(D)$. Let $C = (c_{ij})$ be the $n \times n$ symmetric matrix determined by ℓ, as in Lemma 2, so that $\ell(A) = \text{tr}(AC)$, for all $A \in X$, and $\|C\|_\tau = \|\ell\| = 1$. Hence if the eigenvalues of C are $\{\mu_i\}_{i=1}^n$, then $\|C\|_\tau = \sum_{i=1}^n |\mu_i| = 1$. Direct computation of $\text{tr}(DC)$ yields

$$1 = \ell(D) = \text{tr}(DC) = \sum_{i=1}^n d_{ii} c_{ii}. \tag{2}$$

Hence for some i, $c_{ii} \neq 0$. Let B be the $n \times n$ diagonal matrix, $B = (b_{ij})$ with $b_{ii} = \text{sign}$ c_{ii}, if $c_{ii} \neq 0$, and $b_{ii} = 0$, otherwise.

Now from (1) and (2), noting that $|d_{ii}| \leq 1$, $1 \leq i \leq n$, it is verified that

$$1 = (\ell(D))^2 = \left(\sum_{i=1}^{n} d_{ii}c_{ii}\right)^2 \leq \left(\sum_{i=1}^{n} |d_{ii}|\ |c_{ii}|\right)^2 \leq \left(\sum_{i=1}^{n} |c_{ii}|\right)^2 \leq 1.$$

Hence $\left(\sum_{i=1}^{n} |d_{ii}|\ |c_{ii}|\right)^2 = \left(\sum_{i=1}^{n} |c_{ii}|\right)^2$, from which it follows that for each i, $1 \leq i \leq n$, either $|d_{ii}| = 1$ or $c_{ii} = 0$. Thus from our choice of D it follows that $c_{ii} = 0$ for all $i \neq j$. Hence the equation (1) implies $j_{jj}c_{jj} = 1$. Since $|d_{jj}| = 1$, $c_{jj} = \text{sign } d_{jj}$. Now it is verified, applying the Frobenius equation to the matrix C and noting that $||C||_\tau = 1$, that

$$\sum_{r=1}^{n}\sum_{s=1}^{n} c_{rs}^2 = \sum_{i=1}^{n} \mu_i^2 \leq (\Sigma|\mu_i|)^2 = 1.$$

Since $c_{jj}^2 = 1$, it follows that for all $(r, s) \neq (j, j)$, $c_{rs} = 0$. Thus there exists only one linear functional ℓ, $||\ell|| = 1$, supporting U_n at D, ℓ being the functional determined by the matrix $C = (c_{rs})$, where $c_{jj} = \text{sign } d_{jj}$, and $c_{rs} = 0$, for all $(r, s) \neq (j, j)$, completing the proof of the theorem.

References

[1] R. M. Aron and J. Globevenik, Analytic functions of c_0, Rev. Mat. Univ. Complutense 2(1989), 27–33.

[2] M. M. Day, Normed Linear Spaces, 3rd Ed., Springer–Verlag, New York, 1973.

[3] S. Dineen, Entire functions on c_0, J. Funct. Anal. 52(1983), 201–218.

[4] J. Llavona, Approximation of continuously differentiable functions, Math. Studies, no. 130, North Holland, 1983.

[5] L. Mirsky, An Introduction to Linear Algebra, Oxford University Press, London, 1955.

[6] R. R. Phelps, Convex Functions, Monotone Operators and Differentiability, Lecture Notes in Math. 1364, Springer—Verlag, 1989.

[7] K. Sundaresan and S. Swaminathan, Geometry and Nonlinear Analysis in Banach Spaces, Lecture Notes in Math. 1131, Springer–Verlag, Berlin, 1985.

[8] K. Sundaresan, Geometry of spaces of homogeneous polynomials on Banach lattices, Applied Geometry and Discrete Mathematics, The Victor Klee Festschrift, DIMACS 4(1991), 571–585.

Department of Mathematics
Cleveland State University
Cleveland, Ohio 44115 USA
kondagunta@math.csuohio.edu

LÁSZLÓ SZÉKELYHIDI

Cosine operator functions on inner product spaces

1. Introduction

The cosine functional equation has a long history. It has been studied by several authors under different conditions. Also the operator-valued version has attracted much attention. For details and further references see e.g. [1], [2], [3], [4], [5], [6], [7], [10], [11], [12]. The aim of this paper is to present some results on normal cosine operator functions on Hilbert spaces and on inner product spaces. The technique we use is based on spectral synthesis.

2. Cosine operator functions on abelian groups

Let G be an abelian group and H be a Hilbert space. We denote the set of all bounded linear operators on H by $\mathcal{L}(H)$ and the set of all normal operators on H by $\mathcal{N}(H)$. A mapping $C : G \to \mathcal{L}(H)$ is called a cosine operator function if it satisfies

$$C(x+y) + C(x-y) = 2C(x)C(y) \tag{1}$$

for all x, y in G and $C(0) = I$, the identity of H. The function C is called a normal cosine operator function if its range is contained in $\mathcal{N}(H)$. Our main goal is to show that any normal cosine operator function is the even part of a homomorphism of G into the multiplicative semigroup of $\mathcal{N}(H)$.

THEOREM 1. *Let G be an abelian group, H a Hilbert space and let C be a normal cosine operator function from G into $\mathcal{L}(H)$. Then*

$$C(x) = \frac{1}{2}\Big(M(x) + M(-x)\Big) \tag{2}$$

holds for any x in G, where $M : G \to \mathcal{N}(H)$ is a homomorphism of G into the multiplicative semigroup of $\mathcal{N}(H)$. Conversely, any function C of the form (2) is a normal cosine operator function.

PROOF: The second statement can be verified by direct computation. Indeed, we have

$$C(x+y) + C(x-y) = \frac{1}{2}M(x+y) + \frac{1}{2}M(x-y) + \frac{1}{2}M(-x-y) + \frac{1}{2}M(-x+y) =$$

$$= \frac{1}{2}\Big(M(x) + M(-x)\Big)\Big(M(y) + M(-y)\Big) = 2C(x)C(y).$$

For the proof of the first statement observe that $C(x)$ and $C(y)$ commute for any x and y in G. Indeed, equation (1) implies (by putting $x = 0$) that C is even and hence the left hand side of (1) is symmetric in x and y. By the Fuglede - Putnam theorem [9] this means that the operator algebra generated by the operators $C(x)$ and $C(x)^*$, where x runs through G, is a commutative C^*-algebra with unit $C(0) = I$. Let λ be any multiplicative linear functional of this algebra and we define the mapping $f_\lambda : G \to \mathbf{C}$ by

$$f_\lambda(x) = \lambda(C(x)).$$

By equation (1) and by the properties of λ it follows that f_λ satisfies

$$f_\lambda(x+y) + f_\lambda(x-y) = 2f_\lambda(x)f_\lambda(y)$$

for any x, y in G, further $f_\lambda(0) = 1$.

Let $F \subset G$ be any finitely generated subgroup. The restriction f_λ^F of f_λ to F satisfies

$$f_\lambda^F(x+y) + f_\lambda^F(x-y) = 2f_\lambda^F(x)f_\lambda^F(y)$$

on F and $f_\lambda^F(0) = 1$. It follows that any translate of f_λ^F, moreover any element g of the smallest translation invariant linear space $V(f_\lambda^F)$ which is closed with respect to pointwise convergence and contains f_λ^F, satisfies

$$g(x+y) + g(x-y) = 2g(x)f_\lambda^F(y)$$

for any x, y in F. As f_λ^F is not identically zero, by spectral synthesis on F (see [8], [12]) the space $V(f_\lambda^F)$ contains an exponential m_λ^F. In other words, m_λ^F is a homomorphism of G into the multiplicative group of nonzero complex numbers and it satisfies

$$m_\lambda^F(x+y) + m_\lambda^F(x-y) = 2m_\lambda^F(x)f_\lambda^F(y)$$

for any x, y in F. By the homomorphism property of m_λ^F this latter equation implies

$$f_\lambda^F(y) = \frac{1}{2}\Big(m_\lambda^F(y) + m_\lambda^F(-y)\Big)$$

for any y in F. By the linear independence of different exponentials (see e.g. [12]) it follows that for $F_1 \subset F_2$ we have

$$m_\lambda^{F_1}(x) = m_\lambda^{F_2}(x)$$

or

$$m_\lambda^{F_1}(x) = m_\lambda^{F_2}(-x)$$

for any x in F_1. Obviously we may suppose - by changing the notation, if necessary - that we always have the first relation, that is, the functions $\{m_\lambda^F\}$ ($F \subset G$ is finitely generated) form an inductive set. Let m_λ denote its inductive limit, that is

$$m_\lambda(x) = m_\lambda^F(x)$$

if x is in F. Then m_λ is an exponential and we have

$$f_\lambda(x) = \frac{1}{2}\Big(m_\lambda(x) + m_\lambda(-x)\Big)$$

for all x in G. Now we define the mapping $M : G \to \mathcal{N}(H)$ by

$$\lambda(M(x)) = m_\lambda(x)$$

for any multiplicative linear functional λ on $\mathcal{N}(H)$ and for any x in G. By Gelfand theory, this is a correct definition. Further we have

$$\lambda\Big(M(x+y) - M(x)M(y)\Big) = \lambda\Big(M(x+y)\Big) - \lambda\Big(M(x)\Big)\lambda\Big(M(y)\Big) =$$

$$= m_\lambda(x+y) - m_\lambda(x)m_\lambda(y) = 0,$$

for any multiplicative linear functional λ, hence

$$M(x+y) = M(x)M(y)$$

is valid for any x, y in G. Similarly, we deduce that equation (2) holds for any x in G and the theorem is proved.

3. The noncommutative case

Cosine operator functions have also been studied on noncommutative groups where usually the commutativity of the group is replaced by the condition

(3) $$C(xyz) = C(yxz)$$

for any x, y, z in G (see [6]). Here we show that this so-called Kannappan-condition makes it possible to reduce the problem to the commutative case.

LEMMA 2. *Let G be a group and H a set. The function $C : G \to H$ is constant on the cosets of the commutator subgroup of G if and only if it satisfies equation* (3) *for all x, y, z in G (see [12]).*

PROOF: Suppose that C satisfies the given condition and let G_0 denote the commutator subgroup of G. If a, b belong to the same coset of G_0, then ab^{-1} is in G_0, that is,

$$a = x_1y_1x_1^{-1}y_1^{-1}x_2y_2x_2^{-1}y_2^{-1}\dots x_ky_kx_k^{-1}y_k^{-1}b$$

with some x_i, y_i in G ($i = 1, 2, \dots, k$). Then we have

$$\begin{aligned} C(a) &= C(x_1y_1x_1^{-1}y_1^{-1}x_2y_2x_2^{-1}y_2^{-1}\dots x_ky_kx_k^{-1}y_k^{-1}b) = \\ &= C(y_1x_1x_1^{-1}y_1^{-1}x_2y_2x_2^{-1}y_2^{-1}\dots x_ky_kx_k^{-1}y_k^{-1}b) = \\ &= C(x_2y_2x_2^{-1}y_2^{-1}\dots x_ky_kx_k^{-1}y_k^{-1}b) = \dots = C(b), \end{aligned}$$

that is, C is constant on the cosets of G_0. Conversely, if C is constant on the cosets of G_0, then

$$C(xyz) = C(xyx^{-1}y^{-1}yxz) = C(yxz)$$

for all x, y, z in G, hence our lemma is proved.

Using this lemma we can extend Theorem 1. for noncommutative groups, supposing that C satisfies equation (3).

THEOREM 3. *Let G be a group, H a Hilbert space and let C be a normal cosine operator function from G into $\mathcal{L}(H)$ satisfying condition* (3) *for every x, y, z in G. Then*

$$C(x) = \frac{1}{2}\Big(M(x) + M(x^{-1})\Big) \tag{4}$$

holds for any x in G, where $M : G \to \mathcal{N}(H)$ is a homomorphism of G into the multiplicative semigroup of $\mathcal{N}(H)$. Conversely, any function C of the form (4) *is a normal cosine operator function and it satisfies* (3).

PROOF: By Lemma 2. C is constant on the cosets of G_0, the commutator subgroup of G. On the other hand, G_0 is a normal subgroup and the factor group G/G_0 is commutative. Let $C = C_0 \circ \Phi$ where $\Phi : G \to G/G_0$ is the natural homomorphism. This factorization is possible, because C is constant on the cosets of G_0. It is easy to verify that $C_0 : G/G_0 \to \mathcal{L}(H)$ is a normal cosine operator function hence by Theorem 1. it has the representation

$$C_0(u) = \frac{1}{2}\Big(M_0(u) + M_0(u^{-1})\Big) \tag{5}$$

with some $M_0 : G/G_0 \to \mathcal{N}(H)$ which is a homomorphism of G/G_0 into the multiplicative semigroup of $\mathcal{N}(H)$. Putting $u = \Phi(x)$ with any x in G we have the desired representation for C as obviously $M = M_0 \circ \Phi$ is a homomorphism of G into the multiplicative semigroup of $\mathcal{N}(H)$. The converse of the statement is an easy calculation and the theorem is proved.

4. Inner product spaces

Although the proof of Theorem 1. heavily depends on the Hilbert space structure and Gelfand theory, we can extend it easily to inner product spaces. Let H_0 be any inner product space and let

H denote its completion. Then H is a Hilbert space. If a cosine operator function $C : G \to \mathcal{L}(H_0)$ is given, then obviously $C(x)$ extends to H for any x in G in a unique way, because $C(x)$ is bounded, hence it is uniformly continuous and H_0 is dense in H. Evidently, this extension - which we denote by C too - is a cosine operator function, which is normal, if the original was normal. Hence we get the following corollaries of Theorem 1. and Theorem 3.

COROLLARY 4. *Let G be an abelian group, H_0 an inner product space, H its completion and let C be a normal cosine operator function from G into $\mathcal{L}(H_0)$. Then*

$$C(x) = \frac{1}{2}\Big(M(x) + M(-x)\Big) \tag{5}$$

holds for any x in G, where $M : G \to \mathcal{N}(H)$ is a homomorphism of G into the multiplicative semigroup of $\mathcal{N}(H)$.

COROLLARY 5. *Let G be a group, H_0 an inner product space, H ist completion and let C be a normal cosine operator function from G into $\mathcal{L}(H_0)$ satisfying condition (3) for every x, y, z in G. Then*

$$C(x) = \frac{1}{2}\Big(M(x) + M(x^{-1})\Big) \tag{6}$$

holds for any x in G, where $M : G \to \mathcal{N}(H)$ is a homomorphism of G into the multiplicative semigroup of $\mathcal{N}(H)$.

5. References

[1] Aczél, J., *Lectures on Functional Equations and Their Applications*, Academic Press, New York, London, 1966.

[2] Chojnacki, W., *Group representations of bounded cosine functions*, Technical report of The University of Adelaide, TR94-09, 1994.

[3] Chung, J.K., Kannappan, Pl., Ng, C.T., *A Generalization of the Cosine-Sine Functional Equation on Groups* Linear Algebra Appl., **66**, 1988. pp. 259-277.

[4] Fattorini, H.O., *Uniformly bounded cosine functions in Hilbert space*, Indiana Univ. Math. J., **20**, 1970. pp. 411-425.

[5] Förg-Rob, W., Schwaiger, J., *A generalization of the cosine equation*, Grazer Math. Ber., **315**, 1991. pp. 25-34.

[6] Kannappan, Pl., *The functional equation $f(xy) + f(xy^{-1}) = 2f(x)f(y)$ for groups*, Proc. Amer. Math. Soc., **19**, 1968. pp. 69-74.

[7] Kurepa, S., *Uniformly bounded cosine function in a Banach space*, Math. Balkanica, **2**, 1972. pp. 109-115.

[8] Lefranc, M., *Analyse harmonique dans Z^n*, C.R.Acad.Sci. Paris, **246**, 1958. pp. 1951-1953.

[9] Mlak, W. *Hilbert Spaces and Operator Theory*, Kluwer Academic Publishers, Dordrecht, Boston, London, 1991.

[10] Stetker, H., *Functional equations and spherical functions*, Aarhus Universitet Preprint Series, **No. 18.** 1994.

[11] Székelyhidi, L., *Functional equations on abelian groups*, Acta Math. Acad. Sci. Hungar., **37** 1981. pp. 235-243.

[12] Székelyhidi, L., *Convolution type functional equations on topological abelian groups*, World Scientific Publishing Co. Pte. Ltd., Singapore, New Jersey, London, Hong Kong, 1991.

Department of Mathematics, Janus Pannonius University, Pécs, Hungary